Texts in Algorithmics
Volume 12

Proceedings of the International Workshop on Combinatorial Algorithms 2008

Texts in Algorithmics Series Editor   Costas Iliopoul
                                                       csi@dcs.kcl.ac.u

Proceedings of the International Workshop on Combinatorial Algorithms 2008

Edited by

Mirka Miller
and
Koichi Wada

ISBN 978-1-904987-74-1

College Publications
Scientific Director: Dov Gabbay
Managing Director: Jane Spurr
Department of Computer Science
Strand, London WC2R 2LS, UK

http://www.collegepublications.co.uk

Original cover design by orchid creative   www.orchidcreative.co.uk
Printed by Lightning Source, Milton Keynes, UK

Preface

The Nineteenth International Workshop on Combinatorial Algorithms (IWOCA 2008) was held during September 12-15, 2008, at Chu-nichi Palace, Nagoya, Japan.

Calls for papers were distributed widely around the world. Each paper was refereed by at least two referees, and almost all by three referees. 22 of the papers received were selected for presentation at the workshop. This represents acceptance rate of about 50%. IWOCA 2008 participants and contributors came from Australia, Austria, China, Czech Republic, Finland, Germany, India, Israel, Japan, South Korea, Spain, Taiwan, UK and USA.

After the workshop, a small number of selected papers is to be published in a Special Issue of Mathematics in Computer Science published by Birkhauser/Springer, edited by Mirka Miller and Koichi Wada. The Special Issue will be called Advances in Combinatorial Algorithms.

Details of the conference are available from the website
http://www.iwoca.org/iwoca2008/default.htm

Details of the AWOCA and IWOCA conference series can be found at
http://www.iwoca.org/

We thank our sponsors, the Daiko Foundation, Nagoya, Japan; Special Interest Group of Algorithms (SIGAL), Information Processing Society of Japan; Graph Theory and Applications research laboratory, The University of Newcastle, Australia; and Department of Computer Science and Engineering, Nagoya Institute of Technology, Japan. It would not have been possible to have a successful conference without all their support.

It remains for us to acknowledge and thank all the members of the Program Committee and the Organising Committee for their contribution and commitment to IWOCA 2008 and for their excellent and timely work.

Mirka Miller and Koichi Wada
Editors
September 2008

Organization

Programme Chairs

Mirka Miller, The University of Newcastle, Australia & University of West Bohemia, Czech Republic
Koichi Wada, Nagoya Institute of Technology, Japan

Programme Committee

Martin Baca, Technical University of Kosice, Slovakia
Hajo Broersma, Durham University, UK
Francis Chin, Hong Kong University, Hong Kong
Charlie Colbourn, Arizona State University, USA
Jackie Daykin, Royal Holloway College, UK
Diane Donovan, University of Queensland, Australia
Dalibor Froncek, University Minnesota, USA
Toshihiro Fujito, Toyohashi University of Technology, Japan
Xiao-Dong Hu, Chinese Academy of Sciences, China
Taisuke Izumi, Nagoya Institute of Technology, Japan
Yoshihiro Kaneko, Gifu University, Japan
Jan Kratochvil, Charles University, Czech Republic
Selda Küçükçifçi, Koç University, Turkey
Gadi Landau, University of Haifa, Israel
Thierry Lecroq, University of Rouen, France
Jimmy Lee, Hong Kong University, Hong Kong
Moshe Lewenstein, Bar Ilan University, Israel
Paulette Lieby, NICTA, Australian National University, Australia
Yuqing Lin, The University of Newcastle, Australia
Gonzalo Navarro, University of Chile, Chile
Kunsoo Park, Seoul National University, Korea
Yoan Pinzon, National University of Colombia, Colombia
Andrzej Proskurowski, Oregon State University, USA
Joe Ryan, The University of Newcastle, Australia
Zdenek Ryjacek, University of West Bohemia, Czech Republic
Rinovia Simanjuntak, ITB Bandung, Indonesia
Jamie Simpson, Curtin University, Australia
Jozef Sirãn, University of Auckland, New Zealand
Thanasis Tsakalidis, University of Patras, Greece
Wal Wallis, University of Southern Illinois, USA
Sue Whitesides, McGill University, Canada
Zhiyou Wu, University of Ballarat, Australia
Chee Yap, New York University, USA

Local Organization

Toshihiro Fujito, Toyohashi University of Technology, Japan
Tomio Hirata, Nagoya University, Japan (Co-chair)
Nobuhiro Inuzuka, Nagoya Institute of Technology, Japan (Co-chair)
Taisuke Izumi, Nagoya Institute of Technology, Japan
Yoshihiro Kaneko, Gifu University, Japan
Yoshiaki Katayama, Nagoya Institute of Technology, Japan
Shigeru Masuyama, Toyohashi University of Technology, Japan
Takao Ono, Nagoya University, Japan
Koichi Wada, Nagoya Institute of Technology, Japan
Mitsunori Yagiura, Nagoya Univeristy, Japan

Steering Committee

Costas Iliopoulos, King's College London, UK
Mirka Miller, University of Ballarat, Australia & University of West Bohemia, Czech Republic
Bill Smyth, McMaster University, Canada & Curtin University, Australia

Sponsoring Institutions

Daiko Foundation
1-1-22 Daiko-Minami Higashi-ku
Nagoya, 461-0047, Japan
Special Interest Group of Algorithms(SIGAL)
Information Processing Society of Japan(IPSJ)
Kagaku-Kaikan(Chemistry Hall) 4F
1-5 Kanda-Surugadai, Chiyoda-ku
Tokyo, 101-0062, Japan
Graph Theory and Applications(GTA) research laboratory
The University of Newcastle
University Drive, Callaghan
2308, Australia
Department of Computer Science and Engineering
Nagoya Institute of Technplogy
Gokiso-cho Syowa,
Nagoya, 466-8555, Japan

External Reviewers

Avivit Levy
Fusheng Bai
Maxime Crochemore
Stefan Dantchev
Danny Hermelin
Tomoko Izumi
Lap Chi Lau
George Mertzios
Hirotaka Ono
Sebastian Ordyniak
Arash Rafiey
Hadas Shachnai
Arezou Soleimanfallah
Merce Villanueva
Oren Weimann
Masafumi Yamashita

Table of Contents

Designing Good Random Walks on Finite Graphs

Masafumi Yamashita

Department of Computer Science and Communication Engineering
Kyushu University, Fukuoka, Japan
mak@csce.kyushu-u.ac.jp

Abstract

Random walks on finite graphs are used in many fields. An important application area is Markov Chain Monte Carlo (MCMC), in which a random walk is used to sample data following a given probability distribution. Many algorithms, including the Metropolis-Haystings algorithm, are proposed to compute a transition probability matrix such that its stationary distribution coincides with the given probability distribution. Their main concern is the quick convergence to the stationary distribution. Distributed computing is another application area. We use a random walk to circulate a token or to search for a particular node in a huge network. In these application, the convergence speed is not a crucial issue, but the hitting and the cover time become important measures. Furthermore, the locality of information that is necessary to compute a transition probability is considered to be an important requirement. In this talk, we propose a method to compute a transition probability using local information that achieves an optimal hitting time, and investigate the Metropolis-Haystings algorithm from this viewpoint.

Delivering Multimedia Content in Vehicular Ad Hoc Networks

Stepahn Olariu

Department of Computer Science, Old Dominion University
Norfork, VA 23529-0162 U.S.A.
olariu@cs.odu.edu

Abstract

In the past decade, Vehicular Ad Hoc Networks (VANET), that specialize Mobile Ad Hoc Networks (MANET), to vehicle to vehicle (V2V) and vehicle to infrastructure (V2I) communications have received a great deal of attention in the research community. And with good reason: vehicular communications promise to integrate driving into a ubiquitous and pervasive network that is already redefining the way we live and work.

The potential societal impact of VANET was confirmed by the proliferation of consortia and initiatives involving car manufacturers, government agencies and academia.

While the original impetus for VANET was traffic safety, more recent concerns involve privacy and security. It was recently noticed that allocation of 75MHz spectrum in the 5.9GHz band for Dedicate Short Range Communications (DSRC) in North America opens VANET to multimedia applications including peer-to-peer (P2P) content provisioning and the fast-growing mobile infotainment industry.

This address discusses a number of important open research problem in VANET and shows ways in which multimedia content can de carried by a zero-infrastructure peer to peer system for VANET.

Combinatorial Algorithms in Concorde

Vašek Chvátal

Canada Research Chair in Combinatorial Optimization
Concordia University, Montreal
chvatal@cse.concordia.ca

Abstract

The most efficient known methods for solving particular instances of the (symmetric) traveling salesman problem have evolved from the cutting-plane method developed by Dantzig, Fulkerson, and Johnson in the early nineteen fifties. Here, one has to solve repeatedly the folowing separation problem: Given a real vector x^* with components indexed by edges of a complete graph, find a linear inequality violated by x^* and satisfied by the incidence vectors of all hamiltonian circuits in the complete graph (or return a failure message if no such inequality exists).

Concorde is a computer code developed by Applegate, Bixby, Cook, and the speaker for the symmetric traveling salesman problem and some related network optimization problems. Its TSP solver has been used to obtain optimal solutions of all test instances from Gerhard Reinelt's library called TSPLIB. (The most difficult of these instances, solved in 2006, has 85,900 cities.)

Among the novel separation algorithms introduced in Concorde,two have a particularly combinatorial flavour. One uses an efficient data structure related to the minimum prefix-sum problem; the other uses an efficient data structure (namely, the PQ-tree) related to the consecutive ones property. We will discuss both.

Reference

David L. Applegate, Robert E.Bixby, Vašek Chvátal, and William J.Cook, The Traveling Salesman Problem: A Computational Study, Princeton Series in Applied Mathematics, February 2007.

Automata Approach to Graphs of Bounded Rank-width

Robert Ganian[*] and Petr Hliněný[**]

Faculty of Informatics, Masaryk University
Botanická 68a, 602 00 Brno, Czech Republic
ganian@mail.muni.cz, hlineny@fi.muni.cz

Abstract. Rank-width is a rather new structural graph measure introduced by Oum and Seymour in 2003 in order to find an efficiently computable approximation of clique-width of a graph. Being a very nice graph measure indeed, the only serious drawback of rank-width was that it is virtually impossible to use a given rank-decomposition of a graph for running dynamic algorithms on it. We propose a new independent description of rank-decompositions of graphs using labeling parse trees which is, after all, mathematically equivalent to the recent algebraic graph-expression approach to rank-decompositions of Courcelle and Kanté [WG'07]. We then use our labeling parse trees to build a Myhill-Nerode-type formalism for handling restricted classes of graphs of bounded rank-width, and to directly prove that (an already indirectly known result) all graph properties expressible in MSO logic are decidable by finite automata running on the labeling parse trees.

Keywords: graph, parameterized algorithm, rank-width, clique-width, tree automaton, MSO logic.

1 Introduction

Most graph problems are known to be NP-hard in general, and yet a solution to these is needed for practical applications. One common method to provide such a solution is through restricting the input graph to have a certain structure. Often the input graphs are restricted to have bounded tree-width [19] (or branch-width), but a useful weaker structural restriction has been brought by the notion of *clique-width*, defined by Courcelle and Olariu in [7].

Now, many hard graph problems (in particular all those expressible in MSO logic of adjacency graphs) are solvable in polynomial time [6, 10, 16, 12], as long as the input graph has bounded clique-width and is given in the form of the "decomposition for clique-width", called a *k-expression*. A *k*-expression is an algebraic expression with the following four operations on vertex-labeled graphs using k labels: create a new vertex with label i; take the disjoint union of two labeled graphs; add all edges between vertices of label i and label j; and relabel

---

[*] Supported by the Institute for Theoretical Computer Science ITI, project 1M0545.
[**] Supported by the Czech research grant GAČR 201/08/0308 and project 1M0545.

all vertices with label i to have label j. However, for fixed $k > 3$, it is not known how to find a k-expression of an input graph having clique-width at most k.

Rank-width is another graph complexity measure introduced by Oum and Seymour [18, 8, 17], aiming at providing an $f(k)$-expression of the input graph having clique-width k for some fixed function f in polynomial time. Rank-width is defined (see Section 2) as the branch-width of a so-called *cut-rank* function of graphs. Rank-width turns out to be very useful for algorithms on graphs of bounded clique-width since it can be computed, together with an optimal decomposition, in time $O(n^3)$ on n-vertex graphs of bounded rank-width [15]. Moreover, if rank-width of a graph is k, then its clique-width lies between k and $2^{k+1} - 1$ [18] and a corresponding expression can be constructed from a rank-decomposition of width k.

In view of the previous facts, particularly that clique-width can be up to exponentially larger than rank-width [4], it appears desirable to design efficient algorithms running straight on an optimal rank-decomposition rather than transforming a width-k rank-decomposition into an $f(k)$-expression. Unfortunately, this goal seems practically impossible in a direct way given the rather "strange nature" of a rank-decomposition. Thus one has to look for indirect approaches, say those inspired by a natural geometric link of rank-decompositions of graphs to branch-decompositions of binary matroids [15].

In 2007 Courcelle and Kanté [5] gave an alternative characterization of rank-decompositions of graphs using algebraic terms over multi-coloured graphs. Independently from them, the first author's Master thesis [11] has recently brought the concept of *labeling parse trees* (also called rank-width parse trees) which exactly characterize decompositions of graphs of given rank-width, too. We postpone all the technical definitions till the next section. For now we just note that the latter approach of [11] turns out to be exactly equivalent to the former of [5], though they come from different perspectives.

The aim of this paper is to continue in the labeling parse–tree approach in order to bring a different new view of dynamic algorithms on graphs of bounded rank-width, which is more computer–science oriented (and hence perhaps better understandable among the CS audience) than the algebraic–logic view of Courcelle and Kanté. We make an effort to employ finite tree automata in the task and develop a Myhill–Nerode kind of a characterization of the finite state properties of graphs of bounded rank-width. This viewpoint is not new in other areas, whereas it has been inspired by an analogous handling of bounded tree-width graphs by Abrahamson and Fellows [1] (or [9, Chapter 6]) and of represented matroids of bounded branch-width by the second author [14].

Our characterization then leads to an elementary proof that all MSO expressible graph properties (MS_1, to be precise) are decidable by tree automata over our labeling parse trees, and hence solvable in linear time. We suggest that this new characterization and the related proof techniques may be of independent interest among computer scientists who are working in designing dynamic algorithms for graphs of bounded rank-width (see Section 5 for further discussion).

2 Definitions and Basics

We consider finite simple undirected graphs by default. Here we bring some technical definitions and basic results which are the building blocks of our research.

Branch-width. A set function $f : 2^M \to \mathbb{Z}$ is called *symmetric* if $f(X) = f(M \setminus X)$ for all $X \subseteq M$. A tree is *subcubic* if all its nodes have degree at most 3. For a symmetric function $f : 2^M \to \mathbb{Z}$ on a finite set M, the branch-width of f is defined as follows.

A *branch-decomposition* of f is a pair (T, μ) of a subcubic tree T and a bijective function $\mu : M \to \{t : t$ is a leaf of $T\}$. For an edge e of T, the connected components of $T \setminus e$ induce a bipartition (X, Y) of the set of leaves of T. The *width* of an edge e of a branch-decomposition (T, μ) is $f(\mu^{-1}(X))$. The *width* of (T, μ) is the maximum width over all edges of T. The *branch-width* of f is the minimum of the width of all branch-decompositions of f. (If $|M| \leq 1$, then we define the branch-width of f as $f(\emptyset)$.)

A natural application of this definition is the branch-width of a graph, as introduced by Robertson and Seymour [19] along with better known tree-width, and its natural matroidal counterpart. In that case we use $M = E(G)$, and f the connectivity function of G. There is, however, another interesting application of the aforementioned general notions, in which we consider the vertex set $V(G) = M$ of a graph G as the ground set.

Rank-width. For a graph G, let $\mathbf{A}_G[U, W]$ be the bipartite adjacency matrix of a bipartition (U, W) of the vertex set $V(G)$ defined over the two-element field $GF(2)$ as follows: the entry $a_{u,w}$, $u \in U$ and $w \in W$, of $\mathbf{A}_G[U, W]$ is 1 if and only if uw is an edge of G. The *cut-rank* function $\rho_G(U) = \rho_G(W)$ then equals the rank of $\mathbf{A}_G[U, W]$ over $GF(2)$. A *rank-decomposition* and *rank-width* of a graph G is the branch-decomposition and branch-width of the cut-rank function ρ_G of G on $M = V(G)$, respectively.

The main reason for the popularity of rank-width over clique-width is the fact that there are parameterized algorithms for rank-decompositions [18, 15].

Theorem 2.1 (Hliněný and Oum [15]). *For every fixed t there is an $O(n^3)$-time algorithm that, for a given n-vertex graph G, either finds a rank-decomposition of G of width at most t, or confirms that the rank-width of G is more than t.*

Few rank-width examples. Any complete graph of more than one vertex has clearly rank-width 1 since any of its bipartite adjacency matrices consists of all 1s. It is similar with complete bipartite graphs if we split the decomposition along the parts. We illustrate the situation with graph cycles: while C_3 and C_4 have rank-width 1, C_5 and all longer cycles have rank-width equal 2. A rank-decomposition of, say, the cycle C_5 is shown in Fig. 1. Conversely, every subcubic tree with at least 4 leaves has an edge separating at least 2 leaves on each side, and every corresponding bipartition of C_5 gives a matrix of rank ≥ 2.

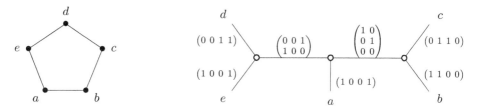

Fig. 1. A rank-decomposition of the graph cycle C_5.

We also mention so-called *distance-hereditary* graphs, i.e. graphs such that the distances in any of their connected induced subgraphs are the same as in the original graph, which have been independently studied, e.g. [3], before. It turns out that distance-hereditary graphs are exactly the graphs of rank-width 1 [17], and this simple fact explains many of their "nice" algorithmic properties.

Labeling parse trees. A (vertex) t-*labeling* of a graph is a mapping $lab : V(G) \to 2^{L_t}$ where $L_t = \{1, 2, \ldots, t\}$ is the set of labels (this notion is equivalent to so-called multicoloured graphs of [5]). Having a graph G with an (implicitly) associated t-labeling lab, we refer to the pair G, lab as to a t-*labeled graph* and use notation $\bar{G}$. Notice that each vertex of a t-labeled graph may have zero, one or more labels. So even an unlabeled graph can be considered as t-labeled with no labels, and every t-labeled graph is also t'-labeled for all $t' > t$.

A t-*relabeling* is a mapping $f : L_t \to 2^{L_t}$. For a t-labeled graph $\bar{G} = (G, lab)$ we define $f(\bar{G})$ as the same graph with a vertex t-labeling $lab' = f \circ lab$. Notice that—since the values of lab are subsets of L_t, or vectors from $\mathrm{GF}(2)^t$—the relabeling f in the composition $f \circ lab$ acts as a linear transformation in the vector space $\mathrm{GF}(2)^t$. Informally, f is applied separately to each label in $lab(v)$ and the outcomes are summed up "modulo 2"; such as for $lab(v) = \{1, 2\}$ and $f(1) = \{1, 3, 4\}$, $f(2) = \{1, 2, 3\}$, we get $lab'(v) = \{2, 4\} = \{1, 3, 4\} \triangle \{1, 2, 3\}$.

Let $\odot$ be a nullary operator creating a single new graph vertex of label $\{1\}$. For relabelings $f_1, f_2, g : L_t \to 2^{L_t}$ let $\oplus[g \mid f_1, f_2]$ be a binary operator over pairs of t-labeled graphs $\bar{G}_1 = (G_1, lab^1)$ and $\bar{G}_2 = (G_2, lab^2)$ defined as follows:

$$(G_1, lab^1) \oplus [g \mid f_1, f_2] (G_2, lab^2) = (H, lab)$$

where the graph H is constructed from the disjoint union $G_1 \dot\cup G_2$ by adding all edges uw, $u \in V(G_1)$ and $w \in V(G_2)$ such that $|lab^1(u) \cap g \circ lab^2(w)|$ is odd, and with the new labeling $lab(v) = f_i \circ lab^i(v)$ for $v \in V(G_i)$, $i = 1, 2$.

A *labeling parse tree* T, see [11, Definition 6.11], is a finite rooted ordered subcubic tree (with the root degree at most 2) such that

- all leaves of T contain the $\odot$ operator, and
- all internal nodes of T contain one of the $\oplus[g \mid f_1, f_2]$ operators.

A parse tree T then *generates* (parses) the graph G which is obtained by successive leaves-to-root applications of the operators in the nodes of T. See Fig. 2.

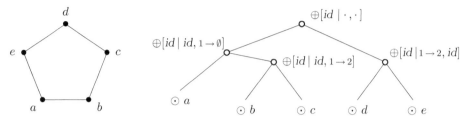

Fig. 2. An example of a labeling parse tree which generates a 2-labeled cycle C_5, with symbolic operators at the nodes (*id* denotes the relabeling preserving all labels).

The following substantial Theorem 2.2 is actually equivalent to [5, Theorem 3.4]. Its independent detailed proof can be found in the first author's Master thesis [11, Chapter 6] (the time complexity bound being implicit there).

Theorem 2.2 (Rank-width parsing theorem [11]). *A graph G has rank-width at most t if and only if (some labeling of) G can be generated by a labeling parse tree using t labels. Furthermore, an optimal rank-decomposition of G can be transformed into a labeling parse tree with t labels in time $O(n^2)$.*

We add a short note that time complexity $O(n^2)$ can be considered "linear" in this case since the size of the graph G can be of order up to n^2. We suggest that this complexity can be improved even to $O(|E(G)|)$ if one carefully reconsiders all the technical details, but that would not be useful in our context in which we use Theorem 2.2 together with Theorem 2.1 to construct an optimal labeling parse tree of a given graph G in parameterized $O(n^3)$ time.

3 Regularity Theorem for Rank-width

The core new contribution of our paper lies in developing a mathematical formalism for easy handling of graph properties which are efficiently solvable on graphs of bounded rank-width. Our formalism is closely tied with the classical Myhill–Nerode regularity theorem in automata theory. As we have already noted above, we are inspired by analogous formalisms used in [1] (graphs of bounded tree-width) and in [14] (matroids of bounded branch-width).

Recalling the notation of labeling parse trees, we shortly write $\oplus[g]$ for $\oplus[g \mid \emptyset, \emptyset]$ where $\emptyset$ stands for the relabeling $L \to \{\emptyset\}$ "forgetting" all vertex labels. Notice that the binary operation $\oplus[g]$ which creates an unlabeled graph from two labeled graphs is not commutative, but its operands can be exchanged together with a suitable modification of g. The role of a specific relabeling g in $\oplus[g]$ is rather technical after all, as the next immediate claim says. Let *id* be the relabeling preserving all labels.

Proposition 3.1. *Let G_1, G_2 be t-labeled graphs generated by labeling parse trees T_1, T_2, and $g : L_t \to 2^{L_t}$ be any relabeling. Then there is a tree T_2^g parsing a t-labeled graph G_2^g (actually unlabeled-equal to G_2) such that*

$$G_1 \oplus[g] G_2 = G_1 \oplus[id] G_2^g.$$

Canonical equivalence. Let Π_t denote the finite set (alphabet) of operators of labeling parse trees with t labels, and let subsequently $P_t \subseteq \Pi_t^{**}$ be the class (language) of all valid labeling parse trees with t labels. If $\mathcal{R}_t$ denotes the class of all unlabeled graphs of rank-width at most t and $\overline{\mathcal{R}}_t$ is the class of all t-labeled graphs parsed by the trees from P_t, then (Theorem 2.2) $G \in \mathcal{R}_t$ if and only if $\bar{G} \in \overline{\mathcal{R}}_t$ for some t-labeling $\bar{G}$ of G.

Let $\mathcal{D}$ be any class of graphs, and $\mathcal{D}_t = \mathcal{D} \cap \mathcal{R}_t$. In analogy to classical theory of regular languages we define a *canonical equivalence* of $\mathcal{D}_t$, denoted by $\approx_{\mathcal{D},t}$, as follows: $\bar{G}_1 \approx_{\mathcal{D},t} \bar{G}_2$ for any $\bar{G}_1, \bar{G}_2 \in \overline{\mathcal{R}}_t$ if and only if, for all $\bar{H} \in \overline{\mathcal{R}}_t$,

$$\bar{G}_1 \oplus [id]\ \bar{H} \in \mathcal{D}_t \quad \Longleftrightarrow \quad \bar{G}_2 \oplus [id]\ \bar{H} \in \mathcal{D}_t.$$

In informal words, the classes of $\approx_{\mathcal{D},t}$ "capture" all information we need to know about a t-labeled subgraph $\bar{G} \in \overline{\mathcal{R}}_t$ to decide membership in $\mathcal{D}$ further on in our parse tree processing.

The previous informal finding can be formalized as follows:

Theorem 3.2 (Rank-width regularity theorem). *Let $t \geq 1$, $\mathcal{D}$ be a graph class, and $\mathcal{D}_t = \mathcal{D} \cap \mathcal{R}_t$. The collection of all those labeling parse trees which generate the members of $\mathcal{D}_t$ is accepted by a finite tree automaton if, and only if, the canonical equivalence $\approx_{\mathcal{D},t}$ of $\mathcal{D}_t$ over $\overline{\mathcal{R}}_t$ is of finite index.*

Sketch of proof. A detailed proof of this statement is contained in [11, Chapter 7]. We only sketch its main ideas due to space restrictions here.

Our starting point is the classical Myhill–Nerode theorem for tree automata. Let Σ^{**} denote the set of all rooted binary trees over a finite alphabet Σ. For a language $\lambda \subseteq \Sigma^{**}$ we can define a congruence $\sim_\lambda$ such that $T_1 \sim_\lambda T_2$ for $T_1, T_2 \in \Sigma^{**}$ if, and only if, $T_1 \diamond_x U \in \lambda \iff T_2 \diamond_x U \in \lambda$ where U runs over all special rooted binary trees over Σ with one distinguished leaf node x, and $T_i \diamond_x U$ results from U by replacing the leaf x with the subtree T_i. Then λ is accepted by a finite tree automaton if and only if $\sim_\lambda$ has finite index.

In our case $\Sigma = \Pi_t$, and λ are the labeling parse trees of the members of $\mathcal{D}_t$. So, to prove our theorem it is enough to show that $\approx_{\mathcal{D},t}$ has infinite index if and only if $\sim_\lambda$ has infinite index.

Suppose the former holds, i.e. there are infinitely many $\bar{G}_k \in \overline{\mathcal{R}}_t$, $k = 1, 2, \ldots$, such that for all indices $i \neq j$ there exists $\bar{H}_{i,j} \in \overline{\mathcal{R}}_t$ for which $\bar{G}_i \oplus [id]\ \bar{H}_{i,j} \in \mathcal{D}_t$ but $\bar{G}_j \oplus [id]\ \bar{H}_{i,j} \notin \mathcal{D}_t$, or vice versa. Let S_k be a labeling parse tree of $\bar{G}_k$, and $Q_{i,j}$ that of $\bar{H}_{i,j}$. We define a new parse tree $U_{i,j}$ such that the root operator is $\oplus [id \mid \emptyset, \emptyset]$, its left son is the distinguished leaf x, and its right subtree is $Q_{i,j}$. Hence the special trees $U_{i,j}$ witness that all the parse trees S_k, $k = 1, 2, \ldots$ belong to distinct classes of $\sim_\lambda$.

Conversely, suppose that the latter holds. So there are infinitely many trees $S_k \in \Pi_t^{**}$, $k = 1, 2, \ldots$, such that for each pair of indices $i \neq j$ there exists $U_{i,j}$ as above for which $S_i \diamond_x U_{i,j} \in \lambda$ but $S_j \diamond_x U_{i,j} \notin \lambda$, or vice versa. We may assume without loss of generality that $S_k \in P_t$ are valid labeling parse trees for all k. Let $\bar{G}_k$ be the graphs parsed by S_k. Using technical [11, Lemma 7.3] and Proposition 3.1, we deduce that there exist graphs $\bar{H}_{i,j}$ such that

- the graph parsed by $S_i \diamond_x U_{i,j}$ is equal up to labeling to $\bar{G}_i \oplus [id] \ \bar{H}_{i,j} \in \mathcal{D}_t$,
- and the graph parsed by $S_j \diamond_x U_{i,j}$ equals up to labeling $\bar{G}_j \oplus [id] \ \bar{H}_{i,j} \notin \mathcal{D}_t$.

This assertion certifies that the graphs $\bar{G}_k$ indeed belong to distinct classes of our canonical equivalence $\approx_{\mathcal{D},t}$. ∎

Remark 3.3. Notice that the arguments used in our proof of Theorem 3.2 *do not* straightforwardly translate from rank-width (and labeling parse trees) to clique-width (and its k-expressions). Quite the opposite, the "only if" direction of this theorem seems not at all provable in the above way since one cannot freely choose the "root" of a k-expression (cf. [11, Lemma 7.3]). We consider that another small reason to favor rank-width over clique-width in CS applications.

3-colourability example. We briefly demonstrate the use of Theorem 3.2 on graph 3-colourability which is a well-known NP-complete problem. Let $\mathcal{C}$ denote the class of all simple 3-colourable graphs. To construct a tree automaton accepting the labeling parse trees of the members of $\mathcal{C}_t = \mathcal{C} \cap \mathcal{R}_t$, it is enough to identify the classes of the canonical equivalence $\approx_{\mathcal{C},t}$. We actually give the finitely many classes $X_0, X_1, X_2, \ldots$ of the following refinement of $\approx_{\mathcal{C},t}$:

- $X_0 = \{G : G \text{ is not 3-colourable}\}$.
- Otherwise, for any t-labeled graph G with a proper 3-colouring χ, we define a vector $c(G, \chi) = (c_\ell : \ell \in 2^{L_t})$ where $c_\ell \subseteq \{1, 2, 3\}$ is the set of χ-colours occurring in the vertices of G labeled by ℓ.
  $X_1, X_2, \ldots, X_{h(t)}$ are then the equivalence classes of $\sim$, where over t-labeled graphs $G_1 \sim G_2$ if and only if it holds $\{c(G_1, \chi) : \chi \text{ is a proper 3-colouring of } G_1\} = \{c(G_2, \chi) : \chi \text{ is a proper 3-colouring of } G_2\}$.

4 Regularity and MSO Definable Properties

From a logic point of view, we consider a graph as a relational structure on the ground set V, with one binary predicate $edge(u, v)$. When the language of MSO logic is applied to such a graph adjacency structure, one gets a descriptional language over graphs commonly abbreviated as MS_1. For an illustration we show an MS_1 expression of the 3-colourability property of a graph:

$$\exists V_1, V_2, V_3 \big[\qquad \forall v \, (v \in V_1 \vee v \in V_2 \vee v \in V_3) \wedge$$
$$\bigwedge\nolimits_{i=1,2,3} \forall v, w \, (v \notin V_i \vee w \notin V_i \vee \neg \, edge(v, w)) \big]$$

It is also common to consider the "counting" version of MSO logic which moreover has predicates $mod_{p,q}(X)$ stating that $|X| \bmod p = q$.

To avoid possible confusion we remark ahead that there is a stronger descriptional language MS_2 of graphs which allows to quantify also over graph edges and their sets, and which is related to graphs of bounded tree-width. There are MS_2 expressible graph properties, e.g. Hamiltonicity, which are not expressible in MS_1, whilst MS_2 properties cannot be efficiently handled in general on graphs of bounded rank-width.

In [6] Courcelle, Makowsky, and Rotics proved that all MS_1 definable graph properties are solvable in linear time, in fact by a tree automaton, running on a given k-expression (k fixed) of the graph. Their indirect proof used MSO interpretation (transduction) of the graphs generated by k-expressions into labeled binary trees. Since a graph class has bounded clique-width if and only if it has bounded rank-width, the results of [6] carry over to graphs of bounded rank-width (with a possible exponential jump in the width parameter).

We, on the other hand, favor the independent direct combinatorial approach to these problems, paralleling [1, 14]:

Theorem 4.1. *Let $t \geq 1$. If $\mathcal{D}$ is a graph class definable in the MS_1 language, then the collection of all those labeling parse trees which generate the members of $\mathcal{D}_t = \mathcal{D} \cap \mathcal{R}_t$ is accepted by a finite tree automaton.*

To prove this statement, in view of Theorem 3.2, it is enough to prove that the associated canonical equivalence $\approx_{\mathcal{D},t}$ is of finite index. However, the latter claim needs a generalization in order to use mathematical induction on the structure of an MS_1 sentence ϕ describing $\mathcal{D}$. This generalization of $\approx_{\mathcal{D},t}$ to $\approx_{\phi,t}^{\circ}$ lies in allowing formulas ϕ with free variables.

Extended canonical equivalence of ϕ. Let $Free(\phi) = Fr(\phi) \cup FR(\phi)$ be the partition of the free variables into those $Fr = Fr(\phi)$ for vertices and those $FR = FR(\phi)$ for vertex sets. We define a *partial equipment signature* of ϕ as a triple $\sigma = (Fr, FR, q)$ where $q : Fr \rightarrow \{0,1\}$. A t-labeled graph G is σ-*partially equipped* if it has distinguished vertices and vertex sets assigned as interpretations of the free variables in σ. Formally, for each $X \in FR$ there is a distinguished subset $S_X \subseteq V(G)$, and for each $x \in Fr$ such that $q(x) = 0$ there is a distinguished vertex $v_x \in V(G)$. Nothing is assigned to variables $x \in Fr$ such that $q(x) = 1$. For σ we define a *complemented* partial equipment signature $\sigma^- = (Fr, FR, q')$ where $q'(x) = 1 - q(x)$ for all $x \in Fr$.

See that if $\bar{H}_1$ is σ-partially equipped and $\bar{H}_2$ is σ^--partially equipped, then $H = \bar{H}_1 \oplus [g] \bar{H}_2$ has a full and consistent interpretation for all the free variables of ϕ (hence this H is a logic model of ϕ). So, we can define equivalence $\approx_{\phi,t}^{\sigma}$ over all t-labeled σ-partially equipped graphs as follows: $\bar{G}_1 \approx_{\phi,t}^{\sigma} \bar{G}_2$ if and only if the following

$$(\bar{G}_1 \oplus [id] \; \bar{H}) \models \phi \iff (\bar{G}_2 \oplus [id] \; \bar{H}) \models \phi$$

holds for all t-labeled σ^--partially equipped graphs $\bar{H}$.

Here we have extended the meaning of $\approx_{\phi,t}^{\sigma}$ in two directions. Firstly, by allowing free variables in ϕ we enlarge the studied universe to partially equipped graphs. Secondly, the universe is further enlarged by allowing all t-labeled underlying graphs – not only those from $\mathcal{R}_t$. Even in this stronger variant we can prove the following key statement which also concludes above Theorem 4.1:

Theorem 4.2. *Let $t \geq 1$ be fixed. Suppose ϕ is a formula in the language MS_1, and σ is a partial equipment signature for ϕ. Then $\approx_{\phi,t}^{\sigma}$ has finite index on the universe of t-labeled σ-partially equipped graphs.*

Proof. We retain the notation introduced above. The induction base is to prove the statement for the atomic formulas in MS_1: $\phi \equiv (v \in W)$, $(v = w)$, $mod_{p,q}(W)$, or $edge(u,v)$. The first three are all rather trivial cases which we skip here, and we focus on the last predicate $edge(u,v)$ (since this one actually "defines" the graph we study).

(4.3) Suppose $\phi \equiv edge(u,v)$. Then the index of $\approx^{\sigma}_{\phi,t}$ is one if $q(u) = q(v) = 1$, two if $q(u) = q(v) = 0$, and 2^t if $q(u) = 0$ and $q(v) = 1$ or vice versa.

In the first case both vertices u,v with a possible edge uv are interpreted in the right-hand graph $\bar{H}$, and hence no matter what $\bar{G}_1$ or $\bar{G}_2$ are, they become equivalent in $\approx^{\sigma}_{\phi,t}$. In the second case both vertices u,v are interpreted in the left-hand graphs $\bar{G}_i$, and hence there are exactly two classes formed by those graphs having and those not having u adjacent to v. It is the third case which interests us: Recalling the definition of our summation operator $\oplus[id]$, we see that all information needed to decide whether some u in the left-hand graph is adjacent to a specific v in the right-hand graph is encoded in the labeling of u, and hence the 2^t possibilities there.

For the inductive step, we consider that a formula ϕ is created from shorter formula(s) in one of the following ways: $\phi \equiv \neg\psi$, $\psi \wedge \eta$, $\exists v\, \psi(v)$, or $\exists W\, \psi(W)$, where $v \in Fr(\psi)$ or $W \in FR(\psi)$ in the latter cases. One may easily express the $\vee$ or $\forall$ symbols using these. The arguments we are going to give in the rest of this proof are not novel, but similar to those used in [1] and merely a translation of the arguments used in [14, Lemma 6.2].

We assume by induction that $\approx^{\pi}_{\psi,t}$ ($\approx^{\rho}_{\eta,t}$) has finite index, where the signature π (ρ) is inherited from σ for ψ (for η, see below the case-by-case details). The first case of $\phi \equiv \neg\psi$ is quite easy to resolve — the equivalence $\approx^{\pi}_{\psi,t}$ is the same as $\approx^{\sigma}_{\phi,t}$. We look at the second case.

(4.4) Suppose $\phi \equiv \psi \wedge \eta$, and let π, ρ denote the restrictions of signature σ to $Free(\psi)$, $Free(\eta)$, respectively. If $\approx^{\pi}_{\psi,t}$ has index p and $\approx^{\rho}_{\eta,t}$ has index r, then $\approx^{\sigma}_{\phi,t}$ has index at most $p \cdot r$.

Consider an arbitrary pair of t-labeled σ-partially equipped graphs $\bar{G}_1 \not\approx^{\sigma}_{\phi,t} \bar{G}_2$, and an associated σ^--partially equipped graph $\bar{H}$ such that $(\bar{G}_1 \oplus[id]\, \bar{H}) \models \phi$ but $(\bar{G}_2 \oplus[id]\, \bar{H}) \not\models \phi$. Then it has to be $(\bar{G}_1 \oplus[id]\, \bar{H}) \models \psi$ (or $\models \eta$) but $(\bar{G}_2 \oplus[id]\, \bar{H}) \not\models \psi$ (or $\not\models \eta$, resp.). Hence it immediately holds that $\bar{G}_1 \not\approx^{\pi}_{\psi,t} \bar{G}_2$ or $\bar{G}_1 \not\approx^{\rho}_{\eta,t} \bar{G}_2$ with the restricted equipments, and so the equivalence classes of $\approx^{\sigma}_{\phi,t}$ are suitable unions of the classes of the "intersection" $\approx^{\pi}_{\psi,t} \cap \approx^{\rho}_{\eta,t}$.

The third case of $\exists v\, \psi(v)$ is technically more complicated, and so we first deal with the similar but easier fourth case of $\exists W\, \psi(W)$.

(4.5) Suppose $\phi \equiv \exists W\, \psi(W)$, and let the signature $\pi = (Fr, FR \cup \{W\}, q)$. If $\approx^{\pi}_{\psi,t}$ has index p, then $\approx^{\sigma}_{\phi,t}$ has index at most $2^p - 1$.

Again consider an arbitrary pair of t-labeled σ-partially equipped graphs $\bar{G}_1 \not\approx^{\sigma}_{\phi,t} \bar{G}_2$, and $\bar{H}$ such that $(\bar{G}_1 \oplus[id]\, \bar{H}) \models \phi$ but $(\bar{G}_2 \oplus[id]\, \bar{H}) \not\models \phi$. We shortly write $\bar{G}[W = S]$ for the π-partially equipped graph obtained from σ-partially equipped $\bar{G}$ by interpreting the variable W as $S \subseteq V(\bar{G})$. Then our assumption about $\bar{G}_1, \bar{G}_2$ means there exist $S_W \subseteq V(\bar{G}_1)$ and $S'_W \subseteq V(\bar{H})$ such that $(\bar{G}_1[W = S_W] \oplus [id] \ \bar{H}[W = S'_W]) \models \psi$, whilst $(\bar{G}_2[W = T_W] \oplus [id] \ \bar{H}[W = S'_W]) \not\models \psi$ for all $T_W \subseteq V(\bar{G}_2)$. Hence $\bar{G}_1[W = S_W] \not\approx^{\pi}_{\psi,t} \bar{G}_2[W = T_W]$.

We now, in search for a contradiction, look at the problem from the other side. Let the equivalence classes of $\approx^{\pi}_{\psi,t}$ over t-labeled π-partially equipped graphs be $\mathcal{C}^1, \mathcal{C}^2, \ldots, \mathcal{C}^p$. For a σ-partially equipped graph $\bar{G}$ we define a nonempty set $Ix(\bar{G}) \subseteq \{1, 2, \ldots, p\}$ as follows: $i \in Ix(\bar{G})$ if and only if $\bar{G}[W = S] \in \mathcal{C}^i$ for some $S \subseteq V(\bar{G})$. If there were 2^p pairwise incomparable σ-partially equipped graphs in the relation $\approx^{\sigma}_{\phi,t}$, then some two of them, say $\bar{G}_1 \not\approx^{\sigma}_{\phi,t} \bar{G}_2$, would receive $Ix(\bar{G}_1) = Ix(\bar{G}_2)$ by the pigeon-hole principle. However, from the argument of the previous paragraph — $\bar{G}_1[W = S_W] \not\approx^{\pi}_{\psi,t} \bar{G}_2[W = T_W]$ for some $S_W \subseteq V(\bar{G}_1)$ and all $T_W \subseteq V(\bar{G}_2)$, we conclude that $j \in Ix(\bar{G}_1) \setminus Ix(\bar{G}_2)$ where j is such that $\bar{G}_1[W = S_W] \in \mathcal{C}^j$. This contradiction proves (4.5).

(4.6) Suppose $\phi \equiv \exists v\, \psi(v)$, and let signatures $\pi = (Fr \cup \{v\}, FR, q_1)$ and $\rho = (Fr \cup \{v\}, FR, q_2)$ where $q_1(v) = 0$ and $q_2(v) = 1$. If $\approx^{\pi}_{\psi,t}$ has index p and $\approx^{\rho}_{\psi,t}$ has index r, then $\approx^{\sigma}_{\phi,t}$ has index at most $2^p \cdot r + 1 - r$.

Notice that a ρ-partial equipment of $\bar{G}$ does not interpret the variable v in $V(G)$, and so σ-partially equipped graph $\bar{G}$ may be viewed also as ρ-partially equipped. Take an arbitrary pair of nonempty t-labeled σ-partially equipped graphs $\bar{G}_1 \not\approx^{\sigma}_{\phi,t} \bar{G}_2$, and $\bar{H}$ such that $(\bar{G}_1 \oplus [id] \ \bar{H}) \models \phi$ but $(\bar{G}_2 \oplus [id] \ \bar{H}) \not\models \phi$. Let $x_v \in V(\bar{G}_1) \cup V(\bar{H})$ be an interpretation of the variable v that satisfies ψ over $\bar{G}_1 \oplus [id] \ \bar{H}$. In particular, ψ is false over $\bar{G}_2 \oplus [id] \ \bar{H}$ here. If $x_v \in V(\bar{H})$, then immediately $\bar{G}_1 \not\approx^{\rho}_{\psi,t} \bar{G}_2$. Otherwise, $x_v \in V(\bar{G}_1)$ and we are in a situation analogous to the first paragraph of (4.5): $(\bar{G}_1[v = x_v] \oplus [id] \ \bar{H}) \models \psi$, whilst $(\bar{G}_2[v = y_v] \oplus [id] \ \bar{H}) \not\models \psi$ for all $y_v \in V(\bar{G}_2)$.

Again, in search for a contradiction, we look at the problem from the other side. If there are $2^p r + 2 - r$ pairwise incomparable σ-partially equipped graphs with respect to $\approx^{\sigma}_{\phi,t}$, then at least $2^p r + 1 - r = (2^p - 1)r + 1$ of those graphs are nonempty, and out of them at least 2^p belong to the same equivalence class of $\approx^{\rho}_{\psi,t}$. Let their set be denoted by $\mathcal{G}$ (Hence for each pair in $\mathcal{G}$, the latter conclusion of the previous paragraph applies). Considering the equivalence classes $\mathcal{C}^1, \mathcal{C}^2, \ldots, \mathcal{C}^p$ of $\approx^{\pi}_{\psi,t}$, we again (as in 4.5) define a nonempty set $Ix(\bar{G}) \subseteq \{1, 2, \ldots, p\}$, for σ-partially equipped $\bar{G}$, by $i \in Ix(\bar{G})$ if and only if $\bar{G}[v = y] \in \mathcal{C}^i$ for some $y \in V(\bar{G})$. Then some pair, say $\bar{G}_1, \bar{G}_2 \in \mathcal{G}$, must satisfy $Ix(\bar{G}_1) = Ix(\bar{G}_2)$ by the pigeon-hole principle. However, that analogously contradicts the latter conclusion of the previous paragraph.

This contradiction proves (4.6), and thus the whole theorem. ∎

5 Concluding Notes

As already mentioned in the introduction, the driving force of our research is to provide a framework for easy design of efficient parameterized algorithms running on a bounded-width rank-decomposition of a graph. In this sense we have provided two directly applicable results in Theorems 2.2 and 4.1. Unfortunately, applicability of Theorem 4.1 is limited to pure decision problems (like 3-colourability), but many practical problems are formulated as optimization ones. (The usual way of transforming optimization problems into decision ones does not work here since MS_1 logic cannot handle arbitrary numbers.)

Nevertheless, there is a known solution. Arnborg, Lagergren, and Seese [2] (while studying graphs of bounded tree-width), and later Courcelle, Makowsky, and Rotics [6] (for graphs of bounded clique-width), specifically extended the expressive power of MSO logic to define so-called *LinEMSO* optimization problems, and consequently shown existence of linear-time algorithms in the respective cases. Briefly saying, *LinEMSO* problems allow, in addition to ordinary MSO expressions, to optimize over and compare between linear evaluation terms.

We can achieve an analogous solution in our framework directly using Theorem 4.2. The basic idea is that, in a dynamic processing of the input parse tree, we can keep track only of suitable "optimal" representatives of the possible interpretations of the free variables in ϕ, per each class of the extended canonical equivalence $\approx^{\sigma}_{\phi,t}$. We illustrate this idea with the next simple example.

Dominating set example. This problem asks for a subset $X \subseteq V(G)$ of the least cardinality such that each vertex not in X is adjacent to some in X. Since it is not a decision question, we cannot hope in a direct application of Theorem 4.1. We, however, can write in MS_1

$$\delta(X) \equiv \forall v \exists w \left[v \in X \vee \left(w \in X \wedge edge(v, w) \right) \right]$$

stating that X is a dominating set in G. And now Theorem 4.2 can be applied.

Let G be a graph of rank-width t, and T its labeling parse tree. We denote by T_x the subtree below a node x of T, and by G_x the subgraph of G parsed by T_x. For any $D \subseteq V(G_x)$, the t-labeled partially equipped graph G_x with interpretation $X = D$ falls into one of the finitely many classes of $\approx^{\sigma}_{\delta,t}$ (where $\sigma = (\emptyset, \{X\}, \emptyset)$). A dynamic algorithm for the dominating set problem has to remember just one representative interpretation $X = D_i$ of the least cardinality from the i-th class of $\approx^{\sigma}_{\delta,t}$, and with knowledge of the associated tree automaton (Theorem 3.2) this information can easily be processed from leaves of T to the root in total linear time.

Non-FPT algorithms for bounded widths. Lastly we note the following interesting phenomenon: for some problems on graphs of bounded width parameters, there are known algorithms which run faster than in general case, but they are not fixed parameter tractable. Among those we mention a (pseudo)polynomial algorithm for the chromatic number of graphs of bounded clique-width [16], or a subexponential algorithm for the Tutte polynomial of graphs of bounded clique-width [13]. Finite automata clearly cannot be applied there. Though, it would be interesting to extend the framework of Theorem 4.2 to also cover the aforementioned situation.

References

1. Abrahamson, K.A. and Fellows, M.R.: Finite Automata, Bounded Treewidth, and Well-Quasiordering. In: Graph Structure Theory, Contemporary Mathematics 147. American Mathematical Society (1993) 539–564
2. Arnborg, S., Lagergren, J., and Seese, D.: Problems easy for Tree-decomposible Graphs. Proc. 15th Colloq. Automata, Languages and Programming. Volume 317 of Lecture Notes in Comput. Sci. Springer, Berlin (1988) 38–51
3. Bandelt, H.-J. and Mulder, H.M.: Distance-hereditary graphs. J. Combin. Theory Ser. B **41**(2) (1986) 182–208
4. Corneil, D.G., Rotics, U.: On the relationship between cliquewidth and treewidth. SIAM J. Comput. **34**(4) (2005) 825–847
5. Courcelle, B. and Kanté, M.M.: Graph Operations Characterizing Rank-Width and Balanced Graph Expressions. In: Graph-theoretic concepts in computer science, Volume 4769 of Lecture Notes in Comput. Sci., Berlin, Springer (2007) 66–75
6. Courcelle, B., Makowsky, J.A., and Rotics, U.: Linear time solvable optimization problems on graphs of bounded clique-width. Theory Comput. Syst. **33**(2) (2000) 125–150
7. Courcelle, B. and Olariu, S.: Upper bounds to the clique width of graphs. Discrete Appl. Math. **101**(1-3) (2000) 77–114
8. Courcelle, B. and Oum, S.: Vertex-minors, monadic second-order logic, and a conjecture by Seese. J. Combin. Theory Ser. B **97**(1) (2007) 91–126
9. Downey, R.G. and Fellows, M.R.: *Parameterized complexity*. Monographs in Computer Science. Springer-Verlag, New York, 1999.
10. Espelage, W., Gurski, F., and Wanke, E.: How to solve NP-hard graph problems on clique-width bounded graphs in polynomial time. In: Graph-theoretic concepts in computer science. Volume 2204 of Lecture Notes in Comput. Sci., Berlin, Springer (2001) 117–128
11. Ganian, R.: Automata formalization for graphs of bounded rank-width. Master thesis. Faculty of Informatics of the Masaryk University, Brno, Czech republic (2008)
12. Gerber, M.U. and Kobler, D.: Algorithms for vertex-partitioning problems on graphs with fixed clique-width. Theoret. Comput. Sci. **299**(1-3) (2003) 719–734
13. Gimenez, O., Hliněný, P. and Noy, M.: Computing the Tutte Polynomial on Graphs of Bounded Clique-Width. In: Graph-theoretic concepts in computer science. Volume 3787 of Lecture Notes in Comput. Sci., Berlin, Springer (2005) 59–68
14. Hliněný, P.: Branch-width, parse trees, and monadic second-order logic for matroids. J. Combin. Theory Ser. B **96**(3) (2006) 325–351
15. Hliněný, P. and Oum, S.: Finding Branch-decomposition and Rank-decomposition. SIAM J. Comput. (2008) to appear
16. Kobler, D. and Rotics, U.: Edge dominating set and colorings on graphs with fixed clique-width. Discrete Appl. Math. **126**(2-3) (2003) 197–221
17. Oum, S.: Rank-width and vertex-minors. J. Combin. Theory Ser. B **95**(1) (2005) 79–100
18. Oum, S. and Seymour, P.: Approximating clique-width and branch-width. J. Combin. Theory Ser. B **96**(4) (2006) 514–528
19. Robertson, N. and Seymour, P.: Graph minors. X. Obstructions to tree-decomposition. J. Combin. Theory Ser. B **52**(2) (1991) 153–190

Notes on generating completely condensed d-neighborhoods

Heikki Hyyrö

Department of Computer Sciences, University of Tampere, Finland.
heikki.hyyro@cs.uta.fi

Abstract. Let $ed(A, B)$ denote the edit distance between string A and B. In this paper we discuss the problem of generating the completely condensed d-beighborhood $CCU_d(P)$, called super condensed d-neigborhood in [4], for a given pattern string P. The set $CCU_d(P)$ consists of those strings A for which $ed(P, A) \leq d$ and no proper substring $A_{i..j}$ of A, for which $A_{i..j} \neq A$, also belongs to $CCU_d(P)$. $CCU_d(P)$ provides a minimal characterization of all strings that are within edit distance d from the pattern P in the sense that if $ed(P, A) \leq d$ for some string A, then $CCU_d(P)$ contains at least one substring of A. Russo and Oliveira [4] discussed how to generate the set $CCU_d(P)$ in $\mathcal{O}(\sigma|P|^2|CCU_d(P)|)$ or $\mathcal{O}(|P||CCU_d(P)|)$ time, where σ is the alphabet size, $|P|$ is the length of P, $|CCU_d(P)|$ is the size of the set $CCU_d(P)$, and the latter complexity corresponds to an optimization, originally due to Myers [3], that uses the KMP failure links [1]. In addition, Russo and Oliveira gave bit-parallel algorithms for generating $CCU_d(P)$. The use of bit-parallelism actually worsened the asymptotic running times, but the methods were still found to be very fast in practice. In this paper we first note that $CCU_d(P)$ can be generated in $\mathcal{O}(\sigma d|P||CCU_d(P)|)$ or $\mathcal{O}(d|CCU_d(P)|)$ time, depending on whether the KMP links are used or not. Building on this, we propose more efficient variants of the bit-parallel method of Russo and Oliveira. Finally, we also present experimental results that show that our algorithms achieve considerable speedups in practice.

Keywords: String algorithms, edit distance, approximate string matching

1 Introduction

We will use the following notation and conventions. Strings consist of a finite sequence of characters from a finite alphabet Σ with alphabet size σ. For a string A, $|A|$ is its length, A_i is its ith character, and $A_{i..j}$ is its substring that spans the characters $A_i \ldots A_j$ for $i \leq j$. Characters are numbered starting from 1, so $A = A_{1..|A|}$. The Levenshtein edit distance [2] between strings A and B is denoted by $ed(A, B)$, and is defined as the minimum number of single-character insertions, deletions and/or substitutions needed in order to transform A into B or vice versa.

Given a pattern string P and an error limit d, the set $U_d(P) = \{A \mid ed(P, A) \leq d\}$ is the d-neighborhood of P. It contains all strings A for which $ed(P, A) \leq d$ and is used e.g. in indexed approximate string matching [3]. Constructing the set $U_d(P)$ transforms the problem of finding strings A for which $ed(P, A) \leq d$ into the simpler and widely supported task of locating exact string matches within a text database.

Myers [3] defined the condensed d-neighborhood of P as the set $CU_d(P) = \{A \mid ed(P, A) \leq d \text{ and } (ed(P, A_{1..j}) > d \text{ when } j < |A|)\}$. If $A \in U_d(P)$ and $A_{1..j} \in U_d(P)$, where $j < |A|$, the set $CU_d(P)$ includes only $A_{1..j}$, as string A is represented by it in the sense that each occurrence of A contains its prefix $A_{1..j}$.

Russo and Oliveira [4] defined a minimal d-neighborhood, which they called "super condensed neighborhood". We find the name "completely condensed neighborhood" to be more descriptive. It is defined as $CCU_d(P) = \{A \mid ed(P, A) \leq d \text{ and } (ed(P, A_{h..j}) > d \text{ when } 1 \leq h \leq j \leq |A| \text{ and } A_{h..j} \neq A)\}$. If both $A \in U_d(P)$ and $A_{h..j} \in U_d(P)$, where $h \geq 2$ or $j < |A|$, then only $A_{h..j}$ is included in $CCU_d(P)$ as each occurrence of A is represented by its substring $A_{h..j}$.

2   Dynamic programming

The basic solution for computing $ed(A, B)$ is to fill an $(|A| + 1) \times (|B| + 1)$ dynamic programming matrix D in which eventually $D[i, j] = ed(A_{1..i}, B_{1..j})$ for $i = 0 \ldots |A|$ and $j = 0 \ldots |B|$. The computation takes $\mathcal{O}(|A||B|)$ time using the well-known Recurrence 1.

Recurrence 1.

When $0 \leq i \leq |A|$ and $0 \leq j \leq |B|$:
$$D[i, j] = \begin{cases} j, \text{ if } i = 0 \\ i, \text{ if } j = 0, \text{ and otherwise} \\ \delta(i, j) + \min\{D[i - 1, j], D[i, j - 1], D[i - 1, j - 1]\}, \end{cases}$$
where $\delta(i, j) = 0$, if $A_i = B_j$, and $\delta(i, j) = 1$, if $A_i \neq B_j$.

Approximate string matching is a problem that is closely related to edit distance computation. Given a pattern string P, a text string T, and an error limit k, the task is to find those indices j where $ed(P, T_{h..j}) \leq k$ for some $h \leq j$. This problem can be solved by using Recurrence 1 with the modification that $D[0, j] = 0$ instead of j. Let us use D' to denote a matrix D that has been filled for P and T using this modified Recurrence 1. Then the relationship $D'[i, j] = \min\{ed(P_{1..i}, T_{h..j}) \mid 1 \leq h \leq j\}$ holds.

The matrix D (and D') have the well-known properties stated in Lemma 1.

Lemma 1. *Let D be a dynamic programming matrix that contains the values $D[i, j] = ed(A_{1..i}, B_{1..j})$ for $0 \leq i \leq |A|$ and $0 \leq j \leq |B|$. Then the following three properties hold for $1 \leq i \leq |A|$ and $1 \leq j \leq |B|$:*

1. $D[i, j] - D[i - 1, j] \in \{-1, 0, 1\}$
2. $D[i, j] - D[i, j - 1] \in \{-1, 0, 1\}$
3. $D[i, j] - D[i - 1, j - 1] \in \{0, 1\}$

```
DFSstep(j)
  1.    ComputeColumn(j)
  2.    If IsMatchColumn(j) = true Then
  3.        AddToNeighborhood(j)
  4.    Else If ExtendsToMatchColumn(j) = true Then
  5.        For a ∈ Σ Do
  6.            A_{j+1} ← a
  7.            DFSstep(j + 1)

Generate-d-Neighborhood(P, d)
  1.    DFS-Step(0)
```

Fig. 1. The high-level procedure for generating a condensed d-neighborhood. The pseudocode assumes that P, d and all other relevant data structures are static and available to all subprocedures without need to pass them explicitly as parameters.

3 The methods of Myers for generating $CU_d(P)$

3.1 The basic method

The scheme of Myers [3] for generating $CU_d(P)$ essentially performs a depth-first search (DFS) in a trie that contains all strings in the string universe Σ^*, ie. each node in the trie has a child for each character $a \in \Sigma$. The high-level process is shown in Fig. 1. The trie does not need to be stored explicitly. Throughout the process, the string $A_{1..j}$ corresponds to the path from the root to the current node in the trie. The subprocedures are as follows in the basic case:

ComputeColumn(j) assumes that the values $D[i, j-1] = ed(P_{1..i}, A_{1..j-1})$ in column $j-1$ are known, and computes the values $D[i, j] = ed(P_{1..i}, A_{1..j})$ for column j.

IsMatchColumn(j) assumes that the values $D[i, j] = ed(P_{1..i}, A_{1..j})$ are available. The procedure returns true if $D[|P|, j] = ed(P, A_{1..j}) \leq d$, which results in calling AddToNeighborhood(j).

AddToNeighborhood(j) adds $A_{1..j}$ to $CU_d(P)$. Note that the DFS backtracks after this, which is correct since $CU_d(P)$ does not contain any string with $A_{1..j} \in CU_d(P)$ as its proper prefix.

ExtendsToMatchColumn(j) assumes that the values $D[i, j] = ed(P_{1..i}, A_{1..j})$ are available. The procedure inspects if $D[i, j] \leq d$ for some $0 \leq i < |P|$. If such an i exists, then the value true is returned as $A_{1..j}$ may be extended into a string that belongs to $CU_d(P)$. Conversely, if $D[i, j] > d$ for all i, then the properties of Lemma 1 guarantee that $D[i, j'] > d$ for all i and $j' > j$, and hence the value false is returned.

Processing a node in the trie corresponds to processing a column j in D. By following the analysis of Myers [3], it can be shown that this basic algorithm processes $\mathcal{O}(\sigma|P||CU_d(P)|)$ nodes, and thus takes overall $\mathcal{O}(X\sigma|P||CU_d(P)|)$ time to generate $CU_d(P)$, where X is the time for processing a node (ie. a column j). If basic dynamic programming is used, $X = \mathcal{O}(|P|)$ and the complexity is $\mathcal{O}(\sigma|P|^2|CU_d(P)|)$.

3.2 Improvement by using the KMP links

One consequence of the property 3 in Lemma 1 is that if $D[i,j] = d$ for some $i < |P|$ in column j, then $D[|P|, j + |P| - i] \leq d$ if and only if $P_{i+h} = A_{j+h}$ for $h = 1 \ldots |P| - i$. Therefore if all values in column j are $\geq d$, one can immediately determine all strings that have $A_{1..j}$ as a prefix and belong to $CU_d(P)$: Each string $B \in CU_d(P)$, where $|B| > j$ and $A_{1..j} = B_{1..j}$, must be of form $A_{1..j}P_{i+1..|P|}$, where $D[i,j] = d$ and the notation means concatenation of $A_{1..j}$ and $P_{i+1..|P|}$. Clearly each such $A_{1..j}P_{i+1..|P|}$ belongs to $U_d(P)$, but not necessarily to $CU_d(P)$: If $D[i,j] = D[i',j] = d$ for $i < i'$ and $P_{i'+1..|P|} = P_{i+1..|P|-i'+i}$, then $A_{1..j}P_{i'+1..|P|}$ is a prefix of $A_{1..j}P_{i+1..|P|}$. Myers [3] showed how to solve this problem with the failure links of the Knuth-Morris-Pratt string matching algorithm [1], which give information about repeating substrings within P. We refer to [3] for further details, but mention that the process requires $\mathcal{O}(|P|)$ time and space for preprocessing the failure links, and that the overall time for checking a contiquous interval of h values $D[i,j]$ in column j in this manner is $\mathcal{O}(h)$, ie. checking $D[i,j]$ for $i = 1 \ldots |P|$ takes $\mathcal{O}(|P|)$ time.

The algorithm of Fig. 1 can be modified to use this improved scheme by replacing some of the procedures with the following versions:

IsMatchColumn2(j) assumes that the values $D[i,j] = ed(P_{1..i}, A_{1..j})$ are available. The procedure returns true if $D[|P|, j] = ed(P, A_{1..j}) \leq d$ or if $D[i,j] \geq d$ for $i = 0 \ldots |P|$, which results in calling AddToNeighborhood2(j).

AddToNeighborhood2(j) adds $A_{1..j}$ to $CU_d(P)$ if $D[|P|, j] \leq d$. Otherwise, the procedure adds to $CU_d(P)$ each string $A_{1..j}P_{i+1..|P|}$, for which $D[i,j] = d$ and it is determined (with the help of the KMP failure links) that there exists no prefix $A_{1..j}P_{i'+1..|P|}$ of $A_{1..j}P_{i+1..|P|}$ for which $D[i',j] = d$.

ExtendsToMatchColumn2(j) assumes that the values $D[i,j] = ed(P_{1..i}, A_{1..j})$ are available. The procedure inspects if $D[i,j] < d$ for some $0 \leq i < |P|$. If such an i exists, then the value true is returned as $A_{1..j}$ may be extended, but not immediately, into one or more strings that belong to $CU_d(P)$.

Myers [3] showed that the improved algorithm processes only $\mathcal{O}(|CU_d(P)|)$ nodes. The overall time complexity is then $\mathcal{O}(|P| + X|CU_d(P)|)$, where X is the time for computing and processing the values $D[i,j]$ in column j. With basic dynamic programming, $X = \mathcal{O}(|P|)$ and the complexity is $\mathcal{O}(|P||CU_d(P)|)$.

3.3 Restricting the computation of $ed(A, B)$

Myers [3] noted that the property 3 in Lemma 1 also allows one to limit the number of cells processed in D, in similar fashion to a single stage in the $\mathcal{O}(dm)$

edit distance computation algorithm of Ukkonen [5]. Since initially $D[i,0] = i$ and $D[0,j] = j$, the conditions $D[d+1+h,h] \geq D[d+1,0] > d$ and $D[h,d+1+h] \geq D[0,d+1] > d$ hold for $h \geq 0$ (within the boundaries of D). The computation may ignore cell values $D[i,j] > d$. Therefore in column j we need to compute $D[i,j]$ only for rows $i = \max\{0, j-d\} \ldots \min\{|P|, j+d\}$, ie. a region of $\mathcal{O}(d)$ cells. This reduces the work X of processing a node to $\mathcal{O}(d)$[1], and the time complexities of Myers' basic and improved method become $\mathcal{O}(\sigma d|P||CU_d(P)|)$ and $\mathcal{O}(|P| + d|CU_d(P)|)$.

4 The methods of Russo and Oliveira for generating $CCU_d(P)$

4.1 The basic scheme

Russo and Oliveira [4] noted that the completely condensed neighborhood $CCU_d(P)$ can be generated by ignoring all cell values $D[i,j] = ed(P_{1..j}, A_{1..j})$ for which $ed(P_{1..j}, A_{1..j}) \leq ed(P_{1..j}, A_{h..j})$ with some $h \geq 2$. Such h can be identified by comparing $D[i,j]$ and $D''[i,j] = \min\{ed(P_{1..i}, T_{h..j}) \mid 2 \leq h \leq j\}$. These latter values correspond to approximate matching between P and the text $A_{2..}$, and it is additionally assumed that $D''[i,j] = D[i,j] + 1$ if $j = 0$. Russo and Oliveira computed such $D''[i,j]$ with an automaton that essentially follows Recurrence 2, which is simple to derive from the approximate string matching variant of Recurrence 1.

Recurrence 2.

When $0 \leq i \leq |P|$ and $0 \leq j \leq |A|$:
$$D''[i,j] = \begin{cases} 0, \text{ if } i = 0 \text{ and } j > 0 \\ i+1, \text{ if } j = 0 \\ i, \text{ if } j = 1, \text{ and otherwise} \\ \delta(i,j) + \min\{D''[i-1,j], D''[i,j-1], D''[i-1,j-1]\}, \end{cases}$$
where $\delta(i,j) = 0$, if $P_i = A_j$, and $\delta(i,j) = 1$, if $P_i \neq A_j$.

The high-level scheme of Myers can now be modified to compute the completely condensed neighborhood $CCU_d(P)$ by computing the values $D[i,j]$ and $D''[i,j]$ in parallel and ignoring all values $D[i,j] \geq D''[i,j]$. The time complexity of generating $CCU_d(P)$ can be analysed in exactly the same manner as before, but now the complexity X of processing a node must include also computing the values $D''[i,j]$ for column j. Russo and Oliveira concluded that the complexities for generating $CCU_d(P)$ are $\mathcal{O}(\sigma|P|^2|CCU_d(P)|)$ and $\mathcal{O}(|P||CCU_d(P)|)$ without and with the KMP failure links, when dynamic programing is used[2].

---

[1] If the string extensions $P_{i+1..|P|}$ are appended to the end of $A_{1..j}$ in the form of a pointer to the ith position in P.

[2] In [4] the complexities of the basic method of Myers are given erroneously without the factor σ.

4.2   A practical bit-parallel method

Russo and Oliveira proposed two practical bit-parallel methods to speedup the computation of the values $D[i,j]$. In their tests, the most practical of these was based on bit-parallel simulation of a non-deterministic finite automaton (NFA) [6]. Let N denote this NFA. N contains $(d+1)\times(|P|+1)$ states that are organized into $d+1$ rows with $|P|+1$ states. The rows are numbered from 0 to d and the states on each row from 0 to $|P|$. After processing the string $A_{1..j}$, state i on row g of N is active if and only if $D[i,j] = ed(P_{1..i}, A_{1..j}) \leq g$, where $0 \leq g \leq d$. Hence N represents all interesting values $D[i,j] \leq d$.

In similar fashion, let N'' be a variant of N for which state i on row g is active if and only if $ed(P_{1..i}, A_{h..j}) \leq g$ for some $h \geq 2$. N'' represents all interesting values $D''[i,j] \leq d$.

Following the bit-parallel approximate string matching algorithm of Wu and Manber[6], Russo and Oliveira [4] represent N as $d+1$ length-$(|P|+1)$ bit-vectors $F_0,\ldots,F_d$. The $(i+1)$th bit of F_g is 1 if and only if state i on row g of N is active. N'' is represented as $d+1$ bit-vectors $S_0,\ldots,S_d$ in the same manner. In addition, a length-$(|P|+1)$ pattern match bit vector PM_a is pre-computed for each character $a \in \Sigma$. The $(i+1)$th bit of PM_a is 1 if and only if $P_i = a$. With this background, Russo and Oliveira implemented the procedures ComputeColumn(j), IsMatchColumn(j) and ExtendsToMatchColumn(j) of Myers' basic method as shown in Fig. 2. The notation is as follows: '&', '|', and '^' and '~' denote bitwise "and", "or", "xor", and "not", respectively, and '<<' and '>>' denote shifting the bit-vector left and right, using zero filling in both directions. With the bit-vectors, the superscript denotes the column for which that vector was computed (e.g. F_g^j denotes F_g for column j, which corresponds to processing $A_{1..j}$).

Now the work X for processing a node without the KMP links is $\mathcal{O}(d\lceil|P|/w\rceil)$, where w is the computer word size[3]. This is often $\mathcal{O}(d)$, as typically in practice $|P| \leq w$. For example Myers' indexed appoximate string matching algorithm [3] generates $CU_d(P)$ for P with length $|P| \approx \log_\sigma(n)$, where n is the size of the database. Already $\sigma = 4$ (e.g. DNA) and a 32-bit computer would suffice for database sizes up to $n = 31^4 \approx 4.6 \cdot 10^{18}$.

The overall time complexities for generating $CCU_d(P)$ become $\mathcal{O}(\sigma d\lceil|P|/w\rceil|P||CC$ and $\mathcal{O}((|P|+d\lceil|P|/w\rceil)|CCU_d(P)|)$ without and with the KMP failure links. In the latter case, the $O(|P|)$ term is for checking the possible extensions for each row $i = 1\ldots|P|-1$.

---

[3] A length-$(|P|+1)$ bit-vector spans $\lceil(|P|+1)/w\rceil$ computer words, and each basic operation on two such bit-vectors takes $\mathcal{O}(\lceil|P|/w\rceil)$ time.

```
ComputeColumnBP(j)
1.      $F_0^j \leftarrow (F_0^{j-1} << 1) \ \& \ PM_{A_j}$
2.      $S_0^j \leftarrow ((S_0^{j-1} << 1) \mid 1) \ \& \ PM_{A_j}$
3.      For $g \leftarrow 1 \ldots d$ Do
4.          $F_g^j \leftarrow ((F_g^{j-1} << 1) \ \& \ PM_{A_j}) \mid F_{g-1}^{j-1} \mid (F_{g-1}^{j-1} << 1) \mid (F_{g-1}^j << 1)$
5.          $S_g^j \leftarrow ((S_g^{j-1} << 1) \ \& \ PM_{A_j}) \mid S_{g-1}^{j-1} \mid (S_{g-1}^{j-1} << 1) \mid (S_{g-1}^j << 1)$

IsMatchColumnBP(j)
1.      If the $(|P|+1)$th bit of $F_d^j$ is 1 and the $(|P|+1)$th bit of $S_d^j$ is 0 Then
2.          return true
3.      Else
4.          return false

ExtendsToMatchColumnBP(j)
1.      $tmp \leftarrow (F_0^j \ \& \ \sim S_0^j) \mid \ldots \mid (F_d^j \ \& \ \sim S_d^j)$
2.      If $tmp \neq 0$ Then
3.          return true
4.      Else
5.          return false
```

Fig. 2. The bit-parallel scheme of Russo and Oliveira for generating $CCU_d(P)$.

5 Improved methods for generating $CCU_d(P)$

5.1 Restricting the computation

Because $D''[0, j] = 0$, Myers' restriction scheme discussed in Section 3.3 is not directly applicable for generating $CCU_d(P)$. We may, however, use the following observations.

Lemma 2. *If* $ed(P_{1..i}, A_{1..j}) \leq d$, *then* $i - d \leq j \leq i + d$.

Proof. This follows from the fact that, depending on whether $i > j$ or $i < j$, we need $i - j$ or $j - i$ operations to make $P_{1..i}$ and $A_{1..j}$ have equal length. □

Lemma 3. *In column* j, *the values* $D''[i, j]$ *need to be computed only for rows* $i = \max\{0, j - 2d\} \ldots \min\{|P|, j + d - 1\}$.

Proof. The limit $i \leq \min\{|P|, j + d - 1\}$ follows from the property 3 of Lemma 1 and the fact that $D''[1, i] = i$ for $i = 0 \ldots |P|$.

Let us define diagonal q of D as the upper-left to lower-right diagonal that contains the cells $D[i, i + q]$. Lemma 2 ensures that $|A| \leq |P| + d$ for each string $A \in U_d(P)$. This guarantees that the cells $D[i, j]$ are not computed further than to column $j = |P| + d$. The cell $D[|P|, |P| + d]$ is on diagonal d. Now consider a cell $D''[i, j]$ that becomes discarded because of the first limit, in which case $i < \max\{0, j - 2d\}$ and $j > 2d$. Such a cell resides on a diagonal $q = j - i > 2d$. Consider a path along the rules of Recurrence 1 from such cell $D[i, j]$ to any cell $D[|P|, j']$, where $j' \leq |P| + d$. The end of such a path resides on a diagonal $q' = j' - |P| \leq |P| + d - |P| = d$, and the path makes at least $q - q' \geq 2d + 1 - d = d + 1$ downward steps when moving from the diagonal q to the diagonal q'. Since each such step corresponds to an insertion or deletion, and thus adds 1 to the distance, the overall distance over the path is $\geq d + 1$. Thus such cells $D[i, j]$ are redundant, as they can never contribute to finding a matching shorter substring $A_{h..j}$ where $j \leq |P| + d$. $\qquad\square$

The relevant region given in Lemma 3 contains at most $j + d - 1 - (j - 2d) + 1 = 3d$ rows in column j. Combining this with the restriction scheme of Section 3.3 enables us to compute and process both the relevant $D[i, j]$ and the relevant $D''[i, j]$ for column j in $\mathcal{O}(d)$ time. This results in $\mathcal{O}(\sigma d|P|\,||CCU_d(P)|)$ and $\mathcal{O}(|P| + d|CCU_d(P)|)$ time complexity for generating $CCU_d(P)$, without and with using the KMP failure links, respectively, when the values are computed using dynamic programming.

5.2 More efficient bit-parallel variants

We first note that the procedures IsMatchColumnBP and ExtendsToMatchColumnBP of Russo and Oliveira [4], shown in Fig. 2, are unnecessarily stringent.

The procedure IsMatchColumnBP does not need to check whether the $(|P| + 1)$th bit of S_d^j is 0, ie. whether $D''[|P|, j] \leq d$. This is because it can be shown that if $D[|P|, j] \leq q$ and $D''[|P|, j] \leq q$, then the high-level DFS would have backtracked already before column j.

The original procedure IsMatchColumnBP checks for all $g = 0 \ldots d$ whether there exists some i for which $D[i, j] = g$ and $D''[i, j] > g$. The purpose of this is to check if there is some cell $D[i, j] < D''[i, j]$ in column j. This condition can be verified by only checking the case $g = d$. It can be shown that if $D[i, j] = g$ and $D''[i, j] > g$ for some $g < d$, then there either is such a row i for also $g = d$, or the DFS would again have backtracked before column j.

A further parallelization for small $|P|$ or d

When looking at the procedure ComputeColumnBP in Fig. 2, it is apparent that the vectors F_g^j and S_g^j are computed in almost identical manner. The only difference is on line 2, where the vector S_0^{j-1} has its first bit set to 1 after shifting left. This similarity leads to the idea of combining the vectors. We will do this in a striped manner. We define a combined vector C_g^j to be a bit-vector of length $2|P| + 2$ whose odd bits represent F_g^j and even bits S_g^j. That is, the $(2i - 1)$th bit of C_g^j holds the value of the ith bit of F_g^j, and the $(2i)$th bit of C_g^j holds the value of the ith bit of S_g^j. To reflect this configuration, we define a corresponding match vector $\overline{PM}_a$ to have its $(2i + 1)$th and $(2i + 2)$th bits set to 1 if $P_i = a$. With this configuration, and taking into account the previous notes about the possibility to streamline the procedures IsMatchColumnBP and ExtendsToMatchColumnBP, the procedures

ComputeColumnCBP, IsMatchColumnCBP and ExtendsToMatchColumnCBP shown in Fig. 3 are straight-forward to verify to be correct. The last procedure uses a mask bit-vector $FMASK$ whose odd bits are 1 and even bits are 0, and it is used to extract set bits from only those positions that correspond to F_g^j.

Using the combined vectors C_g^j may improve the performance up to a constant factor. The scheme is useful mostly in the case where $2|P|+2 \leq w$, that is, when each vector still fits into a single computer word. In this case the amount of work is almost halved. With a 32-bit architecture, the limit is then $|P| \leq 15$. With a 64-bit architecture, which is already quite common, the limit is $|P| \leq 31$. As noted previously, this latter limit is not very strict at least in the context of Myers' indexed approximate string matching method [3].

ComputeColumnCBP(j)
1. $C_0^j \leftarrow ((C_0^{j-1} << 2) \mid 1) \ \& \ \overline{PM}_{A_j}$
2. **For** $g \leftarrow 1 \ldots d$ **Do**
3. $C_g^j \leftarrow ((C_g^{j-1} << 2) \ \& \ \overline{PM}_{A_j}) \mid C_{g-1}^{j-1} \mid (C_{g-1}^{j-1} << 2) \mid (C_{g-1}^{j} << 2)$

IsMatchColumnCBP(j)
1. **If** the $(2|P| + 2)$th bit of C_d^j is 1 **Then**
2.   return true
3. **Else**
4.   return false

ExtendsToMatchColumnCBP(j)
1. $tmp \leftarrow (C_d^j \ \& \ (\sim C_d^j >> 1)) \ \& \ FMASK$
2. **If** $tmp \neq 0$ **Then**
3.   return true
4. **Else**
5.   return false

Fig. 3. A bit-parallel scheme that combines F_g^j and S_g^j into a single vector C_g^j.

The next step is to apply the restriction scheme to the bit-parallel computation. In column j, we need to compute the rows $i = \max\{0, j-d\} \ldots \min\{|P|, j+d\}$ of D and the rows $i = \max\{0, j - 2d\} \ldots \min\{|P|, j+d-1\}$ of D''. We define R_g^j as a restricted combined bit-vector with length $6d$ and again use the striped setup. The $(2i)$th bit of R_g^j corresponds to the $(j + i - 2d)$th bit of F_g^j and the $(2i - 1)$th bit of R_g^j corresponds to the $(j + i - 2d - 1)$th bit of S_g^j. This way R_g^j represents the rows $i = j+1-2d \ldots j+d$ of D and the rows $i = j-2d \ldots j+d-1$ of D''. R_g^j may represent non-existing rows $i > |P|$, but we may simply ignore those. On the other hand, we will make use of the non-existing rows $i < 0$. Let us first define match vectors $\overline{PM}_a^j$ that corresponds to R_g^j. The vector $\overline{PM}_a^j$ has $6d$ bits where the $(2i-2)$th and $(2i-1)$th bits are set if $P_i = a$.

The vector S_0^j should have a set bit corresponding to $D''[0,j] = 0$ for $j = 1\ldots 2d$. This can be enforced for R_g^j by initializing to 1 the bits of R_0^0 that correspond to the non-existing rows $i = -2d\ldots -1$ in column 0, and defining that the rows $-2d\ldots 0$ in the match vectors $\overline{PM}_a^j$ contain 1 bits. This way R_g^0 represents values $D''[i,j] > d$ for each row $i \geq 0$. Then each of the successive $2d$ steps will propagate an active state to the position of R_0^j that corresponds to $D''[0,j]$. Thus the rule $D''[0,j] = 0$ is enforced for columns $j = 1\ldots 2d$.

The start and end points of the row-intervals represented in R_g^j are of form $j+x$, and thus each increment in j shifts the region by one step. The update formula for R_g^j should be adjusted accordingly. To align the vector R_g^j correctly with the vectors R_g^{j-1}, R_{g-1}^{j-1}, the left-shifts of R_g^{j-1} and R_{g-1}^{j-1} in the original formula should be removed. On the other hand, now the previously non-shifted instance of the vector R_{g-1}^{j-1} should be shifted to the right. An implementation that follows these pronciples is shown in Fig. 4. Line 1 of the procedure `IsMatchColumnRBP` uses an auxiliary table $CHECK$. The entry $CHECK[j]$ contains a length-$(6d)$ bit-vector whose ith bit is set to 1 if and only if the ith bit of R_g^j corresponds to row $|P|$ in D.

Using the restricted bit-parallel method results in $\mathcal{O}(\sigma d\lceil d/w\rceil|P||CCU_d(P)|)$ and $\mathcal{O}(d\lceil d/w\rceil|CCU_d(P)|)$ time complexities for generating $CCU_d(P)$ without and with using the KMP failure links. In practice this method is very efficient when $6d \leq w$, as then each R_g^j fits into a single computer word. If $w = 32$, the limit is $d \leq 5$, and if $w = 64$, the limit is $d \leq 10$. Already the smaller limit $d \leq 5$ is usually sufficient for the indexed approximate string matching algorithm of Myers [3].

6 Preliminary test results

We have implemented the discussed bit-parallel methods for generating $CCU_d(P)$ and conducted preliminary tests to find out whether our variants are efficient in practice. The tested methods were the original of Russo and Oliveira (RO), our tuned version of RO that uses simpler `IsMatchColumn` and `ExtendsToMatchColumn` (RO'), our combined bit-parallel method (CBP), and our restricted combined bit-parallel method (RBP).

All code was implemented in C. The test computer was an Intel Celeron M530 with 2 GB RAM, Windows XP operating system and Microsoft Visual C 2003 compiler. The codes were compiled using the "-O2" optimization switch. In correspondence to the results given by Russo and Oliveira [4], our preliminary tests were done without using the KMP links. We used random patterns with $\sigma = 4$ and lengths $|P| = 10$ and $|P| = 10$, and tested for $d = 1\ldots 5$. The results are shown in Table 1 in the form of the time in seconds for generating the set $CCU_d(P)$ 10000 times. It can be seen that our methods achieve considerable speedups over the original RO in these preliminary test cases.

```
ComputeColumnRBP(j)
1.    R_0^j ← R_0^{j-1} & \overline{PM}_{A_j}^j
2.    For g ← 1...d Do
3.        R_g^j ← (R_g^{j-1} & \overline{PM}_{A_j}^j) | (R_{g-1}^{j-1} >> 2) | R_{g-1}^{j-1} | (R_{g-1}^j << 2)

IsMatchColumnRBP(j)
1.    tmp ← R_g^j & CHECK[j]
2.    If tmp ≠ 0 Then
3.        return true
4.    Else
5.        return false

ExtendsToMatchColumnRBP(j)
1.    tmp ← (R_d^j & (∼ R_d^j >> 1)) & FMASK
2.    If tmp ≠ 0 Then
3.        return true
4.    Else
5.        return false
```

Fig. 4. A restricted bit-parallel scheme that computes an $\mathcal{O}(d)$ interval of C_g^j into R_g^j.

method	error limit d ($|P| = 10$)					error limit d ($|P| = 15$)				
	1	2	3	4	5	1	2	3	4	5
RO	0,26	3,27	19,8	33,6	9,1	0,72	20,6	334	3140	13900
RO'	0,25	3,08	19,1	32,2	8,6	0,72	19,2	315	2920	12600
CBP	0,19	2,30	12,1	20,4	5,5	0,52	14,2	236	2000	8750
RBP	0,20	2,53	13,9	21,7	5,5	0,55	15,9	247	2200	9370

Table 1. The time in seconds for generating $CCU_d(P)$ 10000 times.

References

1. D. E. Knuth, J. H. Morris, V. R. Pratt. Fast pattern matching in strings. *SIAM Journal on Computing* 6(1):323–350, 1977.
2. V. Levenshtein. Binary codes capable of correcting deletions, insertions and reversals. *Soviet Physics Doklady* 10(8):707–710, 1966.
3. E. Myers. A sublinear algorithm for approximate keyword searching. *Algorithmica* 12(4/5):345–374, 1994.
4. L. M. S. Russo, A. L. Oliveira. Efficient generation of super condensed neighborhoods. *Journal of Discrete Algorithms* 5(3):501–513, 2007.
5. Algorithms for approximate string matching. *Information and Control* 64:100–118, 1985.
6. S. Wu, U. Manber. Fast text searching allowing errors. *Comm. of the ACM* 35(10):83–91, 1992.

Path factors and parallel knock-out schemes of almost claw-free graphs

Matthew Johnson, Daniël Paulusma, and Chantal Wood

Department of Computer Science,
Durham University,
South Road, Durham, DH1 3LE, U.K.
{matthew.johnson2,daniel.paulusma,chantal.wood}@durham.ac.uk

Abstract. An $H_1, \{H_2\}$-factor of a graph G is a spanning subgraph of G with exactly one component isomorphic to the graph H_1 and all other components (if there are any) isomorphic to the graph H_2. We completely characterise the class of connected almost claw-free graphs that have a $P_7, \{P_2\}$-factor, where P_7 and P_2 denote the paths on seven and two vertices, respectively. We apply this result to parallel knock-out schemes for almost claw-free graphs. These schemes proceed in rounds in each of which each surviving vertex simultaneously eliminates one of its surviving neighbours. A graph is reducible if such a scheme eliminates every vertex in the graph. Using our characterisation we are able to classify all reducible almost claw-free graphs, and we can show that every reducible almost claw-free graph is reducible in at most two rounds. This leads to a quadratic time algorithm for determining if an almost claw-free graph is reducible (which is a generalisation and improvement upon the previous strongest result that showed that there was a $O(n^{5.376})$ time algorithm for claw-free graphs on n vertices).

Keywords: parallel knock-out schemes, (almost) claw-free graphs, perfect matching, factor

1 Introduction

We denote a graph by $G = (V, E)$. An edge joining vertices u and v is denoted by uv. If not stated otherwise a graph is assumed to be finite, undirected and simple. The *neighbourhood* of $u \in V$, that is, the set of vertices adjacent to u is denoted by $N_G(u) = \{v \mid uv \in E\}$, and the *degree* of u is denoted by $\deg_G(u) = |N_G(u)|$. If no confusion is possible, we omit the subscripts. A set $I \subseteq V$ is called an *independent set* of G if no two vertices in I are adjacent to each other, and α denotes the *independence number* of G, the number of vertices in a maximum size independent set of G. See [3] for other basic graph-theoretic terminology.

A graph $(\{u, v_1, v_2, v_3\}, \{uv_1, uv_2, uv_3\})$ is called a *claw* with *claw centre* u and *leaves* v_1, v_2, v_3. A graph is *claw-free* if it does not contain a claw as a induced subgraph. Claw-free graphs form a rich class containing, for example, the class of line graphs and the class of complements of triangle-free graphs. It is a very well-studied graph class, both within structural graph theory and within algorithmic graph theory; see [10] for a survey. We study a generalisation of claw-free graphs, namely *almost claw-free graphs* which were introduced by Ryjáček [22].

Definition 1. *A graph $G = (V, E)$ is almost claw-free if the following two conditions hold:*

1. *The set of all vertices that are claw centres of induced claws in G is an independent set in G.*
2. *For all $u \in V$, either $|N(u)| = 1$ or $N(u)$ contains two vertices v_1, v_2 such that $N(u) \backslash \{v_1, v_2\} \subseteq N(v_1) \cup N(v_2)$.*

Claw-free graphs trivially satisfy the first condition, and they also satisfy the second since otherwise they would contain a vertex with three independent neighbours yielding an induced claw. Hence, every claw-free graph is almost claw-free. It is easy to see that there exist almost claw-free graphs that are not claw-free; see, for example, the graph H in Figure 2.

Several papers have generalised results on claw-free graphs to almost claw-free graphs: see [7, 19, 25] for results on hamiltonicity, shortest walks and toughness. A subgraph $M = (V', E')$ of a graph $G = (V, E)$ is called a *matching* of G if every vertex in M has degree one. It is called a *perfect* matching if $V' = V$. We call G *even* if $|V|$ is even, and *odd* otherwise. Las Vergnas [18] and Sumner [23] have independently proven that every even connected claw-free graph $G = (V, E)$ has a perfect matching. The following theorem by Ryjáček [22] generalises this result to almost claw-free graphs.

Theorem 1 ([22]). *Every even connected almost claw-free graph has a perfect matching.*

For an odd graph $G = (V, E)$, the natural analogue of a perfect matching is a *near-perfect* matching: a matching $M = (V \backslash \{v\}, E')$ for some $v \in V$. In this paper we shall prove the following.

Theorem 2. *Every odd connected almost claw-free graph has a near-perfect matching.*

Jünger, Pulleyblank and Reinelt [14] have shown that odd claw-free graphs have near-perfect matchings so Theorem 2 is an extension of this result to *almost* claw-free graphs. In fact, our main result, Theorem 3, is much stronger and more general, but we require some further preliminaries before we can state it.

To capture both even and odd graphs, the notion of a (near-)perfect matching has been generalised in various ways. We consider two such generalisations for almost claw-free graphs, namely *path factors* and *parallel knock-out numbers*, which we relate to each other.

In Section 2, we completely characterise the class of connected almost claw-free graphs that have a spanning subgraph with exactly one component isomorphic to a path on seven vertices while all other components form a matching. In Section 4 we sketch the proof of this result and present a polynomial algorithm for finding such a subgraph, but first we apply this result in Section 3 to parallel knock-out schemes for almost claw-free graphs.

These schemes proceed in rounds in each of which each surviving vertex simultaneously eliminates one of its surviving neighbours. A graph is *reducible* if such a scheme eliminates every vertex in the graph. Using our characterisation we are able to classify all reducible almost claw-free graphs, and we can show that every reducible almost claw-free graph is reducible in at most two rounds. This leads to a quadratic time algorithm for determining if an almost claw-free graph is reducible. This is a generalisation and improvement upon the $O(n^{5.376})$ time algorithm for n-vertex claw-free graphs given by Broersma et al. in [6]. Although, in general, determining if a graph is reducible is an NP-complete problem, the new technique that uses (path) factors for this problem might be promising for other graph classes as well. We discuss this in Section 5.

2 Path factors

Let $\mathcal{H} = \{H_1, H_2, \ldots, \}$ be a family of graphs. An $\mathcal{H}$-factor of a graph G is a spanning subgraph of G with each component isomorphic to a graph in $\{\mathcal{H}\}$. Let P_n denote the path on n vertices. A *path factor* of a graph G is a $\{P_1, P_2, \ldots\}$-factor of G. Path factors generalise perfect matchings, which are $\{P_2\}$-factors. Path factors have been the subject of considerable study: see, for example, [24] for a characterisation of bipartite graphs with a $\{P_3, P_4, P_5\}$-factor and [15, 16] for a characterisation of general graphs with a $\{P_3, P_4, P_5\}$-factor. A more recent result [20] shows that the square of any graph on at least six vertices has a $\{P_3, P_4\}$-factor. Connected claw-free graphs with minimum degree d have a $\{P_{d+1}, P_{d+2}, \ldots\}$-factor [1]. In general, obtaining good characterisations of graph classes with path factors might be difficult as it is shown in [11] that the problem of deciding if a given graph has a H-factor is NP-complete for any fixed H with $|V_H| \geq 3$. For a more general survey on factors see [21].

We are interested in another class of path factors. Let H_1, H_2 be graphs. Then an $H_1, \{H_2\}$-*factor* of a graph G is a spanning subgraph of G with exactly one component isomorphic to H_1 and all other components (if there are any) isomorphic to H_2. The components are called H_1-*components* and H_2-*components*. A $P_2, \{P_2\}$-factor of a graph corresponds to a perfect matching, and a $P_1, \{P_2\}$-factor corresponds to a near-perfect matching.

In order to state our main result, we must define two families $\mathcal{F}$ and $\mathcal{G}$ of connected almost claw-free graphs. For an integer $k \geq 0$, let the graph F_k be obtained from the complete graph on $k + 1$ vertices $x_0, \ldots, x_k$ by adding a vertex y_i and an edge $x_i y_i$ for $i = 1, \ldots, k$ (note there is no vertex y_0). We say that x_0 is the *root* of F_k. Note that each graph F_k is claw-free. In particular, F_0 is isomorphic to P_1 and F_1 is isomorphic to P_3. For integers $k, \ell \geq 1$, let $F_{k,\ell}$ denote the graph obtained from two vertex-disjoint copies of F_k and F_ℓ after removing their roots and adding a new vertex x^* adjacent to precisely those vertices to which the roots were adjacent in F_k, F_ℓ. We call x^* the *root* of $F_{k,\ell}$. Note that each graph $F_{k,\ell}$ is claw-free. In particular, $F_{1,1}$ is isomorphic to P_5. Finally, for integers $k, \ell \geq 1$, let $F'_{k,\ell}$ denote the graph obtained from $F_{k,\ell}$ with root x^* after adding two new vertices y and z with y adjacent to z and z also adjacent to all vertices in $N_{F_{k,\ell}}(x^*)$. We call x^* the *root* of $F'_{k,\ell}$. Since z is the (only) centre of an induced claw, $F'_{k,\ell}$ is not claw-free. However, it is easy to check that each $F'_{k,\ell}$ is almost claw-free. Let $\mathcal{F} = \{F_0, F_k, F_{k,\ell}, F'_{k,\ell} \mid k, \ell \geq 1\}$. See Figure 1 for some examples of graphs that belong to this family. Let C_n denote the cycle on n vertices. For

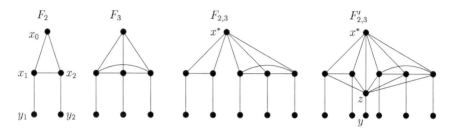

Fig. 1. The graphs $F_2, F_3, F_{2,3}$, and $F'_{2,3}$.

$k \geq 0$, the graph G_k is obtained from F_k by adding two new vertices a and b that are adjacent to the root of F_k and to each other. Note that G_0 is isomorphic to C_3; see Figure 2 for some other examples. The family $\mathcal{G}$ contains the graphs G_k, $k \geq 0$, and also all other connected graphs on five vertices that have a $C_3, \{P_2\}$-factor. There are eleven such graphs which are depicted in Figure 3 together with the graph G_1. Note that each graph in $\mathcal{G}$ is claw-free and contains a $C_3, \{P_2\}$-factor. Let $H = (\{u_1, u_2, u_3, u_4, u_5\}, \{u_1u_2, u_1u_3, u_1u_4, u_2u_4, u_3u_4, u_4u_5\})$ be the almost claw-free graph in Figure 2. Note that the only connected almost claw-free graphs on five vertices not in $\mathcal{G}$ are $F_2, F_{1,1}, C_5$, and H.

Theorem 3 below states that to check whether a graph G on n vertices has a $P_7, \{P_2\}$-factor can be done by checking whether or not $G \in \mathcal{F} \cup \mathcal{G} \cup \{C_5, H\}$ (and this can clearly be done in time $O(|V|^2)$). The theorem also states that finding such a factor takes $O(|V|^{3.5})$ time. This is a major improvement upon the trivial brute-force algorithm that checks for every 7-tuple of vertices $\{v_1, \ldots, v_7\}$ whether the graph obtained after removing $\{v_1, \ldots v_7\}$ contains a perfect matching.

Theorem 3. *Let $G = (V, E)$ be an odd connected almost claw-free graph. If $G \notin \mathcal{F} \cup \mathcal{G} \cup \{C_5, H\}$ then G has a $P_7, \{P_2\}$-factor, which we can find in $O(|V|^{3.5})$ time.*

Note that Theorem 3 implies Theorem 2. We prove Theorem 3 in Section 4. There we describe an algorithm that computes a $P_7, \{P_2\}$-factor in $O(|V|^{3.5})$ time. The running time of the algorithm on an input graph $G = (V, E)$ depends on the running time of a subalgorithm that is performed $O(|V|)$ times and that finds a perfect matching in at most two subgraphs of G and then attempts to transform these perfect matchings into a $P_7, \{P_2\}$-factor of G. As such a transformation already requires $\Omega(|V|^2)$ time for some almost claw-free graphs, we did not aim to bring down the running time of the $O(|V|^{0.5}|E|) = O(|V|^{2.5})$ time algorithm of Blum that computes a maximum matching for general graphs [2].

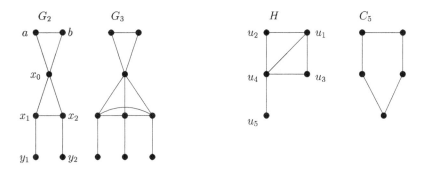

Fig. 2. The graphs G_2, G_3, H and C_5.

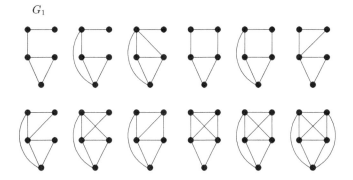

Fig. 3. All connected 5-vertex graphs with a $C_3, \{P_2\}$-factor.

3 Parallel knock-out schemes

3.1 Definitions and Observations

In this section we continue the study on *parallel knock-out schemes* for finite undirected simple graphs begun in [17] and continued in [4–6]. Such a scheme proceeds in rounds. In the first round each vertex in the graph selects exactly one of its neighbours, and then all the selected vertices are eliminated simultaneously. In subsequent rounds this procedure is repeated in the subgraph induced by those vertices not yet eliminated. The scheme continues until there are no vertices left, or until an isolated vertex is obtained (since an isolated vertex will never be eliminated).

More formally, for a graph $G = (V, E)$, a *KO-selection* is a function $f : V \to V$ with $f(v) \in N(v)$ for all $v \in V$. If $f(v) = u$, we say that vertex v *fires at* vertex u, or that vertex u *is knocked out* by vertex v. For a KO-selection f, we define the corresponding *KO-successor* of G as the subgraph of G that is induced by the vertices in $V \setminus f(V)$; if G' is the KO-successor of G we write $G \rightsquigarrow G'$. Note that every graph without isolated vertices has at least one KO-successor. A graph G is called *KO-reducible*, if there exists a *KO-reduction scheme*, that is, a finite sequence

$$G \rightsquigarrow G^1 \rightsquigarrow G^2 \rightsquigarrow \cdots \rightsquigarrow G^r,$$

where G^r is the null graph $(\emptyset, \emptyset)$. A single step in this sequence is called a *round*, and the parallel knock-out number of G, $\text{pko}(G)$, is the smallest number of rounds of any KO-reduction scheme. If G is not KO-reducible, then $\text{pko}(G) = \infty$.

Note that $\text{pko}(P_1) = \text{pko}(P_3) = \text{pko}(P_5) = \infty$, as in each case there is at least one isolated vertex after the first round of any parallel knock-out scheme, and $\text{pko}(P_{2k}) = 1$, for $k \geq 1$, and $\text{pko}(C_k) = 1$, for $k \geq 3$, as we can define a first round firing along the perfect matching and cycle edges, respectively. Finally, $\text{pko}(P_{2k+1}) = 2$ for $k \geq 3$. To see this, consider a KO-reduction scheme for a path $p_1 p_2 \cdots p_{2k+1}$ such that in the first round p_{2i-1} and p_{2i} fire at each other for $i = 1, \ldots, k - 2$, p_{2k-3} fires at p_{2k-4}, p_{2k-2} fires at p_{2k-3}, p_{2k-1} fires at p_{2k}, and p_{2k} and p_{2k+1} fire at each other. Then, after round one, p_{2k-2} and p_{2k-1} are the only two vertices left and they fire at each other in round two. This yields the following observation which explains our interest in $P_7, \{P_2\}$-factors; note that the reverse implication is not true.

Observation 4 *Let G be a graph. If G has a perfect matching or a $C_k, \{P_2\}$-factor for some $k \geq 3$, then $\text{pko}(G) = 1$. If G has a $P_{2k+1}, \{P_2\}$-factor for some $k \geq 3$, then $\text{pko}(G) \leq 2$.*

The paper [6] shows that a KO-reducible n-vertex graph G has

$$\text{pko}(G) \leq \min\left\{ -\frac{1}{2} + \sqrt{2n - \frac{7}{4}}, \; \frac{1}{2} + \sqrt{2\alpha - \frac{7}{4}} \right\},$$

(recall that α is the independence number). This bound is asymptotically tight due to the existence of a family of graphs in [4] whose knock-out numbers grow proportionally to the square root of the number of vertices (and to the square root of the independence number as these graphs are bipartite). KO-reducible claw-free graphs, however, can be knocked out in at most two rounds [4]. Connected claw-free graphs with minimum degree $d \geq 5$ have a $\{P_6, P_7, \ldots\}$-factor [1]: this implies they are KO-reducible in at most two rounds by Observation 4. Using Theorem 3 we can strengthen and generalise the result on parallel knock-out numbers for claw-free graphs to almost claw-free graphs. First, note that every graph $F \in \mathcal{F}$ is not KO-reducible as in the first round of any KO-reduction scheme all neighbours of the root x of F must fire at their neighbour of degree one, and vice versa. So, in the next round, x would be the only remaining vertex which is not possible in a KO-reduction scheme. We find that $\text{pko}(H) = 2$ as u_1 can fire at u_2, while u_2 and u_3 fire at u_4, and u_4 and u_5 fire at each other in the first round, and then u_1 and u_3 fire at each other in the second round. By Observation 4, $\text{pko}(G) = 1$ if $G \in \mathcal{G} \cup \{C_5\}$. If G is an even connected almost claw-free graph, then G has a perfect matching by Theorem 1 and consequently $\text{pko}(G) = 1$ by Observation 4. Hence we have the following result.

Corollary 1. *Let G be a connected almost claw-free graph. Then G is KO-reducible if and only if $pko(G) \leq 2$ if and only if $G \notin \mathcal{F}$.*

Note that odd paths on at least seven vertices are examples of (almost) claw-free graphs with parallel knock-out number two. We observe that Corollary 1 restricted to claw-free graphs states that a connected claw-free graph G is KO-reducible if and only if $pko(G) \leq 2$ if and only if G is not isomorphic to some F_k or $F_{k,\ell}$. This characterisation of claw-free graphs is new.

3.2 Running Times

In [4], a polynomial time algorithm is given that determines the parallel knock-out number of any tree. For general bipartite graphs, however, the problem of finding the parallel knock-out number is NP-hard [5]. In fact, even the problem of deciding if $pko(G) \leq 2$ for a given bipartite graph G is NP-complete. On the positive side, a polynomial time algorithm for finding a KO-reduction scheme for general claw-free graphs was presented in [6]. Corollary 1 provides us with an $O(|V|^2)$ algorithm for checking if an *almost* claw-free graph $G = (V, E)$ is KO-reducible as it takes $O(|V|^2)$ time to verify that each component of G does not belong to $\mathcal{F}$. This is a considerable improvement upon the polynomial time algorithm for claw-free graphs in [6] which we briefly describe now as its running time was not previously analysed.

The algorithm first checks if $pko(G) = 1$ by determining if G has a *[1,2]-factor* (a spanning subgraph in which every component is either a cycle or an edge). The problem of deciding if $G = (V, E)$ contains a [1,2]-factor is a folklore problem appearing in many standard books on combinatorial optimisation. It is solved as follows. Let $V = \{v_1, v_2, \ldots, v_n\}$. Define the *product graph* of G as the bipartite graph $G' = (V', E')$ with vertex set $V' = \{u_1, u_2, \ldots, u_n, w_1, w_2, \ldots, w_n\}$ in which $u_i w_j \in E'$ and $u_j w_i \in E'$ if and only if $v_i v_j \in E$. A [1,2]-factor in G corresponds to a perfect matching in G'. The fastest known algorithms for checking if a bipartite graph $G = (V, E)$ has a perfect matching have running time $O(|V|^{0.5}|E|)$ [9, 12] or $O(|V|^{2.376})$ [13].

If $pko(G) \neq 1$, the algorithm checks if $pko(G) = 2$ by using a result (also proved in [6]) that any connected claw-free graph G with $pko(G) = 2$ allows a KO-reduction scheme in which only two vertices x, y remain in the second round such that

1. x knocks out a vertex w in the first round that is not knocked out by any other vertex and that fires at a vertex that is knocked out by some other vertex as well.
2. y knocks out a vertex in the first round that is knocked out by some other vertex as well.

The algorithm simply checks all possibilities for x, y, w. After guessing these three vertices, it checks if the remaining graph has parallel knock-out number one. Thus the algorithm of [6] takes $O(|V|^{5.376})$ time if we use the algorithm of [13] and $O(|V|^{3.5}|E|)$ time if we use the algorithms in [9, 12] for finding a perfect matching in a bipartite graph. (We have not examined if the algorithms in [9, 12, 13] can be improved if the bipartite graph under consideration is the product graph of a claw-free graph.) Note that our new algorithm *finds* a KO-reduction scheme for the class of almost claw-free graphs in $O(|V|^{3.5})$ time. This can be seen as follows. We first check in $O(|V|^2)$ time if our input graph $G = (V, E)$ that is almost claw-free belongs to $\mathcal{G} \cup \{C_5, H\}$. If so, then we can immediately deduce a KO-reduction scheme. We then check in $O(|V|^2)$ time if G belongs to $\mathcal{F}$. If so, then $\text{pko}(G) = \infty$. If not then G contains a $P_7, \{P_2\}$-factor which we can find in $O(|V|^{3.5})$ time by Theorem 3. This $P_7, \{P_2\}$-factor immediately provides us with a KO-reduction scheme of G.

We summarise what we have proved:

Corollary 2. *Let $G = (V, E)$ be an almost claw-free graph. Deciding whether G is KO-reducible or has $\text{pko}(G) \leq 2$, respectively, can be done in $O(|V|^2)$ time. The problem of finding a KO-reduction scheme for G can be done in $O(|V|^{3.5})$ time.*

4 Proof of Theorem 3

4.1 Definitions and Lemmas

In this section we prove Theorem 3 after first introducing some additional notation and preliminary results. The subgraph of a graph $G = (V, E)$ induced by set $U \subseteq V$ is denoted by $G[U]$. A set $U \subseteq V$ is a *dominating set* of G if each vertex in V is in U or adjacent to a vertex in U. If $U = \{u\}$ we call u a *dominating vertex* of G and if $U = \{u_1, u_2\}$ we call u_1 and u_2 a *dominating pair*. Note that condition 2 of Definition 1 is equivalent to: "for all $v \in V$, $G[N(v)]$ must contain a dominating vertex or dominating pair". We denote the set of vertices in a graph G that have degree i by V_i and all vertices that have degree at least i by $V_{\geq i}$. We denote by $V_{\geq 2}'$ the subset of $V_{\geq 2}$ containing vertices that do not have neighbours of degree 1. For convenience, we sometimes use the notation $|G|$ to denote the number of vertices in G.

The following fact is a complicating factor in the proof of Theorem 3: removing a vertex x from an almost claw-free graph does not automatically result in a new almost claw-free graph. Note that claw-free graphs do satisfy such a property. An example is the almost claw-free graph H: if we remove u_1 from H then we obtain a claw, which does not satisfy condition 2 of Definition 1. Hence, one of the conditions in Lemma 5 below, namely that $G[V \backslash \{x\}]$ is almost claw-free, is not satisfied by every almost claw-free graph (if it were, then Lemma 5 alone would imply Theorem 3). The next lemma tells us about the structure of a graph obtained by removing a single vertex from an almost claw-free graph.

Lemma 1. *Let x be a vertex of an almost claw-free graph $G = (V, E)$ such that $G[V \backslash \{x\}]$ is not almost claw-free. Let Y be the subset of $V \backslash \{x\}$ such that $G[N(y) \backslash \{x\}]$ does not contain a dominating pair. Then the following holds:*

(i) Y is an independent set with $|Y| \in \{1, 2\}$.
(ii) Each $y \in Y$ is adjacent to x.

(iii) *For each $y \in Y$ there exist vertices $a, b \in N(x)$ and $c \notin N(x) \cup \{x\}$ such that y is the centre of an induced claw with edges ya, yb, yc.*

Proof. Let x be a vertex of an almost claw-free graph $G = (V, E)$ and let $G' = G[V \setminus \{x\}]$. Suppose G' is not almost claw-free. If G' violates condition 1 of Definition 1, then G would violate this condition as well. Hence G' violates condition 2 of Definition 1. Then there exists a vertex y^*, such that $G'[N_{G'}(y^*)] = G[N(y^*) \setminus \{x\}]$ does not contain a dominating pair. As G is almost claw-free, x is in any dominating pair of $G[N(y^*)]$. Then $y^* \in Y$ and $xy^* \in E$. This proves $|Y| \geq 1$ and (ii).

Let x, c be a dominating pair of $G[N(y)]$ for some $y \in Y$. Since $G[N(y) \setminus \{x\}]$ does not contain a dominating pair, x has a neighbour $a \in N(y) \setminus \{x, c\}$ not adjacent to c. Because $\{a, c\}$ is not a dominating pair of $G[N(y) \setminus \{x\}]$, x has a neighbour $b \in N(y) \setminus \{a, x, c\}$ neither adjacent to a nor to c. We note that y is the centre of an induced claw in G with edges ya, yb, yc. Then, by condition 1 of Definition 1, x is not the centre of an induced claw. We then deduce that $xc \notin E$. This proves (iii).

Because each $y \in Y$ is the centre of an induced claw, Y is an independent set of G due to condition 1 of Definition 1. To finish the proof of (i), suppose $Y = \{y_1, \ldots, y_r\}$ with $r \geq 3$. Because $\{y_1, y_2, y_3\}$ is an independent set in $G[N(x)]$, we then find that x is the centre of an induced claw with edges xy_1, xy_2, xy_3. We already observed x is not the centre of an induced claw. Hence we conclude that $r \leq 2$. This completes the proof of Lemma 1. □

The following lemmas are used in the proof of Theorem 3. Due to page restrictions only the proofs of Lemma 2 and 5 are given (in Section 4.3). The proof of Lemma 6 uses Lemma 1.

Lemma 2. *If $G = (V, E)$ is an odd connected almost claw-free graph not in $\mathcal{F} \cup \mathcal{G} \cup \{C_5, H\}$, then $|V| \geq 7$, $V'_{\geq 2} \neq \emptyset$, and all vertices in $V'_{\geq 2}$ have a neighbour in $V'_{\geq 2}$.*

Lemma 3. *Let $G = (V, E) \notin \mathcal{G}$ be a connected almost claw-free graph with a $C_3, \{P_2\}$-factor. Then G has a $P_7, \{P_2\}$-factor. Moreover, given a $C_3, \{P_2\}$-factor of G, there is an algorithm that finds a $P_7, \{P_2\}$-factor of G in $O(|V|^2)$ time.*

Lemma 4. *Let $G = (V, E)$ with $|V| \geq 7$ be a connected almost claw-free graph that has a $C_5, \{P_2\}$-factor or an $H, \{P_2\}$-factor. Then G has a $P_7, \{P_2\}$-factor. Moreover, given a $C_5, \{P_2\}$-factor or $H, \{P_2\}$-factor of G, there is an algorithm that finds a $P_7, \{P_2\}$-factor of G in $O(|V|^2)$ time.*

Lemma 5. *Let $G = (V, E) \notin \mathcal{F} \cup \mathcal{G} \cup \{C_5, H\}$ be an odd connected almost claw-free graph. If $G[V \setminus \{x\}]$ is almost claw-free for some $x \in V'_{\geq 2}$, then G has a $P_7, \{P_2\}$-factor. Moreover, given such a vertex x, there is an algorithm that finds a $P_7, \{P_2\}$-factor of G in $O(|V|^{2.5})$ time.*

Lemma 6. *Let $G = (V, E)$ be an odd connected almost claw-free graph not in $\mathcal{F} \cup \mathcal{G}$ such that $G[V \setminus \{x\}]$ is not almost claw-free for all $x \in V'_{\geq 2}$. Then, for each $x \in V'_{\geq 2}$, there exist two vertices $\{c, y\}$ with $y \in N(x)$ and $c \in N(y) \cap V_1$ such that $G^* = G[V \setminus \{c, y\}]$ is either in $\mathcal{G} \cup \{C_5, H\}$ or else G^* is an odd connected almost claw-free graph not in $\mathcal{F}$ such that $G^*[V_{G^*} \setminus \{x\}]$ is almost claw-free.*

4.2   The Algorithm

We restate Theorem 3 before presenting the algorithm that provides a proof.

Theorem 3 *Let $G = (V, E)$ be an odd connected almost claw-free graph. If $G \notin \mathcal{F} \cup \mathcal{G} \cup \{C_5, H\}$ then G has a $P_7, \{P_2\}$-factor, which we can find in $O(|V|^{3.5})$ time.*

Outline of the algorithm. Let $G = (V, E)$ be an odd connected almost claw-free graph. Suppose $G \notin \mathcal{F} \cup \mathcal{G} \cup \{C_5, H\}$. We show how to find a $P_7, \{P_2\}$-factor of G in $O(|V|^{3.5})$ time.

Step 1. Determine the set $V'_{\geq 2}$.

This takes time $O(|V|^2)$ time, and, by Lemma 2, the set is nonempty. (In fact Lemma 2 says more than this as it is used in the proofs of later lemmas.)

Step 2. For each vertex $x \in V'_{\geq 2}$, run the algorithm of Lemma 5.

If $G[V \setminus \{x\}]$ is almost claw-free, then, by Lemma 5, we will find a $P_7, \{P_2\}$-factor of G. If, after trying all possible choices for x, we still have not found a $P_7, \{P_2\}$-factor of G, then we know that $G[V \setminus \{x\}]$ is not almost claw-free for all $x \in V'_{\geq 2}$. Step 2 takes time $|V'_{\geq 2}| O(|V|^{2.5}) = O(|V|^{3.5})$.

Step 3. Choose an arbitrary vertex $x \in V_{\geq 2}$. Find all edges cy where $c \in V_1$, $y \in N(x)$ and $N(y) \setminus \{c\}$ is dominated by x.

After Step 3 we have obtained a set of p edges $c_1 y_1, \ldots, c_p y_p$ with $c_i \in N(y) \cap V_1$ and $y_i \in N(x)$ with $N(y_i) \setminus \{c_i\} \subseteq N(x)$ for each $i = 1, \ldots, p$. Note that $p \leq |V|$. Step 3 takes time $O(|V|^3)$.

Step 4. For each i, consider the graph $G_i^* = G[V \setminus \{c_i, y_i\}]$. Check whether $G_i^* \in \mathcal{G} \cup \{C_5, H\}$.

Step 4a. If $G_i^* \in \mathcal{G}$, then find a $C_3, \{P_2\}$-factor of G_i^* (this is easy). Extend this factor with the P_2-component $c_i y_i$ to obtain a $C_3, \{P_2\}$-factor of G. Use the algorithm of Lemma 3 to obtain a $P_7, \{P_2\}$-factor of G.

We can use the algorithm of Lemma 3 since $G \notin \mathcal{G}$. Step 4a takes time $O(|V|^2)$.

Step 4b. If G_i^* is isomorphic to C_5 or H, then find a $C_5, \{P_2\}$-factor or $H, \{P_2\}$-factor of G (by adding the edge $c_i y$). Then use the algorithm of Lemma 4 to find a $P_7, \{P_2\}$-factor of G.

Step 4b takes time $O(|V|^2)$. If we have still not found a $P_7, \{P_2\}$-factor of G at the end of Step 4, then we have taken $p \cdot O(|V|^2) = O(|V|^3)$ time to find that $G_i^* \notin \mathcal{G} \cup \{C_5, H\}$ for each i.

Step 5. Apply the algorithm of Lemma 5 to G_i^* and x for each i.

By Lemma 6, there must exist an i such that $G_i^* \notin \mathcal{F} \cup \mathcal{G} \cup \{C_5, H\}$ and both G_i^* and $G_i^*[V_{G_i^*} \setminus \{x\}]$ are almost claw-free. Hence we obtain a $P_7, \{P_2\}$-factor of some G_i^* in $p \cdot O(|V|^{2.5}) = O(|V|^{3.5})$ time. We extend this $P_7, \{P_2\}$-factor to a $P_7, \{P_2\}$-factor of G by adding the P_2-component $c_i y_i$. □

4.3 Proofs

Proof of Lemma 2. Let $G = (V, E)$ be an odd connected almost claw-free graph not in $\mathcal{F} \cup \mathcal{G}$. We first prove the following claim.

Claim 1. Each vertex in V has at most one neighbour in V_1.

We first prove Claim 1. Let $u \in V$ have two neighbours u' and u'' in V_1. As $G \notin \mathcal{F}$, we know that G is not isomorphic to $F_1 = P_3$. Hence u has a neighbour $v \notin \{u', u''\}$. Thus each dominating set of $G[N(u)]$ contains u', u'' and at least one other vertex. This violates condition 2 of Definition 1. So we have proven Claim 1.

We now prove that $|V| \geq 7$ and $V'_{\geq 2} \neq \emptyset$. If $V = V_0$ then G is isomorphic to $F_0 = P_1$. This contradicts our assumption that $G \notin \mathcal{F}$. If $V = V_1$ then G is isomorphic to P_2. This contradicts our assumption that $|V|$ is odd. If $|V| = 3$ then G is isomorphic to $P_3 = F_1$ or to $C_3 = G_0$, which both violate the assumption that $G \notin \mathcal{F} \cup \mathcal{G}$. Hence $V_{\geq 2} \neq \emptyset$ and $|V| \geq 5$. Suppose $|V| = 5$. If G has a $C_3, \{P_2\}$-factor then $G \in \mathcal{G}$ by definition. The only four remaining connected almost claw-free graphs on five vertices are $F_2, F_{1,1}, C_5$, and H. All these four graphs are excluded. Hence $|V| \geq 7$. Suppose $V'_{\geq 2} = \emptyset$, that is, all vertices in $V_{\geq 2}$ are adjacent to a vertex in V_1. By Claim 1, each vertex in $V_{\geq 2}$ has exactly one neighbour in V_1. This means that G has a perfect matching and contradicts the assumption that G is odd. Hence we find that $V'_{\geq 2} \neq \emptyset$.

We now prove the second statement of the lemma by contradiction. Suppose x is a vertex in $V'_{\geq 2}$ such that $N(y) \cap V_1 \neq \emptyset$ for all $y \in N(x)$. We first show by contradiction that $V = \{x\} \cup N(x) \cup N'(x)$, where $N'(x) = \{z' \in V_1 \mid z' \in N(z) \text{ for some } z \in N(x)\}$ denotes the set of degree one neighbours of vertices in $N(x)$. If $V \neq \{x\} \cup N(x) \cup N'(x)$ then there exists a vertex $w \in N(x)$ that has a neighbour w^* not in $\{x\} \cup N(x) \cup N'(x)$. Let w' be the neighbour of w in V_1 (so $w' \in N'(x)$). Note that $\{w', w^*, x\}$ is an independent set in $G[N(w)]$. Due to condition 2 in Definition 1, $G[N(w)]$ must have a dominating pair. Hence w^* and x must have a common neighbour z in $G[N(w)]$. Then $z \in V_{\geq 2} \cap N(x)$, and z must have a neighbour z' in V_1. Thus w is the centre of an induced claw in G with edges ww^*, ww', wx, and z is the centre of an induced claw in G with edges zw^*, zx, zz'. This is in contradiction to condition 1 of Definition 1, as z and w are adjacent. Hence we may indeed conclude that $V = \{x\} \cup N(x) \cup N'(x)$.

We now need to distinguish two cases (both turn out to be impossible).

Case 1. x has a neighbour y that is adjacent to all vertices in $N(x) \setminus \{y\}$.

Let y' be the neighbour of y in V_1. As $x \in V'_{\geq 2} \subseteq V_{\geq 2}$, we have $|N(x) \setminus \{y\}| \geq 1$. Suppose $G[N(x) \setminus \{y\}]$ is connected. If $G[N(x) \setminus \{y\}]$ is not a complete graph, then $G[N(x)\backslash\{y\}]$ contains two non-adjacent vertices s and t. Let $P = u_1u_2\cdots u_p$ be a shortest (and consequently induced) path from $s = u_1$ to $t = u_p$ in $G[N(x)\backslash\{y\}]$. Then $p \geq 3$ and $u_1u_3 \notin E$. By our assumption, u_2 has a neighbour u_2' in V_1. Hence, y is the centre of an induced claw with edges yy', yu_1, yu_3, and u_2 is the centre of an induced claw with edges u_2u_2', u_2u_1, u_2u_3. However, y is adjacent to u_2. This is not possible as condition 1 of Definition 1 is violated. Hence we find that $G[N(x)\backslash\{y\}]$, and consequently, $G[N(x)]$ is a complete graph. Recall that $V = \{x\} \cup N(x) \cup N'(x)$. By Claim 1 and our assumption on x, every vertex in $N(x)$ has exactly one neighbour in $N'(x)$. This would mean that G is isomorphic to $F_{|N(x)|}$, which contradicts our assumption that $G \notin \mathcal{F}$. Hence, $G[N(x)\backslash\{y\}]$ is not connected.

Let $D_1, \ldots D_q$ be the $q \geq 2$ components of $G[N(x)\backslash\{y\}]$. Suppose $q \geq 3$. Then x is the centre of an induced claw in G with edges xd_i for some $d_i \in V_{D_i}$ for $i = 1, 2, 3$. Also y is the centre of an induced claw with edges yd_i for $i = 1, 2, 3$. As $xy \in E$, we get a contradiction with condition 1 of Definition 1. Hence $q = 2$.

We claim that both D_1 and D_2 are complete graphs. If, say, D_1 is not complete, then D_1 contains two vertices a and b with $ab \notin E$. Let $c \in D_2$. Then x and y are adjacent centres of induced claws with edges xa, xb, xc and ya, yb, yc respectively. By condition 1 of Definition 1, this is not possible. Hence D_1 and D_2 are complete graphs. Recall that $V = \{x\} \cup N(x) \cup N'(x)$. Then G is isomorphic to $F'_{|D_1|,|D_2|}$. This contradicts our assumption that $G \notin \mathcal{F}$. We conclude that Case 1 cannot occur.

Case 2. $N(x)$ does not contain a vertex adjacent to all vertices in $N(x)$.

By condition 2 of Definition 1, $N(x)$ contains a dominating pair y_1 and y_2. First suppose $y_1y_2 \in E$. By our assumption, y_1 is not adjacent to some vertex $z_1 \in N(x)$, and y_2 is not adjacent to some vertex $z_2 \in N(x)$. As y_1, y_2 form a dominating pair, we deduce that y_1z_2 and y_2z_1 are edges of G. Let y_1' be the neighbour of y_1 in V_1 and let y_2' be the neighbour of y_2 in V_1. Then y_1 is the centre of an induced claw in G with edges y_1y_1', y_1y_2, y_1z_2, and y_2 is the centre of an induced claw in G with edges y_2y_1, y_2y_2', y_2z_1. This violates condition 1 of Definition 1, because y_1 and y_2 are adjacent. Hence we find that $y_1y_2 \notin E$.

Let $D_1, \ldots, D_p$ denote the components of $G[N(x)]$. Suppose $p \geq 3$. We may without loss of generality assume $\{y_1, y_2\} \subseteq V_{D_1} \cup V_{D_2}$. Then $\{y_1, y_2\}$ does not dominate D_i for $i \geq 3$. Hence $p \leq 2$. Suppose $p = 1$ and let $P = u_1u_2\cdots u_r$ be a shortest (and consequently induced) path from $u_1 = y_1$ to $u_r = y_2$ in $G[N(x)]$. Let u_i' be the neighbour of u_i in V_1 for $i = 1, \ldots, r$. As $y_1y_2 \notin E$ and P is an induced path, we find that $r \geq 3$. Suppose $r \geq 4$. Then u_2, u_3 are adjacent centres of induces claws in G with edges u_2u_1, u_2u_2', u_2u_3 and u_3u_2, u_3u_3', u_3u_4 respectively. As this is not possible by condition 1 of Definition 1, we find that $r = 3$. Because u_2 cannot be a dominating vertex of $G[N(x)]$ due to our Case 2 assumption, there exists a vertex $z \in N(x)$ not adjacent to u_2. Since $\{u_1, u_3\} = \{y_1, y_2\}$ is a dominating pair of $G[N(x)]$, we have u_1z or u_3z in E. We may without loss of generality assume $u_1z \in E$. Then u_1 and u_2 are adjacent centres of induced claws in G with edges u_1u_1', u_1u_2, u_1z and u_2u_1, u_2u_2', u_2u_3, respectively. This is not possible due to condition 1 of Definition 1.

Hence $p = 2$. We assume without loss of generality that y_1 belongs to D_1 and y_2 to D_2 (if y_1, y_2 are in the same component, say D_1, they will not dominate the vertices in D_2). Suppose D_1 is not a complete graph. Then there exist vertices a, b in D_1 with $ab \notin E$. Let y_1' be the neighbour of y_1 in V_1. Then x and y_1 are adjacent centres of induced claws with edges xa, xb, xy_2 and y_1a, y_1b, y_1y_1' respectively. By condition 1 of Definition 1, this is not possible. Hence D_1, and by the same arguments, D_2 are complete graphs. Recall that $V = \{x\} \cup N(x) \cup N'(x)$. Hence G is isomorphic to $F_{|D_1|,|D_2|}$. This contradicts our assumption that $G \notin \mathcal{F}$. We conclude that Case 2 does not occur. This completes the proof of Lemma 2. $\qquad \square$

Proof of Lemma 5. Let $G = (V, E)$ be an odd connected almost claw-free graph that is not in $\mathcal{F} \cup \mathcal{G} \cup \{C_5, H\}$. Assume that $G[V \setminus \{x\}]$ is almost claw-free for some $x \in V_{\geq 2}'$. Denote the components of $G[V \setminus \{x\}]$ by $Q_1, \ldots, Q_l$. If $l \geq 3$, then $G[N(x)]$ does not have a dominating pair. This is not possible by condition 2 of Definition 1. Hence $l \leq 2$. We distinguish two subcases.

Case 1. $l = 1$ or $l = 2$ but Q_1 and Q_2 are both even.

We first compute a perfect matching M of $G[V \setminus \{x\}]$ as follows. Suppose $l = 1$. Since $|V|$ is odd, Q_1 is even. Since Q_1 is almost claw-free and connected as well, by Theorem 1, Q_1 has a perfect matching. We define M as the perfect matching that we compute in $O(|V|^{0.5}|E|) = O(|V|^{2.5})$ time by Blum's algorithm [2]. Suppose $l = 2$. Since Q_1 and Q_2 are even, almost claw-free and connected, both Q_1 and Q_2 have a perfect matching, due to Theorem 1. We can compute these perfect matchings M_1 and M_2, respectively, in $O(|V|^{2.5})$ time by Blum's algorithm and define $M := (V_{M_1} \cup V_{M_2}, E_{M_1} \cup E_{M_2})$.

Below we show how we can obtain a $P_7, \{P_2\}$-factor of G from M in $O(|V|^2)$ extra time.

By Lemma 2, x has a neighbour $y \in V_{\geq 2}'$. We can find y in $O(|V|^2)$ time. Let $ay \in E_M$. If $ax \in E$, then G has a $C_3, \{P_2\}$-factor with components $axya$ and the remaining matching edges of M. Since $G \notin \mathcal{G}$, we use Lemma 3 to find a $P_7, \{P_2\}$-factor of G in $O(|V|^2)$ extra time. Suppose $ax \notin E$. As $x \in V_{\geq 2}$, x is adjacent to some vertex $z \neq y$. Since y does not have degree one neighbours, a has at least two neighbours.

Suppose a has a neighbour $b \notin \{y, z\}$. Since $ax \notin E$, $b \neq x$. Let $bc \in E_M$. If $c = z$, we obtain a $C_5, \{P_2\}$-factor L of G with components $abzxya$ and the remaining edges in M. By Lemma 2, $|V| \geq 7$, and we find a $P_7, \{P_2\}$-factor of G in $O(|V|^2)$ extra time, due to Lemma 4. Hence $c \neq z$. Note that $c \notin \{a, b, x, y\}$ either. Let $zd \in E_M$. Then $d \notin \{a, b, c, x, y, z\}$. Hence we have found a $P_7, \{P_2\}$-factor of G with components $dzxyabc$ and the remaining edges in M. We can check this case in $O(|V|^2)$ time.

In the remaining case, a has exactly two neighbours, namely y and z. Again, let $dz \in M$. If $dx \in E$, then again we find a $C_3, \{P_2\}$-factor of G, and consequently, we find a $P_7, \{P_2\}$-factor of G in $O(|V|^2)$ extra time, due to Lemma 3. Suppose $dx \notin E$. Note that $ad \notin D$ since $N(a) = \{y, z\}$. Hence z is the centre of induced claw with edges za, zd, zx. By condition 2 of Definition 1, there exists a vertex p adjacent to z and at least two vertices in $\{a, x, d\}$, and consequently, to at least one vertex in $\{a, d\}$.

First assume $p = y$ (meaning that $yz \in E$). If $yd \in E$, then G contains two adjacent centres, namely y, z, of induced claws with edges ya, yd, yx and za, zd, zx, respectively. This is not possible due to condition 1 of Definition 1. Hence $yd \notin E$. However, then $G[\{a, d, x, y, z\}]$ is isomorphic to H. Recall that $|V| \geq 7$. Then, by Lemma 4, we find a $P_7, \{P_2\}$-factor of G in $O(|V|^2)$ extra time.

Now suppose $p \neq y$. Let $pq \in E_M$. Note that $q \notin \{a, d, p, x, y, z\}$. Assume that p is adjacent to a. We find a path $qpayxzd$ on seven vertices in G. This path together with the remaining edges in M forms a $P_7, \{P_2\}$-factor of G. If $ap \notin E$, then $dp \in E$ and we find a path $qpdzxya$ on seven vertices in G. So, also in this case, which we can check in $O(|V|^2)$ time, we have found a $P_7, \{P_2\}$-factor of G. This finishes Case 1.

Case 2. $l = 2$ but either Q_1 or Q_2 is odd.

As $|V|$ is odd, we find that both $|Q_1|$ and $|Q_2|$ are odd, and consequently $G_1 = G[V_{Q_1} \cup \{x\}]$ and $G_2 = G[V_{Q_2} \cup \{x\}]$ are even. Then G_1 and G_2 are almost claw-free, as otherwise G would not be almost claw-free. Since G_1 and G_2 are almost claw-free and connected as well, they have a perfect matching M_1, M_2, respectively, due to Theorem 1. By Using Blum's algorithm [2], we can find M_1 and M_2 in $O(|V|^{0.5}|E|) = O(|V|^{2.5})$ time. Let xu_1 be an edge in M_1 and xu_2 an edge in M_2. Since $x \in V'_{\geq 2}$, u_1 and u_2 are in $V_{\geq 2}$ by definition. Let $u_1^* \neq x$ be a neighbour of u_1 (in Q_1) and let $u_2^* \neq x$ be a neighbour of u_2 (in Q_2). Let $w_i u_i^* \in E_{M_i}$ for $i = 1, 2$. We note that $|\{u_1, u_1^*, u_2, u_2^*, w_1, w_2, x\}| = 7$. Hence we found a $P_7, \{P_2\}$-factor of G with components $w_1 u_1^* u_1 x u_2 u_2^* w_2$ and the remaining edges in M_1 and M_2. This finishes Case 2 and completes the proof of Lemma 5. □

5 Conclusions

We completely characterised the class of connected almost claw-free graphs that have a $P_7, \{P_2\}$-factor. Using this characterisation we were able to classify all KO-reducible almost claw-free graphs, and we could show that every reducible almost claw-free graph is reducible in at most two rounds. This lead to a quadratic time algorithm for determining if an almost claw-free graph is KO-reducible.

The following open questions are interesting. Can we characterise all (almost) claw-free graphs that have a $P_{2k+1}, \{P_2\}$-factor for $k \geq 4$? Let $K_{1,r}$ denote the *star* on $r + 1$ vertices, that is, the complete bipartite graph with partition classes X and Y with $|X| = 1$ and $|Y| = r$. Can we characterise all KO-reducible $K_{1,r}$-free graphs for $r \geq 4$? This already seems to be a difficult question for $r = 4$, since there exist $K_{1,4}$-free graphs with parallel knock-out number equal to three. Hence, the family of forbidden subgraphs seems considerably more difficult to characterise.

References

1. K. Ando, Y. Egawa, A. Kaneko, K. Kawarabayashi, and H. Matsuda (2002). Path factors in claw-free graphs, *Discrete Mathematics 243*, 195-200.
2. N. Blum (1990). A new approach to maximum matching in general graphs, *Proceedings of the 17th International Colloquium on Automata, Languages and Programming (ICALP 1990)*, Lecture Notes in Computer Science 443, 586–597.
3. J.A. Bondy and U.S.R. Murty (1976). *Graph Theory with Applications*. Macmillan, London and Elsevier, New York.
4. H.J. Broersma, F.V. Fomin, R. Královič, and G.J. Woeginger (2007). Eliminating graphs by means of parallel knock-out schemes, *Discrete Applied Mathematics 155*, 92–102.
5. H.J. Broersma, M. Johnson, D. Paulusma, and I.A. Stewart (2008). The computational complexity of the parallel knock-out problem, *Theoretical Computer Science 393*, 182–195.
6. H.J. Broersma, M. Johnson, and D. Paulusma (2008). Upper bounds and algorithms for parallel knock-out numbers, to appear in *Theoretical Computer Science*.
7. H.J. Broersma, Z. Ryjáček, and I. Schiermeyer (1996). Toughness and hamiltonicity in almost claw-free graphs, *Journal of Graph Theory 21*, 431–439.
8. O. Favaron, E. Flandrin, H. Li, and Z. Ryjáček (1996). Shortest walks in almost claw-free graphs, *Ars Combinatoria 42*, 223–232.
9. T. Feder and R. Motwani (1995). Clique partitions, graph compression and speeding-up algorithms, *Journal of Computer and System Sciences 51*, 261–272.
10. R. Faudree, E. Flandrin, and Z. Ryjáček (1997). Claw-free graphs-a survey, *Discrete Mathematics 164*, 87–147.
11. P. Hell and D.G. Kirkpatrick (1983). On the complexity of general graph factor problems, *SIAM Journal on Computing 12*, 601-609.
12. J. Hopcroft and R.M. Karp (1973). An $n^{5/2}$ algorithm for maximum matchings in bipartite graphs, *SIAM Journal on Computing 2*, 225–231.
13. O.H. Ibarra and S. Moran (1981). Deterministic and probabilistic algorithms for maximum bipartite matching via fast matrix multiplication, *Information Processing Letters 13*, 12–15.
14. S. Jünger, W.R. Pulleyblank, and G. Reinelt (1985). On partitioning the edges of graphs into connected subgraphs, *Journal of Graph Theory 9*, 539–549.
15. A. Kaneko (2003). A necessary and sufficient condition for the existence of a path factor every component of which is a path of length at least two, *Journal of Combinatorial Theory, Series B 88*, 195-218.
16. M. Kano, G.Y. Katona, and Z. Király (2004). Packing paths of length at least two, *Discrete Mathematics 283*, 129-135.
17. D.E. Lampert and P.J. Slater (1998). Parallel knockouts in the complete graph, *American Mathematical Monthly 105*, 556–558.
18. M. Las Vergnas (1975). A note on matchings in graphs, *Cahiers Centre Études Recherche Oprationelle 17*, 257-260.
19. M. Li (2001). Hamiltonian Cycles in Almost Claw-Free Graphs, *Graphs and Combinatorics 17*, 687–706.
20. X. Li and Z. Zhang (2008). Path factors in the square of a tree, *Graphs and Combinatorics 24*, 107–111.
21. M.D. Plummer (2007). Graph factors and factorization: 1985-2003: A survey, *Discrete Mathematics 307*, 791–821.
22. Z. Ryjáček (1994). Almost claw-free graphs, *Journal of Graph Theory 18*, 469–477.
23. D.P. Sumner (1976). 1-factors and antifactor sets, *Journal of the London Mathematical Society 13*, 351-359.
24. H. Wang (1994). Path factors of bipartite graphs, *Journal of Graph Theory 18*, 161-167.
25. M. Zhan (2002). Neighborhood intersections and Hamiltonicity in almost claw-free graphs, *Discrete Mathematics 243*, 171–185.

Construction of Extremal Graphs

Jianmin Tang[1], Yuqing Lin [1], and Mirka Miller [1,2]

[1] School of Electrical Engineering and Computer Science
The University of Newcastle, NSW 2308, Australia
[2] *Department of Mathematics*
University of West Bohemia, Pilsen, Czech Republic
jianmin.tang@studentmail.edu.au
{yuqing.lin,mirka.miller}@newcastle.edu.au
http://www.newcastle.edu.au

Abstract. By *extremal number* $ex(n;t) = ex(n;\{C_3, C_4, \ldots, C_t\})$ we denote the maximum *size* (that is, number of edges) in a graph of order n and girth at least $g \geq t + 1$. The set of all graphs of order n, containing no cycle of length $\leq t$, and of size $ex(n;t)$, is denoted by $EX(n;t) = EX(n;\{C_3, C_4, \ldots, C_t\})$; such graphs are called *extremal graphs*. In 1975, Erdős mentioned the problem of determining the extremal number $ex(n;4)$ of a graph of order n and girth at least 5. In this paper, we consider a generalized version of this problem, for $t \geq 5$. In particular, we prove that $ex(29;6) = 45$. We also improve upper bounds of $ex_u(n;t)$, for some particular values of n and t. Furthermore, we generate graphs in which the number of edges improves the current lower bounds of $ex(n;6)$ when $n \leq 40$.

Key words: extremal graph, extremal number, cages.

1 Introduction

Throughout this paper, only undirected simple graphs without loops or multiple edges are considered. Unless stated otherwise, we follow [2] for terminology and definitions.

The *vertex set* (respectively, *edge set*) of a graph G is denoted by $V(G)$ (respectively, $E(G)$). The *order* (respectively, *size*) of a graph G is equal to $n = n(G) = |V(G)|$ (respectively, $e = e(G) = |E(G)|$). The set of all the vertices adjacent to a vertex v is called the *neighbourhood* of v, denoted by $N(v)$. The *degree* of a vertex v is $deg(v) = |N(v)|$. We denote by $\delta(G)$ the minimum degree of G and by $\Delta(G)$ the maximum degree of G. A graph is G called *k-regular* when all the vertices in G have the same degree k. The *distance* $d(u,v)$ of two vertices u and v in $V(G)$ is the length of a shortest path between u and v. The longest distance between any two vertices in a graph G is called the *diameter* of G, denoted by D. We also use the notion of *distance between a vertex v and a set of vertices*

F, written $d(v, F)$, which is the distance from the vertex v to a closest vertex in F. The set $N_r(v) = \{w \in V : d(w, v) = r\}$ denotes the *neighbourhood of the vertex v at distance r*. For $F \subset V$, the *neighbourhood of F excluding F itself, at distance r* is denoted by $N_r(F) = \{w \in V \setminus F : d(w, F) = r\}$. When $r = 1$, we write $N(v)$ and $N(F)$, instead of $N_1(v)$ and $N_1(F)$.

The length of a shortest cycle in a graph G is called the *girth* and is denoted by $g = g(G)$. A k-regular graph with girth g is called a (k, g)-*graph*. Recall that a $(k; g)$-graph is called a $(k; g)$-*cage* if it has the least possible number of vertices.

For a graph G, let $C(G)$ denote the set of integers which correspond to the lengths of cycles in G. By the *extremal number* $ex(n; t) = ex(n; \{C_3, C_4, \ldots, C_t\})$ we denote the maximum number of edges in a graph of order n and girth at least $g \geq t + 1$, and by $EX(n; t) = EX(n; \{C_3, C_4, C_5, \ldots, C_t\})$ we denote the set of all graphs of order n, girth at least $t + 1$, having the number of edges equal to $ex(n; t)$; these graphs are also called *extremal graphs*.

In 1975, Erdős [4] mentioned the problem of determining the value of the extremal number, $ex(n; 4)$, the maximum number of edges in a graph of order n with girth at least 5. He conjectured that $ex(n; 4) = (1/2 + o(1))^{3/2} n^{3/2}$. The current best known result [5] is

$$\frac{1}{2\sqrt{2}} \leq \lim_{n \to \infty} \sup \frac{ex(n; 4)}{n^{3/2}} \leq \frac{1}{2}.$$

It is known that $ex(n; 3) = \lfloor n^2/4 \rfloor$, and the extremal graph is $K_{\lfloor n/2 \rfloor, \lceil n/2 \rceil}$.

In [11] Garnic *et al.* developed algorithms, combining hill-climbing and backtracking techniques, to generate graphs with order up to 201 for $ex(n; 4)$, and Wang *et al.* [8] used simulated annealing to generate graphs for several values of n and t; these results provide constructive lower bounds for $ex(n; 4)$. However, in general, the current best lower bounds of $ex(n; t)$ are still far from the current best upper bounds.

In [1], Alon *et al.* found the exact value of the smallest number of vertices $n = n(\bar{d}; g)$ in a graph of girth g and an average degree $\bar{d}$, that is, $\bar{d} = \frac{1}{n} \sum_{x \in V(G)} \deg(x)$.

Theorem 1 *[1] Let $g \geq 3$ and $\bar{d} > 2$. Then*

$$n(\bar{d}, g) = \begin{cases} 1 + \bar{d} \sum_{i=0}^{\frac{g-3}{2}} (\bar{d} - 1)^i & \text{if } g \text{ is odd.} \\ 2\bar{d} \sum_{i=0}^{\frac{g-2}{2}} (\bar{d} - 1)^i & \text{if } g \text{ is even.} \end{cases}$$

Therefore, the theoretical upper bound of $ex(n; t)$, denoted by $ex_u(n; t)$, is obtained by

$$ex_u(n; t) \leq \lfloor \bar{d}n/2 \rfloor. \tag{1}$$

We know that, for particular values of girth and order, there do exist graphs with the largest possible number of edges and minimum degree 1. For example, there are graphs in $EX(11; 4)$ with degree sequences $\{4_1, 3_9, 1_1\}$, but also $\{3_{10}, 2_1\}$ or $\{4_1, 3_8, 2_2\}$. Furthermore, in general, it is believed that the degrees are not far from the average degree [8]. This observation relates the problem of constructing extremal graphs of the family $EX(n; t)$ to the problem of constructing cages. Cages were introduced by Tutte [9]. A $(k; g)$-cage is a k-regular graph with girth g and the smallest possible number of vertices. In most cases, if extremal graphs have a regular degree, then these extremal graphs are cages. For instance, the $(3; 6)$-cage has order 16, regular degree 3, and girth 6; it is a graph in $EX(16; 6)$. However, as pointed out in many papers, these two classes of graphs are not the same. For example, the $(5; 5)$-cage on $n = 30$ vertices has 75 edges, while there exists a graph G in $EX(30; 4)$ with 76 edges, and the degree sequence $\{6_6, 5_{20}, 4_4\}$ [8].

In this paper, we consider a generalized version of determining the value of extremal number of extremal graphs, for $t \geq 5$. In particular, we prove that $ex(29; 6) = 45$. Additionally, some upper bounds of $ex(n; 6)$ and $ex(n; 7)$, when $n \leq 40$, are improved in this paper. Furthermore, we generate some graphs in which the number of edges improves the current lower bounds of $ex(n; 6)$ when $n \leq 40$.

2 Proofs

Since the maximum degree is at least as big as the average degree, we have

$$\Delta(G) \geq \lceil \bar{d} \rceil = \lceil 2e(G)/n(G) \rceil. \tag{2}$$

We have constructed a graph of order $n = 29$ and girth $g = 7$ with size 45, that is, $ex_l(29; 6) = 45$ (see Table 2). We will prove that $ex(29; 6) = 45$.
Firstly, we have proved that $ex_u(29; 6) \leq 46$. Then we assume that there exists a graph $G \in EX(29; 6)$ with 46 edges. Then $\delta(G) = 3$ and $\Delta(G) = 4$ and in the graph G there exist 5 vertices of degree 4. Based on our assumption that $ex(29; 6) = 46$, we obtain

Proposition 1 *If a graph $G \in EX(29; 6)$ and $e(G) = 46$ then the eccentricity of any special vertex is 3.*

Proof The reason for Proposition (1) is that the graph G can be drawn in Figure 1 as it starts from a special vertex, say x, such that $n(G) \geq 1 + |N(x)| + |N_2(x)| + |N_3(x)| = 29$. ∎

By Proposition 1, we obtain

Corollary 1 *If a graph $G \in EX(29; 6)$ and $e(G) = 46$ then every vertex of degree 3 is within 3 steps from any other special vertex.*

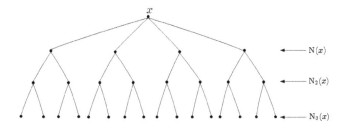

$N(x)$

$N_2(x)$

$N_3(x)$

Fig. 1. The eccentricity of any special vertex is 3.

Making use of Proposition 1 and Corollary 1, we prove that

Theorem 2 *Let a graph* $G \in EX(29; 6)$ *then* $e(G) = ex(29; 6) = 45$.

3 Current lower bounds of $ex(n; t)$

We construct extremal graphs for any value of girth by using the *hybrid simulated annealing and genetic algorithm* (HSAGA). For more details of HSAGA, see the paper [10]. We have modified our algorithm, and then using this algorithm with different values of parameters, such as the *cooling rate* and *crossover rate*, some new results for $t = 6$ have been obtained.

Tables 1, 2 and 3 give the current lower bounds of $ex(n; 5)$, $ex(n; 6)$ and $ex(n; 7)$, for $n \leq 40$. In these tables, n represents the order of graphs; $(ex_l(n; t); \mathcal{D})$ means our new lower bounds, and $\mathcal{D}$ is the degree sequence of a corresponding generated graph. In addition, $ex_u(n; t)$ represents the theoretical upper bounds of extremal graphs. The value of this lower bound $ex_l(n; t)$ is written in bold font, whenever a lower bound of $ex_l(n; t)$ reaches the theoretical upper bound of $ex(n; t)$, Furthermore, the value of new lower bounds and upper bounds are written in italic font. For instance, in Table 1, $ex_l(26; 5) = ex_u(26; 5) = 52$, and its degree sequence is 4_{26}, in fact, this graph that we obtained is a $(4; 6)$-cage. In another example from Table 2, we generated a graph of order $n = 39$ with girth $g = 7$ and size 65, which is the new lower bound of $ex(39; 6)$; and the degree sequence of this graph is $\mathcal{D} = \{4_{15}, 3_{26}, 2_2\}$.

n	$(ex_l(n;t); \mathcal{D})$	$ex_u(n;t)$	n	$(ex_l(n;t); \mathcal{D})$	$ex_u(n;t)$
1			**21**	$(\mathbf{36}; \{4_{10}, 3_{10}, 2_1\})$	36
2			**22**	$(\mathbf{39}; \{4_{12}, 3_{10}\})$	39
3			**23**	$(\mathbf{42}; \{4_{15}, 3_8\})$	42
4			**24**	$(\mathbf{45}; \{4_{18}, 3_6\})$	45
5			**25**	$(\mathbf{48}; \{4_{21}, 3_4\})$	48
6	$(\mathbf{6}; \{2_6\})$	6	**26**	$(\mathbf{52}; \{4_{26}\})$	52
7	$(\mathbf{7}; \{3_1, 2_5, 1_1\})$	7	**27**	$(\mathbf{53}; \{5_1, 4_{24}, 3_1, 2_1\})$	53
8	$(\mathbf{9}; \{3_2, 2_6\})$	9	**28**	$(\mathbf{56}; \{4_{28}\})$	56
9	$(\mathbf{10}; \{3_2, 2_7\})$	10	**29**	$(\mathbf{58}; \{5_2, 4_{26}, 2_1\})$	58
10	$(\mathbf{12}; \{3_4, 2_6\})$	12	**30**	$(\mathbf{61}; \{5_4, 4_{24}, 3_2\})$	61
11	$(\mathbf{14}; \{3_6, 2_5\})$	14	**31**	$(\mathbf{64}; \{5_6, 4_{23}, 3_2\})$	64
12	$(\mathbf{16}; \{3_8, 2_4\})$	16	**32**	$(\mathbf{67}; \{5_6, 4_{26}\})$	67
13	$(\mathbf{18}; \{3_{10}, 2_3\})$	18	**33**	$(\mathbf{70}; \{5_{10}, 4_{21}, 3_2\})$	70
14	$(\mathbf{21}; \{3_{14}\})$	21	**34**	$(\mathbf{74}; \{4_{22}, 3_{12}\})$	74
15	$(\mathbf{22}; \{3_{14}, 2_1\})$	22	**35**	$(\mathbf{77}; \{5_{15}, 4_{19}, 3_1\})$	77
16	$(\mathbf{24}; \{3_{16}\})$	24	**36**	$(\mathbf{81}; \{5_{18}, 4_{18}\})$	81
17	$(\mathbf{26}; \{4_3, 3_{12}, 2_2\})$	26	**37**	$(\mathbf{84}; \{5_{15}, 4_{21}, 3_1\})$	84
18	$(\mathbf{29}; \{4_4, 3_{14}\})$	29	**38**	$(\mathbf{88}; \{5_{24}, 4_{14}\})$	88
19	$(\mathbf{31}; \{4_6, 3_{12}, 2_1\})$	31	**39**	$(\mathbf{92}; \{5_{28}, 4_{11}\})$	92
20	$(\mathbf{34}; \{4_8, 3_{12}\})$	34	**40**	$(\mathbf{96}; \{5_{32}, 4_8\})$	96

Table 1. Current lower bounds $ex_l(n;5)$ and current upper bounds $ex_u(n;5)$ for $n \leq 40$.

n	$(ex_l(n;t); \mathcal{D})$	$ex_u(n;t)$	n	$(ex_l(n;t); \mathcal{D})$	$ex_u(n;t)$
1			**21**	$(\mathbf{29}; \{3_{16}, 2_5\})$	29
2			**22**	$(\mathbf{31}; \{3_{18}, 2_4\})$	31
3			**23**	$(\mathbf{33}; \{3_{20}, 2_3\})$	33
4			**24**	$(\mathbf{36}; \{3_{24}\})$	36
5			**25**	$(\mathbf{37}; \{3_{24}, 2_1\})$	37
6			**26**	$(\mathbf{39}; \{4_2, 3_{22}, 2_2\})$	39
7	$(\mathbf{7}; \{2_7\})$	7	**27**	$(\mathbf{41}; \{5_1, 4_{24}, 3_1, 2_1\})$	41
8	$(\mathbf{8}; \{2_8\})$	8	**28**	$(\mathbf{43}; \{5_2, 4_{23}, 3_2, 2_1\})$	43
9	$(\mathbf{9}; \{3_1, 2_7, 1_1\})$	9	**29**	$(\mathit{45}; \{4_6, 3_{20}, 2_3\})$	$\mathit{45}$
10	$(\mathbf{11}; \{3_2, 2_8\})$	11	**30**	$(\mathit{47}; \{4_4, 3_{26}\})$	$\mathit{48}$
11	$(\mathbf{12}; \{3_3, 2_7, 1_1\})$	12	**31**	$(\mathit{49}; \{4_5, 3_{26}\})$	$\mathit{51}$
12	$(\mathbf{14}; \{3_4, 2_8\})$	14	**32**	$(\mathit{51}; \{4_7, 3_{24}, 2_1\})$	54
13	$(\mathbf{15}; \{3_5, 2_7, 1_1\})$	15	**33**	$(\mathit{53}; \{4_9, 3_{22}, 2_2\})$	56
14	$(\mathbf{17}; \{3_6, 2_8\})$	17	**34**	$(\mathbf{55}; \{4_8, 3_{26}\})$	58
15	$(\mathbf{18}; \{4_1, 3_4, 2_{10}\})$	18	**35**	$(\mathbf{58}; \{4_{11}, 3_{24}\})$	61
16	$(\mathbf{20}; \{3_8, 2_8\})$	20	**36**	$(\mathbf{59}; \{4_{11}, 3_{24}, 2_1\})$	63
17	$(\mathbf{22}; \{3_{10}, 2_7\})$	22	**37**	$(\mathit{61}; \{4_{24}, 3_{12}, 2_1\})$	65
18	$(\mathbf{23}; \{4_1, 3_8, 2_9\})$	23	**38**	$(\mathit{63}; \{4_{14}, 3_{22}, 2_2\})$	68
19	$(\mathbf{25}; \{3_{12}, 2_7\})$	25	**39**	$(\mathbf{65}; \{4_{15}, 3_{22}, 2_2\})$	70
20	$(\mathbf{27}; \{3_{14}, 2_6\})$	27	**40**	$(\mathbf{67}; \{4_{14}, 3_{26}\})$	73

Table 2. Current lower bounds $ex_l(n;6)$ and current upper bounds $ex_u(n;6)$ for $n \leq$ 40.

n	$(ex_l(n;t); \mathcal{D})$	$ex_u(n;t)$	n	$(ex_l(n;t); \mathcal{D})$	$ex_u(n;t)$
1			**21**	$(\mathbf{27}; \{3_{12}, 2_9\})$	27
2			**22**	$(\mathbf{29}; \{3_{14}, 2_8\})$	29
3			**23**	$(\mathbf{30}; \{3_{15}, 2_7, 1_1\})$	30
4			**24**	$(\mathbf{32}; \{3_{16}, 2_8\})$	32
5			**25**	$(\mathbf{34}; \{3_{18}, 2_7\})$	34
6			**26**	$(\mathbf{36}; \{3_{20}, 2_6\})$	36
7			**27**	$(\mathbf{38}; \{3_{22}, 2_5\})$	38
8	$(\mathbf{8}; \{2_8\})$	8	**28**	$(\mathbf{40}; \{3_{24}, 2_4\})$	40
9	$(\mathbf{9}; \{2_9\})$	9	**29**	$(\mathbf{42}; \{3_{26}, 2_3\})$	42
10	$(\mathbf{10}; \{3_2, 2_6, 1_2\})$	10	**30**	$(\mathbf{45}; \{3_{30}\})$	45
11	$(\mathbf{12}; \{3_2, 2_9\})$	12	**31**	$(\mathbf{46}; \{3_{30}, 2_1\})$	46
12	$(\mathbf{13}; \{3_2, 2_{10}\})$	13	**32**	$(\mathbf{47}; \{3_{30}, 2_2\})$	47
13	$(\mathbf{14}; \{3_3, 2_9, 3_3\})$	14	**33**	$(\mathbf{49}; \{4_2, 3_{28}, 2_3\})$	49
14	$(\mathbf{16}; \{3_4, 2_{10}\})$	16	**34**	$(\mathbf{51}; \{4_3, 3_{28}, 2_3\})$	51
15	$(\mathbf{18}; \{3_6, 2_9\})$	18	**35**	$(\mathbf{53}; \{4_5, 3_{26}, 2_4\})$	53
16	$(\mathbf{19}; \{3_7, 2_8, 1_1\})$	19	**36**	$(\mathbf{55}; \{5_1, 4_5, 3_{25}, 2_5\})$	55
17	$(\mathbf{20}; \{3_7, 2_9, 1_1\})$	20	**37**	$(56; \{4_{10}, 3_{26}, 2_1\})$	*58*
18	$(\mathbf{22}; \{3_8, 2_{10}\})$	22	**38**	$(58; \{4_{11}, 3_{26}, 2_1\})$	*60*
19	$(\mathbf{24}; \{3_{10}, 2_9\})$	24	**39**	$(60; \{4_7, 3_{28}, 2_4\})$	63
20	$(\mathbf{25}; \{3_{10}, 2_{10}\})$	25	**40**	$(62; \{4_4, 3_{36}\})$	65

Table 3. Current lower bounds $ex_l(n; 7)$ and current upper bounds $ex_u(n; 7)$ for $n \leq 40$.

4 Future work

Notice that in Tables 1, 2 and 3, there exist a few values of n and t, where the gaps between current $ex_l(n;t)$ and $ex_u(n;t)$ are small. In the future, we plan to work on these small gaps, in order to decrease the size of these gaps. Furthermore, we will aim to prove some exact values of extremal number, for several particular values of n and t.

References

1. N. Alon, S. Hoory and N. Linial, The Moore bound for irregular graphs, *Graphs Combin.* 18 (2002) 53-57.
2. G. Chartrand and L. Lesniak, *Graphs and digraphs*, Third edition, Chapman and Hall, London, 1996.
3. M. Downs, R.J. Gould, J. Mitchem and F. Saba, $(D;n)$-cages, *Congr. Number.* 32 (1981) 179 − 193.
4. P. Erdős, Some recent progress on extremal problems in graph theory, *Congr. Numer.* 14 (1975) 3-14.
5. D.K. Garnick and N.A. Nieuwejaar, Non-isomorphic extremal graphs with three-cylces or four-cycles, *J. Combin. Math. Combin. Comput.* 12 (1993) 33 − 56.
6. Y.H.H. Kwong, D.K. Garnick and F. Lazebnik, Extremal graphs without three-cycles or four-cycles, *J. Graph Theory* 17 (5) (1993) 633-645.
7. F. Lazebnik and P. Wang, On the structure of extremal graphs of high girth, *J. Graph Theory* 26 (1997)
8. P. Wang, G.W. Dueck and S. MacMillan, Using simulated annealing to construct extremal graphs, *Discrete Mathematics* 235 (2001) 125-135.
9. W.T. Tutte, A family of cubical graphs. *Proc. Cambridge Philos. Soc.*, (1947), 459 − 474.
10. J. Tang, Y. Lin, C. Balbuena and M. Miller, Calculating the extremal number $ex(n; \{C_3, C_4, \ldots, C_t\})$. *Journal of Discrete Applied Mathematics*, In Press, Corrected Proof, Feb 2008.
11. Y.H.H. Kwong, D.K. Garnick, F. Lazebnik, Extremal graphs without three-cycles or four-cycles, *J. Graph Theory* 17 (5) (1993) 633-645.

Partitioning Bispanning Graphs into Spanning Trees

Matthias Baumgart

Department of Computer Science, Technische Universität München
Boltzmannstraße 3, D-85748 Garching bei München, Germany
baumgart@in.tum.de

Abstract. Given a weighted bispanning graph $\mathcal{B} = (V, P, Q)$ consisting of two edge-disjoint spanning trees P and Q such that $w(P) < w(Q)$ and Q is the only spanning tree with weight $w(Q)$, it is conjectured that there are $|V| - 1$ spanning trees with pairwise different weight where each of them is smaller than $w(Q)$. This conjecture due to Mayr and Plaxton is proven for bispanning graphs restricted in terms of the underlying weight function and the structure of the bispanning graphs. Furthermore, a slightly stronger conjecture is presented and proven for the latter class.

Keywords: Bispanning graph, edge-disjoint spanning trees, tree graph, number of spanning trees

1 Introduction

Let $G = (V, E)$ be a weighted graph. One of the fundamental problems in computer science and graph theory is to compute a minimum spanning tree (MST) of G, i.e., an acyclic spanning subgraph of minimum weight in G. The history of minimum spanning tree algorithms dates back at least to Borůvka [3] in 1926. The most popular textbook algorithms are those by Kruskal [9] and Prim [11].

Depending on the order in which the vertices are visited, it is possible to obtain different minimum spanning trees provided that there is more than one MST. If there are different minimum spanning trees then it is possible to transform any MST T into another MST by performing exactly one edge swap which means that we remove one edge of T and insert one of the remaining edges.

In this context, the so-called tree graph of a graph G can be defined. The vertex set of this graph is the set of all spanning trees of G with an edge between two spanning trees if and only if they are related by an edge swap. Regarding this tree graph, various questions arise. Some of them were discussed by Kano [7] who proposed four conjectures concerning distances (regarding the number of edge swaps) between different spanning trees in tree graphs motivated by a paper by Kawamoto, Kajitani and Shinoda [8].

One of Kano's conjectures was that any minimum spanning tree can be transformed into a kth smallest spanning tree (a so-called k-MST) by at most $k - 1$ edge swaps. This conjecture was proven by Mayr and Plaxton [10]. Furthermore, they formulated a new conjecture which unified Kano's remaining conjectures.

The present paper addresses this new conjecture and its equivalent formulation that each weighted bispanning graph $\mathcal{B} = (V, P, Q)$ such that $w(P) < w(Q)$ and Q is the only spanning tree with weight $w(Q)$ has at least $|V| - 1$ spanning trees with pairwise different weights strictly less than $w(Q)$. We show that this is true if the spanning tree P is the only spanning tree of weight $w(P)$ or if the given bispanning graph has no minor isomorphic to the complete graph on four vertices K_4.

This paper is organized as follows. In Section 2, we give some definitions and introduce different conjectures concerning tree graphs. The content of the subsequent sections is the analysis of bispanning graphs. The first step of this analysis is the restriction on special weight functions in Section 3. Afterwards, we turn our main focus onto the structure of bispanning graphs. In Section 4, we use techniques from matroid theory, and in Section 5 we show that it suffices to consider weighted bispanning graphs that are 2-vertex-connected and 3-edge-connected.

2 Preliminaries

Throughout this paper, we assume that graphs $G = (V, E)$ are always connected undirected graphs where multiple edges are allowed. We denote the number of vertices by n and the number of edges by m, respectively. Let $w \colon E \to \mathbb{R}$ be a weight function. For any subset $E' \subseteq E$ we define the weight of E', denoted by $w(E')$, as the sum of the weights of all edges in E', that is, $w(E') = \sum_{e \in E'} w(e)$.

A *spanning tree* T of G is any subset of E for which the graph $G = (V, T)$ is acyclic and connected. We denote by $\mathcal{T}(G)$ the set of all spanning trees of G. Given a graph $G = (V, E)$ and a weight function $w \colon E \to \mathbb{R}$, we denote by $\mathcal{W}(G)$ the set of different weights of spanning trees of G and by $\mathcal{W}_i(G)$ the ith smallest element of $\mathcal{W}(G)$. Analogously, we denote by $\mathcal{T}_i(G)$ the set of spanning trees T where $w(T) = \mathcal{W}_i(G)$. We define the order $\mathrm{ord}(G, T)$ of a spanning tree T with respect to G as the number $i \in \mathbb{N}$ such that $T \in \mathcal{T}_i(G)$. We denote the number of spanning trees with weight $w(T)$ by $\sigma(G, T)$, that is, $\sigma(G, T) = |\mathcal{T}_{\mathrm{ord}(G,T)}(G)|$.

Let $G = (V, E)$ be a graph, T a spanning tree of G and $f \in E \setminus T$. We denote by $Cyc(T, f)$ the *fundamental cycle* of G defined by f with respect to T. Given a pair of distinct edges e, f such that $e \in Cyc(T, f)$, we define (e, f) to be a single *edge swap*. We denote by $L_k(G, T)$ the set of all those spanning trees T' of G such that T can be transformed into T' by at most k edge swaps.

The following four conjectures were proposed by Kano [7].

Conjecture 1. If T is a 1-MST of G then $L_{i-1}(G, T)$ contains an i-MST for all $1 \le i \le |\mathcal{W}(G)|$.

Conjecture 2. If T is an i-MST in $L_i(G, T)$ then T is an i-MST of G.

Conjecture 3. If T is an i-MST of G then T is an i-MST in $L_{i-1}(G, T)$.

Conjecture 4. Let $G(i, j)$ denote the graph with vertex set $\mathcal{T}_i(G)$ where an edge exists between each pair of i-MSTs T and T' if T can be transformed into T' by at most j edge swaps in G. Then $G(i, i)$ is connected.

Conjecture 1 was proven by Mayr and Plaxton [10]. Moreover, they gave the following conjecture and showed that this one implies Conjectures 2 through 4.

Conjecture 5. If T is a j-MST of G then $L_{i-1}(G, T)$ contains an i-MST for all $1 \leq j < i \leq |\mathcal{W}(G)|$.

In their proof, Mayr and Plaxton used so-called *bispanning graphs*.

Definition 1. *A graph* $G = (V, E)$ *is called a* bispanning graph *if E is the union of two (edge) disjoint spanning trees P and Q. We denote a bispanning graph by the triple* $\mathcal{B} = (V, P, Q)$.

More precisely, Mayr and Plaxton showed that Conjecture 1 holds if and only if there is no weighted bispanning graph $\mathcal{B} = (V, P, Q)$ such that $1 = \mathrm{ord}(\mathcal{B}, P) < \mathrm{ord}(\mathcal{B}, Q) < n$ and $\sigma(\mathcal{B}, Q) = 1$ holds. Indeed there is no such bispanning graph. Furthermore, they proved that under the assumption that there is no weighted bispanning graph $\mathcal{B} = (V, P, Q)$ such that $\mathrm{ord}(\mathcal{B}, P) < \mathrm{ord}(\mathcal{B}, Q) < n$ and $\sigma(\mathcal{B}, Q) = 1$ hold, Conjecture 5 is a theorem, too. Hence, we would be done by proving the following conjecture implying Conjectures 2 through 5.

Conjecture 6. Let $\mathcal{B} = (V, P, Q)$ be a weighted bispanning graph with $\mathrm{ord}(\mathcal{B}, P) < \mathrm{ord}(\mathcal{B}, Q)$ and $\sigma(\mathcal{B}, Q) = 1$. Then it holds that $\mathrm{ord}(\mathcal{B}, Q) \geq n$.

A powerful tool are contractions and deletions of edges. Let $G = (V, E)$ be a graph and $e, f \in E$ be two edges. We denote by $G[e, f]$ the graph we obtain by contracting the edge e and deleting the edge f. The following lemma is very useful. For a proof, we refer the reader to [10].

Lemma 1. *Let* $G = (V, E)$ *be a weighted graph, T a spanning tree of G, and $e \in E$. If $e \notin T$, consider the graph $G' = G[\emptyset, e]$ and $T' = T$. Otherwise, let $G' = G[e, \emptyset]$ and $T' = T[e, \emptyset]$. In either case, the following statements hold:*

1. T' *is a spanning tree of* G'.
2. $\mathrm{ord}(G', T') \leq \mathrm{ord}(G, T)$.
3. $\sigma(G', T') \leq \sigma(G, T)$.

3 Assuming singularity of both spanning trees

In this section, we will prove Conjecture 6 under the assumption that the spanning tree P is also unique, i.e., the weight function is additionally restricted to satisfy $\sigma(\mathcal{B}, P) = 1$.

Theorem 1. *Let* $\mathcal{B} = (V, P, Q)$ *be a weighted bispanning graph with* $\mathrm{ord}(\mathcal{B}, P) < \mathrm{ord}(\mathcal{B}, Q)$, $\sigma(\mathcal{B}, Q) = 1$, *and* $\sigma(\mathcal{B}, P) = 1$. *Then it holds that* $\mathrm{ord}(\mathcal{B}, Q) \geq n$.

Proof. This Theorem is proven by induction over the number of vertices of $\mathcal{B}$. Clearly, if $n = 2$ then $\mathcal{B}$ consists of two parallel edges with distinct weights, thus $\mathrm{ord}(\mathcal{B}, Q) \geq 2$.

We assume $n > 2$ and consider an arbitrary symmetric exchange (p, q) with $p \in P$ and $q \in Q$, that is, $P \setminus \{p\} \cup \{q\}$ and $Q \setminus \{q\} \cup \{p\}$ are spanning trees. Since $\sigma(\mathcal{B}, Q) = 1$ (or $\sigma(\mathcal{B}, P) = 1$), the edges p and q must have different weight. Now, we distinguish different cases.

1. If $w(q) < w(p)$ holds then we consider the bispanning graph $\mathcal{B}' = \mathcal{B}[q,p]$. The elements of $\mathcal{W}(\mathcal{B}')$ in strictly increasing order are (for the sake of readability, we associate by a spanning tree T also its weight $w(T)$)

$$(A_1', \ldots, A_\alpha', P', B_1', \ldots, B_\beta', Q', C_1', \ldots, C_\gamma')$$

with $P' = P[q,p]$ and $Q' = Q[q,p]$. By Lemma 1, it holds that $\sigma(\mathcal{B}', Q') = 1$, thus, applying the induction hypothesis, we obtain $\operatorname{ord}(\mathcal{B}', Q') \geq n - 1$. Observe that each spanning tree of $\mathcal{B}'$ together with the edge q forms a spanning tree of $\mathcal{B}$. Thus, there are at least $n - 2$ spanning trees with the distinct weights

$$(A_1, \ldots, A_\alpha, P' + w(q), B_1, \ldots, B_\beta)$$

with $A_i = A_i' + w(q)$, $1 \leq i \leq \alpha$, and $B_j = B_j' + w(q)$, $1 \leq j \leq \beta$ such that each of these weights is strictly smaller than $w(Q)$. Since $\sigma(\mathcal{B}, P) = 1$, none of these spanning tree weights can map into the weight $w(P)$, thus, we obtain $\operatorname{ord}(\mathcal{B}, Q) \geq n$. This situation is illustrated in Figure 1(a).

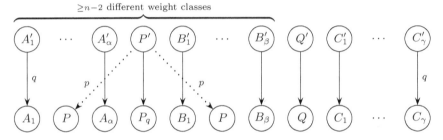

(a) Constructing new classes of spanning trees by adding the edge q (solid lines) to the classes of $\mathcal{B}[q,p]$. Depending on whether $w(q) < w(p)$ or $w(p) < w(q)$ (together with $w(P') < w(Q')$) hold, we get either the right spanning tree P or the left one.

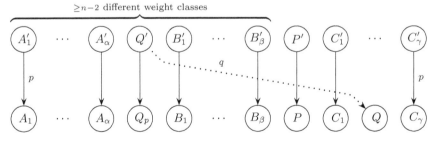

(b) Constructing new classes of spanning trees by adding the edge p (solid lines) to the classes of $\mathcal{B}[p,q]$.

Fig. 1. Mapping of spanning tree weight classes.

2. We assume $w(p) < w(q)$ and

    (a) $w(P) - w(p) < w(Q) - w(q)$. Again, we consider the bispanning graph $\mathcal{B}' = \mathcal{B}[q, p]$ and follow the same line as seen in case 1 (see Figure 1(a)).

    (b) $w(Q) - w(q) < w(P) - w(p)$. We contract p and delete q, that is, we consider the bispanning graph $\mathcal{B}' = \mathcal{B}[p, q]$. The increasing sequence of weights from $\mathcal{W}(\mathcal{B}')$ is

$$(A_1', \ldots, A_\alpha', Q', B_1', \ldots, B_\beta', P', C_1', \ldots, C_\gamma') \ .$$

Now, it holds that $\sigma(\mathcal{B}', P') = 1$. Thus, by induction hypothesis, we have $\operatorname{ord}(\mathcal{B}', P') \geq n - 1$. Combining these spanning trees with the contracted edge p, we get at least $n - 1$ different spanning trees with weights

$$(A_1, \ldots, A_\alpha, Q' + w(p), B_1, \ldots, B_\beta, P)$$

with $A_i = A_i' + w(p)$, $1 \leq i \leq \alpha$, and $B_j = B_j' + w(p)$, $1 \leq j \leq \beta$. Since $w(P) < w(Q)$, we obtain $\operatorname{ord}(\mathcal{B}, Q) \geq n$ and the theorem follows. This situation is illustrated in Figure 1(b). $\qquad\square$

4   Strongly base orderable matroids

In the previous section, we proved that Conjecture 6 holds for weighted bispanning graphs $\mathcal{B} = (V, P, Q)$ restricted to weight functions $w \colon E \to \mathbb{R}$ with the property that both spanning trees, P and Q, have unique weight. An opposed approach is to analyze specially structured bispanning graphs $\mathcal{B} = (V, P, Q)$ where the weight function is only required to satisfy $w(P) < w(Q)$ (equivalent to $\operatorname{ord}(\mathcal{B}, P) < \operatorname{ord}(\mathcal{B}, Q)$) and $\sigma(\mathcal{B}, Q) = 1$. For this reason, we will now introduce the concept of matroids.

Definition 2. *A pair $(E, \mathcal{I})$ consisting of a finite set E and a nonempty family $\mathcal{I}$ of subsets of E is called a* matroid *if $\mathcal{I}$ satisfies*

    (i) $\emptyset \in \mathcal{I}$,
    (ii) $I_1 \in \mathcal{I}$ and $I_2 \subseteq I_1$ imply $I_2 \in \mathcal{I}$, and
    (iii) $I_1, I_2 \in \mathcal{I}$ and $|I_1| < |I_2|$ imply $I_1 \cup \{x\} \in \mathcal{I}$ for some $x \in I_2 \setminus I_1$.

The following proposition is well known from graph theory.

Proposition 1. *Let $G = (V, E)$ be a graph and let $\mathcal{I}$ be the family of all subsets of E such that each $I \in \mathcal{I}$ is a forest of G. Then $M = (E, \mathcal{I})$ is a matroid which is called the* cycle matroid *of G.*

A subset I of E is called *independent* if $I \in \mathcal{I}$, and *dependent* otherwise. An independent subset B of E is called a *base* if there is no subset B' of E such that $B \subset B'$. The bases of the cycle matroid of a connected graph G are the spanning trees of G.

    In the following, we consider bispanning graphs $\mathcal{B} = (V, P, Q)$ where the cycle matroid of $\mathcal{B}$ is strongly base orderable.

Definition 3. *A matroid $M = (E, \mathcal{I})$ is called* strongly base orderable *if there exists a bijection $\varphi \colon B \to B'$ for each two bases B, B' such that for each subset X of B the set $(B \setminus X) \cup \varphi(X)$ is a base, too.*

Lemma 2. *Let $M = (E, \mathcal{I})$ be a strongly base orderable matroid. Then for each two bases B and B' there exists a bijection $\varphi \colon B \to B'$ such that for all subsets X of B the sets $(B \setminus X) \cup \varphi(X)$ and $(B' \setminus \varphi(X)) \cup X$ are bases.*

Proof. By the definition of strongly base orderable matroids, there exists a bijection $\varphi \colon B \to B'$ such that for each $X \subseteq B$ the set $(B \setminus X) \cup \varphi(X)$ is a base. Let X be a subset of B of minimal cardinality such that $(B' \setminus \varphi(X)) \cup X$ is not a base, that is, $(B' \setminus \varphi(X)) \cup X$ contains exactly one cycle C since otherwise X is not minimal. Moreover, because of this minimality, we have $X \subset C$. Let $X' = C \setminus X$ and let $\tilde{X} = \varphi^{-1}(X') \subseteq B$ be the elements of B that map onto an element of X'. Clearly, it holds that $X \cap \tilde{X} = \emptyset$. Furthermore, the set $(B \setminus \tilde{X}) \cup \varphi(\tilde{X})$ contains the cycle C in contradiction to the property of M to be strongly base orderable.

Note that the bases B and B' need not to be disjoint. In this case, the relation for all elements in $B \cap B'$ is the identity. However, in our analysis, we only need the special case of disjoint bases denoted by P and Q.

Theorem 2. *Let $\mathcal{B} = (V, P, Q)$ be a weighted bispanning graph with $\operatorname{ord}(\mathcal{B}, P) < \operatorname{ord}(\mathcal{B}, Q)$ and $\sigma(\mathcal{B}, Q) = 1$. Let $M = (P \cup Q, \mathcal{I})$ be the cycle matroid of $\mathcal{B}$. If M is strongly base orderable then it holds that $\operatorname{ord}(\mathcal{B}, Q) \geq n$.*

Proof. Since M is strongly base orderable, there exists a bijection $\varphi \colon Q \to P$ such that for each subset Q' of Q the set $(Q \setminus Q') \cup \varphi(Q')$ is a spanning tree. Let $w \colon (P \cup Q) \to \mathbb{R}$ be the weight function of $\mathcal{B}$. Clearly, it holds that $w(Q') \neq w(\varphi(Q'))$ for each $Q' \subseteq Q$ since otherwise we have $\operatorname{ord}(\mathcal{B}, Q) > 1$. Let $\delta(q) = w(\varphi(q)) - w(q)$ be the difference of the weights of an edge q and its image $\varphi(q)$ with respect to the weight function w. Thus, the function δ measures the increase of the spanning tree weight after changing the edge q by its image $\varphi(q)$. In general, we define for a subset Q' of Q

$$\delta(Q') = \sum_{q \in Q'} \delta(q) = \sum_{q \in Q'} \big(w(\varphi(q)) - w(q) \big) . \tag{1}$$

If we choose $Q' = Q$, equation (1) implies

$$\delta(Q) = \sum_{q \in Q} \big(w(\varphi(q)) - w(q) \big) = w(P) - w(Q) < 0 \iff w(P) < w(Q) .$$

Now, we arrange the elements of Q in such a way that $\delta(q_1) \leq \ldots \leq \delta(q_{n-1})$ holds and consider the $n - 1$ different sets $Q_i = \{q_1, \ldots, q_i\}$ for $1 \leq i \leq n - 1$. Clearly, for each $1 \leq i \leq n - 1$, the set $(Q \setminus Q_i) \cup \varphi(Q_i)$ is a spanning tree since M is a strongly base orderable matroid. Furthermore, each set Q_i satisfies

$$\delta(Q_i) = \sum_{q \in Q_i} \delta(q) < 0$$

that is, if we remove the edges Q_i from Q and add the edges $\varphi(Q_i)$, the weight of the resulting spanning tree is smaller than $w(Q)$. Thus, if we show that these spanning tree weights are distinct, the claim follows. Let i and j be two indices such that $\delta(Q_i) = \delta(Q_j)$. In this case, it holds that

$$\delta(Q') = \sum_{q \in Q'} \delta(q) = \sum_{q \in Q'} \left(w(\varphi(q)) - w(q) \right)$$
$$= \sum_{q \in Q_j} \left(w(\varphi(q)) - w(q) \right) - \sum_{q \in Q_i} \left(w(\varphi(q)) - w(q) \right)$$
$$= \delta(Q_j) - \delta(Q_i) = 0$$

which implies that the weight of the spanning tree $(Q \setminus Q') \cup \varphi(Q')$ is equal to $w(Q)$ contradicting $\sigma(\mathcal{B}, Q) = 1$. Hence, there are at least $n - 1$ spanning trees with distinct weight such that each of them is smaller than $w(Q)$ implying $\mathrm{ord}(\mathcal{B}, Q) \geq n$ which proves the theorem. □

Observe that because of Lemma 2, we only counted spanning trees T of a bispanning graph $\mathcal{B} = (V, P, Q)$ such that the remaining edges $(P \cup Q) \setminus T$ also form a spanning tree. There are strong indications that these spanning trees are sufficient to prove Conjecture 6. This approach will be pursued in the next section.

A questions which arises now is how to distinguish bispanning graphs whose cycle matroid is strongly base orderable from bispanning graphs that do not have this property? The following two theorems are well known from matroid theory and give an answer to this question. For a proof, we refer the reader to [2, 13].

Theorem 3. *Transversal matroids are strongly base orderable.*

Theorem 4. *Let $G = (V, E)$ be a finite graph. Then its cycle matroid $M = (E, \mathcal{I})$ is transversal if and only if G contains no minor isomorphic to K_4 or C_k^2 $(k > 2)$.*

Since a bispanning graph cannot contain a minor isomorphic to C_k^2 $(k > 2)$, we have to focus on graphs which contain a K_4 minor.

Corollary 1. *Let $\mathcal{B} = (V, P, Q)$ be a finite bispanning graph. Then its cycle matroid is transversal if and only if $\mathcal{B}$ contains no minor isomorphic to K_4.*

5 Partitioning bispanning graphs

In this section, we merge the ideas of the previous two sections. In Section 3, we have seen that Conjecture 6 holds under the assumption that the weight function is required to satisfy $\sigma(\mathcal{B}, P) = 1$. In general, the weight of spanning tree P is not unique even if the number of vertices is small. The main observation in this case is that given a spanning tree $T \neq P$ where $w(T) = w(P)$ the remaining edges $E \setminus T$ contain at least one cycle. To avoid the problem of $\sigma(\mathcal{B}, P) > 1$, we introduce the concept of partitioning bispanning graphs into spanning trees which was already indicated in Section 4. This approach leads to a somewhat stronger conjecture compared to Conjecture 6.

Definition 4. *Let $\mathcal{B} = (V, P, Q)$ be a bispanning graph. A spanning tree T of $\mathcal{B}$ is called a* partition spanning tree *if and only if its complement $E \setminus T$ is a spanning tree, too.*

Let $\mathcal{B} = (V, P, Q)$ be a bispanning graph. We denote by $\mathcal{T}'(\mathcal{B})$ the set of all partition spanning trees of $\mathcal{B}$. Given a weight function $w\colon (P \cup Q) \rightarrow \mathbb{R}$, we denote by $\mathcal{W}'(\mathcal{B})$ the set of different weights of spanning trees of $\mathcal{B}$ and by $\mathcal{W}'_i(\mathcal{B})$ the ith smallest element of $\mathcal{W}'(\mathcal{B})$. Moreover, $\mathcal{T}'_i(\mathcal{B})$ is the set of partition spanning trees T where $w(T) = \mathcal{W}'_i(\mathcal{B})$. We define the order $\mathrm{ord}'(\mathcal{B}, T)$ of a partition spanning tree T with respect to $\mathcal{B}$ as the number $i \in \mathbb{N}$ such that $T \in \mathcal{T}'_i(\mathcal{B})$. The number of partition spanning trees with weight $w(T)$ is denoted by $\sigma'(\mathcal{B}, T)$, that is, $\sigma'(\mathcal{B}, T) = |\mathcal{T}'_{\mathrm{ord}'(\mathcal{B},T)}(\mathcal{B})|$.

Conjecture 7. Let $\mathcal{B} = (V, P, Q)$ be a weighted bispanning graph with $\mathrm{ord}(\mathcal{B}, P) < \mathrm{ord}(\mathcal{B}, Q)$ and $\sigma(\mathcal{B}, Q) = 1$. Then it holds that $\mathrm{ord}'(\mathcal{B}, Q) \geq n$.

If Conjecture 7 holds then it implies immediately Conjecture 6 since $\mathrm{ord}'(\mathcal{B}, T) \leq \mathrm{ord}(\mathcal{B}, T)$ for all partition spanning trees T of $\mathcal{B}$.

5.1 Strictly 2-edge-connected bispanning graphs

In this section, we consider strictly 2-edge-connected bispanning graphs and show that each of these bispanning graphs can be reduced to some 3-edge-connected bispanning graph under the assumption that it is only necessary to count partition spanning trees.

Theorem 5. *Let $\mathcal{B} = (V, P, Q)$ be a weighted bispanning graph with edge connectivity $\lambda = 2$, $\mathrm{ord}(\mathcal{B}, P) < \mathrm{ord}(\mathcal{B}, Q)$, and $\sigma(\mathcal{B}, Q) = 1$. Then there are two edges $p \in P$ and $q \in Q$ such that $\mathrm{ord}'(\mathcal{B}[q, p], Q[q, p]) < \mathrm{ord}'(\mathcal{B}, Q)$.*

Proof. Since $\lambda = 2$, there exists a cut $(V', V \setminus V')$ in G with exactly two edges between V' and $V \setminus V'$. Clearly, one of these edges belongs to P and the other one belongs to Q since otherwise either P or Q is not a spanning tree. We denote by p the edge which belongs to P and by q the edge which belongs to Q, respectively. Now, we consider the bispanning graph $\mathcal{B}' = \mathcal{B}[q, p]$ and observe that each partition spanning tree of this graph can be combined either with p or with q yielding a partition spanning tree of $\mathcal{B}$. Depending on the weight of p and q the set of different partition spanning tree weights of $\mathcal{B}'$ in increasing order is either

$$(A'_1, \ldots, A'_\alpha, P', B'_1, \ldots, B'_\beta, Q', C'_\alpha, \ldots, C'_1) \tag{2}$$

or

$$(A'_1, \ldots, A'_\mu, Q', B'_1, \ldots, B'_\nu, P', C'_\mu, \ldots, C'_1) \tag{3}$$

where, for the sake of readability, we associate with a tree T also its weight $w(T)$.

If (2) holds, we combine each partition spanning tree with the edge q resulting in $\alpha + \beta + 1$ partition spanning trees with distinct weight where each of them is smaller than $w(Q)$. Since none of these weights can map into $w(P)$, there are $\alpha + \beta + 2$ implying $\mathrm{ord}'(\mathcal{B}', Q') < \mathrm{ord}'(\mathcal{B}, Q)$.

In the case that (3) holds, we combine each partition spanning tree with the edge p. Since $w(P' \cup \{p\}) < w(Q' \cup \{q\})$, we arrive at $\mathrm{ord}'(\mathcal{B}', Q') < \mathrm{ord}'(\mathcal{B}, Q)$ and the claim follows. □

Analogous to Theorem 5, it is possible to omit multiple edges even if the edge connectivity of $\mathcal{B}$ is greater than two. The consequence is that we only have to consider 3-edge-connected bispanning graphs without multiple edges to prove Conjecture 7.

5.2 Breaking up bispanning graphs with articulation vertices

In the previous subsection, we have seen that we can turn our attention to 3-edge-connected bispanning graphs. Now, we want to show that it is also sufficient to consider only 2-vertex-connected bispanning graphs, i.e., graphs that do not contain any articulation vertex. Given a graph $G = (V, E)$, we call a vertex $v \in V$ an articulation vertex (or cut vertex) if the removal of v and its adjacent edges will disconnect the graph.

In the following theorem, we assume that the bispanning graph $\mathcal{B}$ contains exactly one articulation vertex. This is no restriction since it is possible to decompose each bispanning graph into its 2-vertex-connected components and inductively apply the theorem.

Theorem 6. *Let $\mathcal{B} = (V, P, Q)$ be a weighted bispanning graph with $\mathrm{ord}(\mathcal{B}, P) < \mathrm{ord}(\mathcal{B}, Q)$ and $\sigma(\mathcal{B}, Q) = 1$ which consists of exactly one articulation vertex $v \in V$. Let $\mathcal{B}_1 = (V_1, P_1, Q_1)$ and $\mathcal{B}_2 = (V_2, P_2, Q_2)$ be the two 2-vertex-connected components. Then both components are weighted bispanning graphs and if Conjecture 7 holds for both of them, Conjecture 7 holds for $\mathcal{B}$.*

In the proof of Theorem 5, we combined partition spanning trees with the edges p and q. Now we have to combine partition spanning trees of $\mathcal{B}_1$ with partition spanning tree of $\mathcal{B}_2$. Since this is somewhat more difficult, we take a look at the following two lemmas because they will simplify the proof of Theorem 6.

Lemma 3. *Let $X = (a_1, \ldots, a_\alpha, p_1, b_1, \ldots, b_\beta, q_1)$ and $Y = (c_1, \ldots, c_\mu, p_2, d_1, \ldots, d_\nu, q_2)$ be strictly increasing sequences of numbers. Let S be the set of all possible sums of two elements $x \in X$ and $y \in Y$. Then there are at least $\alpha + \beta + \mu + \nu + 2$ distinct $s \in S$ such that $s < q_1 + q_2$.*

Proof. We consider the chain of distinct sums $(a_1 + c_1) < (a_1 + c_2) < \ldots < (a_1 + c_\mu) < (a_1 + p_2) < (a_1 + d_1) < (a_1 + d_2) < \ldots < (a_1 + d_\nu) < (a_1 + q_2) < (a_2 + q_2) < \ldots < (a_\alpha + q_2) < (p_1 + q_2) < (b_1 + q_2) < (b_2 + q_2) < \ldots < (b_\beta + q_2) < (q_1 + q_2)$. Hence, there are $\alpha + \beta + \mu + \nu + 2$ distinct sums smaller than $q_1 + q_2$. □

Lemma 3 will help us to prove the case if $\mathrm{ord}'(\mathcal{B}_1, P_1) < \mathrm{ord}'(\mathcal{B}_1, Q_1)$ and $\mathrm{ord}'(\mathcal{B}_2, P_2) < \mathrm{ord}'(\mathcal{B}_2, Q_2)$ holds. On the other hand, the next lemma will help us to analyze the remaining case that either $\mathrm{ord}'(\mathcal{B}_1, Q_1) < \mathrm{ord}'(\mathcal{B}_1, P_1)$ or $\mathrm{ord}'(\mathcal{B}_2, Q_2) < \mathrm{ord}'(\mathcal{B}_2, P_2)$ holds. The case $\mathrm{ord}'(\mathcal{B}_1, Q_1) < \mathrm{ord}'(\mathcal{B}_1, P_1)$ and $\mathrm{ord}'(\mathcal{B}_2, Q_2) < \mathrm{ord}'(\mathcal{B}_2, P_2)$ is impossible since it implies $w(Q) < w(P)$.

Lemma 4. *Let $X = (a_1, \ldots, a_\alpha, q_1, b_1, \ldots, b_\beta, p_1)$ and $Y = (c_1, \ldots, c_\mu, p_2, d_1, \ldots, d_\nu, q_2)$ be strictly increasing sequences of numbers with the following restrictions. Let E_X, E_Y, F_X, and F_Y defined depending on X and Y as*

$$E_X = \{p_1 - x \mid x \in X \setminus \{p_1\}\} \qquad E_Y = \{q_2 - y \mid y \in Y \setminus \{q_2\}\}$$
$$F_X = \{b_i - q_1 \mid 1 \le i \le \beta\} \qquad F_Y = \{d_j - p_2 \mid 1 \le j \le \nu\} \ .$$

We assume X and Y satisfy $F_X \subseteq E_X$, $F_Y \subseteq E_Y$, $F_X \cap F_Y = \emptyset$, and $p = p_1 + p_2 < q_1 + q_2 = q$. If $p = x + y$ holds if and only if $x = p_1$ and $y = p_2$ where $x \in X$ and $y \in Y$ then the set of all possible sums S of two elements $x \in X$ and $y \in Y$ consists of at least $\alpha + \beta + \mu + \nu + 2$ distinct elements $s \in S$ such that $s < q_1 + q_2$.

Proof. We consider the chain of pairwise different sums $(a_1 + c_1) < (a_1 + c_2) < \ldots < (a_1 + c_\mu) < (a_1 + p_2) < (a_2 + p_2) < \ldots < (a_\alpha + p_2) < (q_1 + p_2) < (b_1 + p_2) < (b_2, p_2) < \ldots < (b_\beta + p_2) < (p_1 + p_2)$. Obviously, all of these sums are less than $q_1 + q_2$, that is, we have already found $\alpha + \beta + \mu + 2$ distinct sums.

Now, we consider the sums that are formed by q_1 and d_j with $1 \le j \le \nu$. Obviously, these sums are distinct where each of them is greater than $q_1 + p_2$ and smaller than $q_1 + q_2$. Since $p = p_1 + p_2$ is unique, they can only conflict with some pair $b_i + p_2$ in the chain given above. Therefore, we assume $q_1 + d_j = b_i + p_2 \iff d_j - p_2 = b_i - q_1$ for arbitrarily chosen $1 \le i \le \beta$ and $1 \le j \le \nu$. But this is a contradiction to our assumption $F_X \cap F_Y = \emptyset$ since $b_j - q_1 \in F_X$ and $d_i - p_2 \in F_Y$. This proves the lemma. $\square$

Proof (Theorem 6). As described above, we have to distinguish between the following cases. Either it holds that

$$\text{ord}'(\mathcal{B}_1, P_1) < \text{ord}'(\mathcal{B}_1, Q_1) \text{ and } \text{ord}'(\mathcal{B}_2, P_2) < \text{ord}'(\mathcal{B}_2, Q_2) \tag{4}$$

or it holds that

$$\text{ord}'(\mathcal{B}_1, P_1) < \text{ord}'(\mathcal{B}_1, Q_1) \text{ and } \text{ord}'(\mathcal{B}_2, P_2) > \text{ord}'(\mathcal{B}_2, Q_2) \ . \tag{5}$$

In the first case (4), the ordered sequences of all partition spanning tree weights of $\mathcal{B}_1$ and $\mathcal{B}_2$ are

$$X = (A_1, \ldots, A_\alpha, P_1, B_1, \ldots, B_\beta, Q_1, \tilde{A}_\alpha, \ldots, \tilde{A}_1) \text{ and}$$
$$Y = (C_1, \ldots, C_\mu, P_2, D_1, \ldots, D_\nu, Q_2, \tilde{C}_\mu, \ldots, \tilde{C}_1) \ .$$

Note that $\sigma(\mathcal{B}_1, Q_1) = \sigma(\mathcal{B}_2, Q_2) = 1$ since otherwise $\sigma(\mathcal{B}, Q) > 1$. If Conjecture 7 holds then we have $\text{ord}'(\mathcal{B}_1, Q_1) \ge n_1$ and $\text{ord}'(\mathcal{B}_2, Q_2) \ge n_2$ with $n_1 = |V_1|$ and $n_2 = |V_2|$. Clearly, if we combine a partition spanning tree of $\mathcal{B}_1$ with an partition spanning tree of $\mathcal{B}_2$ we get a partition spanning tree of $\mathcal{B}$. The number of vertices in $\mathcal{B}$ is $|V| = n = n_1 + n_2 - 1$. Thus, if we can construct $n_1 + n_2 - 2$ partition spanning trees with distinct weights where each weight is smaller than $w(Q)$, we are done. Applying Lemma 3 we obtain $\alpha + \beta + 1 \ge n_1 - 1$

and $\mu + \nu + 1 \geq n_2 - 1$. Hence, it holds that $\alpha + \beta + \mu + \nu + 2 \geq n_1 + n_2 - 2$ which implies $\mathrm{ord}'(\mathcal{B}, Q) \geq n_1 + n_2 - 1 = n$.

All combinations are illustrated in Figure 2 where it is easy to see that all of them lead to partition spanning trees of different weights since there are no crossing lines.

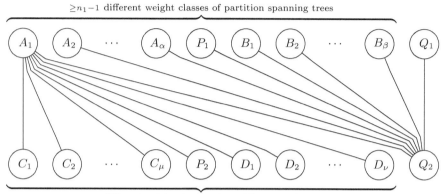

Fig. 2. Combinations if $\mathrm{ord}'(\mathcal{B}_1, P_1) < \mathrm{ord}'(\mathcal{B}_1, Q_1)$ and $\mathrm{ord}'(\mathcal{B}_2, P_2) < \mathrm{ord}'(\mathcal{B}_2, Q_2)$.

Assume now that (5) holds. Again, we consider the ordered sequences of all partition spanning tree weights of $\mathcal{B}_1$ and $\mathcal{B}_2$ which are

$$X = (A_1, \ldots, A_\alpha, Q_1, B_1, \ldots, B_\beta, P_1, \tilde{A}_\alpha, \ldots, \tilde{A}_1)$$
$$Y = (C_1, \ldots, C_\mu, P_2, D_1, \ldots, D_\nu, Q_2, \tilde{C}_\mu, \ldots, \tilde{C}_1) \ .$$

Analogous to the previous case, we have $\sigma(\mathcal{B}_1, Q_1) = \sigma(\mathcal{B}_2, Q_2) = 1$ as well as $\mathrm{ord}'(\mathcal{B}_2, Q_2) \geq n_2$. Because of symmetry, we obtain $\mathrm{ord}'(\mathcal{B}_1, P_1) \geq n_1$ and $\alpha + \beta + 1 \geq n_1 - 1$. On the other hand, it holds that $\mu + \nu + 1 \geq n_2 - 1$. Let E_X, E_Y, F_X, and F_Y be defined as follows (for the sake of readability, we write T instead of $w(T)$)

$$\begin{array}{ll} E_X = \{P_1 - x \mid x \in X \setminus \{P_1\}\} & E_Y = \{Q_2 - y \mid y \in Y \setminus \{Q_2\}\} \\ F_X = \{B_i - Q_1 \mid 1 \leq i \leq \beta\} & F_Y = \{D_j - P_2 \mid 1 \leq j \leq \nu\} \ . \end{array}$$

It holds that $F_X \subseteq E_X$ and $F_Y \subseteq F_Y$ (again because of symmetry). Assuming $F_X \cap F_Y \neq \emptyset$, there exist indices $1 \leq i \leq \beta$ and $1 \leq j \leq \nu$ such that $B_i - Q_1 = D_j - P_2 = Q_2 - D_{n-j} \iff Q_1 + Q_2 = B_i + D_{n-j}$ resulting in a contradiction to $\sigma(\mathcal{B}, Q) = 1$. Hence, it holds that $F_X \cap F_Y = \emptyset$. Furthermore, P is the only partition spanning tree with weight $w(P)$. Therefore, we can apply Lemma 4 and construct $\alpha + \beta + \mu + \nu + 2 \geq n_1 + n_2 - 2$ partition spanning trees with distinct weights smaller than $w(Q)$ (see Figure 3). Hence, we arrive at $\mathrm{ord}'(\mathcal{B}, Q) \geq n_1 + n_2 - 1 = n$ proving the theorem. $\qquad \square$

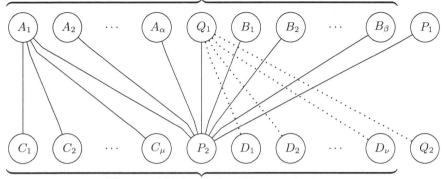

Fig. 3. Combinations if $\mathrm{ord}'(\mathcal{B}_1, Q_1) < \mathrm{ord}'(\mathcal{B}_1, P_1)$ and $\mathrm{ord}'(\mathcal{B}_2, P_2) < \mathrm{ord}'(\mathcal{B}_2, Q_2)$.

5.3 Partitioning the complete graph on four vertices

We have seen that we can separately analyze the 2-vertex-connected and 3-edge-connected bispanning components of an arbitrary bispanning graph. The following theorem is well known from graph theory [6].

Theorem 7. *A 2-connected simple graph in which the degree of every vertex is at least 3 has a minor isomorphic to the complete graph K_4.*

Hence, each of these 2-vertex-connected and 3-edge-connected bispanning graphs $\mathcal{B} = (V, P, Q)$ has a K_4 minor and therefore their cycle matroids are *not* strongly base orderable. Nevertheless, Conjecture 7 holds even if the given weighted bispanning graph $\mathcal{B} = (V, P, Q)$ is isomorphic to the complete graph on four vertices with a weight function satisfying $\mathrm{ord}(\mathcal{B}, P) < \mathrm{ord}(\mathcal{B}, Q)$ and $\sigma(\mathcal{B}, Q) = 1$.

Theorem 8. *Let $\mathcal{B} = (V, P, Q)$ be a weighted bispanning graph on four vertices such that $\mathrm{ord}(\mathcal{B}, P) < \mathrm{ord}(\mathcal{B}, Q)$ and $\sigma(\mathcal{B}, Q) = 1$. Then it holds $\mathrm{ord}'(\mathcal{B}, Q) \geq 4$.*

Proof. If $\mathcal{B} = (V, P, Q)$ contains any multiple edges then its cycle matroid is strongly base orderable and we are done. Now, we consider the complete graph K_4 and observe that there exists up to isomorphism only one assignment of the edges to two disjoint spanning trees. According to this assignment, there are 12 different partition spanning trees where at least three of them have weight less than $w(Q)$. For further details, we refer the interested reader to [1]. □

Since Conjecture 7 is at least as strong as Conjecture 6 (in fact we only consider a smaller class of spanning trees), we obtain the following corollary.

Corollary 2. *Let $\mathcal{B} = (V, P, Q)$ be a weighted bispanning graph on four vertices such that $\mathrm{ord}(\mathcal{B}, P) < \mathrm{ord}(\mathcal{B}, Q)$ and $\sigma(\mathcal{B}, Q) = 1$. Then it holds $\mathrm{ord}(\mathcal{B}, Q) \geq 4$.*

6 Summary

In this paper, we discussed a conjecture of Mayr and Plaxton [10] and its equivalent formulation viz. that all weighted bispanning graphs $\mathcal{B} = (V, P, Q)$ satisfying $\text{ord}(\mathcal{B}, P) < \text{ord}(\mathcal{B}, Q)$ and $\sigma(\mathcal{B}, Q) = 1$ have $|V| - 1$ distinct spanning trees with pairwise different weights strictly smaller than $w(Q)$. We have shown that this conjecture is true when we restrict ourselves to special weight functions (see Section 3) or to specially structured bispanning graphs (see Section 4). Furthermore, we formulated a slightly stronger conjecture where we count only so-called partition spanning trees of bispanning graphs. Under this stronger conjecture, it is sufficient to analyze 2-vertex-connected and 3-edge-connected bispanning graphs. The complete graph K_4, which is the smallest graph of this class, has the desired property. Unfortunately, there only exists a proof which is a tedious case analysis. It remains open to find a shorter proof which might be extended to other 2-vertex-connected and 3-edge-connected bispanning graphs.

References

1. M. Baumgart. Partitioning bispanning graphs into spanning trees. Technical Report TUM-I0813, Institut für Informatik, Technische Universität München, 2008. http://www.in.tum.de/forschung/pub/reports/2008/TUM-I0813.ps.gz
2. J. A. Bondy. Transversal matroids, base-orderable matroids and graphs. *The Quarterly Journal of Mathematics*, 23(1):81–89, 1972.
3. O. Borůvka. O jistém problému minimálním. *Práca Moravské Přírodověcké Společnosti*, 3:37–58, 1926. (in Czech).
4. R.A. Brualdi. Comments on bases in dependence structures. *Bulletin of the Australian Mathematical Society*, 1:161–167, 1969.
5. T.H. Brylawski. Some properties of basic families of subsets. *Discrete Mathematics*, 6:333–341, 1973.
6. G.A. Dirac. A property of 4-chromatic graphs and some remarks on critical graphs. *Journal of the London Mathematical Society*, 27:85–92, 1952.
7. M. Kano. Maximum and k-th maximal spanning trees of a weighted graph. *Combinatorica*, 7(2):205–214, 1987.
8. T. Kawamoto, Y. Kajitani, and S. Shinoda. On the second maximal spanning trees of a weighted graph. *Transactions of the IECE of Japan*, 61-A:988–995, 1978. (in Japanese).
9. J.B. Kruskal. On the shortest spanning subtree of a graph and the traveling salesman problem. *Proceedings of the American Mathematical Society* volume 7, 48–50, 1956.
10. E.W. Mayr and C.G. Plaxton. On the spanning trees of weighted graphs. *Combinatorica*, 12(4):433–447, 1992.
11. R.C. Prim. Shortest connection networks and some generalizations . *The Bell System Technical Journal*, 36:1389–1401, 1957.
12. A. Schrijver. *Combinatorial Optimization*. Springer Berlin, Heidelberg, New York, 2003.
13. D. Welsh. *Matroid Theory*. Academic Press, 1976.

On Irreducibility of Maximal Cliques*

Tao-Ming Wang[1][**], Peter Che Bor Lam[1], Jun-Lin Kuo[1], and Feng-Rung Hu[2]

[1] Department of Mathematics
Tunghai University
Taichung, 40704, Taiwan, ROC
{wang,cblam,jkuo}@thu.edu.tw
[2] Department of Mathematical Education
National Taichung Education University
Taichung, 40306, Taiwan, ROC
fengrung@math.ntcu.edu.tw

Abstract. An edge maximal clique covering of a graph is a set of maximal cliques that contains every edge. In 1990, W. D. Wallis and G.-H. Zhang introduced the concept of essential maximal cliques, which are maximal cliques containing an edge that is not in any other maximal clique. They studied the graphs for which the set of all maximal cliques forms an edge maximal clique covering of minimum size, namely maximal clique irreducible graphs. In 2003, T.-M. Wang introduced and studied the class of graphs for which the set of all essential maximal cliques forms an edge maximal clique covering of minimum size, namely weakly maximal clique irreducible graphs. On the other hand, the Helly property of the set of maximal cliques is related to the maximal clique irreducibility. A graph is clique-Helly if the set of all maximal cliques satisfies the Helly property, and it is called hereditary clique-Helly if every induced subgraph is clique-Helly. In 1993, E. Prisner showed that hereditary clique-Helly graphs and hereditary maximal clique irreducible graphs are the same, and in general clique-Helly graphs and maximal clique irreducible graphs are mutually exclusive. In this article, we survey on (weakly) maximal clique irreducible line graphs and the line graphs with a unique set of maximal cliques covering all the edges. In particular we show that all line graphs have a unique set, up to isomorphisms, of maximal cliques covering all the edges. Finally we characterize the graphs which are clique-Helly but not hereditary clique-Helly, and characterize the graphs which are maximal clique irreducible but not hereditary maximal clique irreducible.

1 Introduction and Background

In this paper, all graphs considered are finite, undirected, and without multiple edges or loops. If G and H are two graphs, then G is said to be H-free if it

---
* Supported partially by the National Science Council under Grants NSC 96-2115-M-029-001 and NSC 96-2115-M-029-007
** The corresponding author

contains no induced subgraph isomorphic to H. For terminology not defined here, please see [16].

A clique in a graph G is a subset Q of the vertex set $V(G)$ such that every two vertices in Q are adjacent. The complete subgraph induced by a clique is also called a clique. A clique is maximal if it is not properly contained in any other clique. An edge (maximal respectively) clique covering of a graph G is a set of (maximal respectively) cliques that contains every edge in G. We denote by $cc(G)$ the least number of cliques in an edge clique covering of a graph G. A maximal clique containing an edge which is not in any other maximal clique is of our interest:

Definition 1. *Let $M(G)$ denote the set of all maximal cliques in a graph G, and $m(G) = |M(G)|$, the number of maximal cliques in G. We call a maximal clique $Q \in M(G)$ essential if there exists an edge e in Q that is not in any other maximal clique in G, and in-essential otherwise. Also let $EM(G)$ denote the set of all essential maximal cliques in G, and $em(G) = |EM(G)|$, the number of essential maximal cliques in G.*

In [8] we have the following observations for edge clique covering and edge maximal clique covering of graphs:

Proposition 1. *For any graph G,*

1. $EM(G) \subseteq q \subseteq M(G)$, *where q is an edge maximal clique covering of G;*
2. $em(G) \le cc(G) \le m(G)$;
3. $em(G) = cc(G)$ *if and only if $EM(G)$ is an minimum edge clique covering;*
4. $cc(G) = m(G)$ *if and only if $M(G)$ is an minimum edge clique covering;*
5. $cc(G) = m(G)$ *if and only if $M(G) = EM(G)$;*
6. $cc(G) = m(G) \Rightarrow em(G) = cc(G) \Rightarrow$ *there exists a unique minimum edge maximal clique covering up to isomorphisms.*

A graph G satisfying $cc(G) = m(G)$, or equivalently each maximal clique is essential, is said to be maximal clique irreducible. The concept was introduced and studied in 1990 by W.D. Wallis and G.-H. Zhang[13].

On the other hand, a graph is clique-Helly if the set of all maximal cliques satisfies the Helly property, meaning that the total intersection of any pairwise intersecting sub-collection is nonempty. A graph is called hereditary clique-Helly if every induced subgraph is clique-Helly. In 1993, E. Prisner[11] studied the relationship between clique-Helly graphs and maximal clique irreducible graphs. He showed that in particular hereditary clique-Helly graphs and hereditary maximal clique irreducible graphs are the same, and in general clique-Helly graphs and maximal clique irreducible graphs are mutually exclusive, by giving examples.

In this paper, we first study maximal clique irreducible line graphs, weakly maximal clique irreducible line graphs, as two special classes of the line graphs with a unique set of maximal cliques covering all the edges. We study the properties of the graphs G with $m(G) = cc(G)$, the properties of graphs with $em(G) = cc(G)$, and also the properties of the graphs G with unique edge maximal clique covering (of minimum size). Then we make use of them to characterize, up to isomorphisms, all line graphs which have a unique set of maximal cliques covering all the edges. Finally, we give a characterization of the graphs which are clique-Helly but not hereditary clique-Helly, and also a characterization of the graphs which are irreducible but not hereditary irreducible. These characterizations make E. Prisner's results more clear and the relationship between above two notions is clarified.

2   Maximal Clique Irreducible Graphs

It was noticed by Opsut and Roberts [10] in 1981 that, any interval graph is maximal clique irreducible. Therefore by the property 6 in the above Proposition 1, the class of interval graphs serves as a good example satisfying all three conditions. That is, interval graphs have a unique set of maximal cliques covering all edges, in particular.

W.D. Wallis and G.-H. Zhang in [13] have obtained a result that if a graph G is not maximal clique irreducible, then G must contain an induced subgraph isomorphic to F_1, F_2, F_3, or F_4. (Please see Figure 1) It is equivalent to the following characterization theorem 3, and we restate it in a forbidden subgraph characterization format. Let's start with a lemma and its application first:

Lemma 1. *For any in-essential maximal clique Q of a graph, there must exist an induced triangle in Q such that the three edges belong to three different maximal cliques (on at least three vertices) respectively other than Q itself.*

Proof. Let $Q_1, Q_2, ..., Q_m$ be a collection of least number of maximal cliques other than Q which covers all the edges of Q. Without loss of generality, there must exist an edge $uv \in E(Q \cap Q_1)$, but uv is not in $E(Q_2 \cup ... \cup Q_m)$, otherwise the size of the collection can be reduced. Now pick a vertex $w \in Q - Q_1$, say $w \in Q_2$, then we claim the induced triangle on these three vertices u, v, and w is the desired one in Q. The reason is that $uw \in E(Q_2)$, and $vw \in E(Q_3)$ (without loss of generality), where w is not in Q_1, v is not in Q_2, and u is not in Q_3. Note that if $v \in Q_2$ or $u \in Q_3$, then uv will be in Q_2 or Q_3, a contradiction.

Q.E.D.

Note that every maximal K_2 is always essential. Therefore, if a graph G has an in-essential maximal clique Q, then $|Q| \geq 3$, and G has at least four maximal cliques (including Q) on three or more vertices, by the above Lemma 1. In other words,

Proposition 2. *For a graph G, if the number of maximal cliques on three or more vertices are less than four, then we have $cc(G) = m(G)$, $cc(G) = em(G)$, and there exists a unique minimum edge maximal clique covering for G.*

Again this class of graphs provides with examples with a unique minimum edge maximal clique covering, as interval graphs do.

Moreover, W.D. Wallis and G.-H. Zhang showed in [13] the following characterization of hereditary class of maximal clique irreducible graphs, and it can be derived directly by the above Lemma 1.

Proposition 3. *For a graph G, for each $A \subseteq V(G)$ the induced subgraph G_A satisfies $cc(G_A) = m(G_A)$ if and only if G is F_1-free, F_2-free, F_3-free and F_4-free. (see Figure 2)*

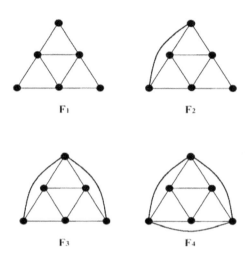

Fig. 1. F_1, F_2, F_3 and F_4 (Ocular Graphs)

A clique-Helly graph is a graph whose maximal cliques obey the Helly properties. In [11], E. Prisner showed that the hereditary class of maximal clique irreducible graphs is exactly the same with the hereditary clique-Helly graphs. A strongly chordal graph is a graph which is chordal and every even cycle of length at least 6 has a strong chord, meaning a chord joining vertices whose distance along the cycle is odd [7]. A graph has hereditary property P if every induced subgraph has the property P. E. Prisner also concluded in [11] that strongly chordal graphs are hereditary clique-Helly, hence hereditary maximal clique irreducible. Therefore, again the class of strongly chordal graphs provide us with more general examples that are weakly maximal irreducible, and examples with a unique edge maximal clique covering.

More generally, we have proved in [14] the following characterization of hereditary class of maximal clique irreducible graphs:

Proposition 4. *For a graph G, for each $A \subseteq V(G)$, the induced subgraph G_A satisfies $em(G_A) = cc(G_A)$ if and only if G is G_1-free, G_2-free,......, and G_{19}-free. (Please see Figure 2)*

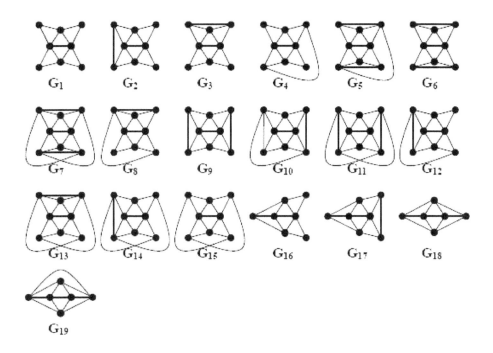

Fig. 2. $G_1,......, G_{19}$

Also an immediate corollary is as follows:

Proposition 5. *For a graph G, for each $A \subseteq V(G)$, the induced subgraph G_A has a unique minimum edge maximal clique covering if and only if G is G_1-free, G_2-free,......, and G_{19}-free. (see Figure 2)*

In particular, forbidding these 19 subgraphs $G_1, G_2, ..., G_{19}$ gives a sufficient condition for graphs having a unique minimum edge maximal clique covering, and hence a polynomial time algorithm for recognizing such class of graphs as well. While we have many classes of graphs which are (weakly) maximal clique irreducible and with a unique edge maximal clique covering as shown above, the question remains open for recognizing the general class of graphs with a unique minimum edge maximal clique covering or (weakly) maximal clique irreducible. In next section, we will focus on line graphs for the characterization problems.

3 Maximal Clique Irreducible Line Graphs

Using the above characterizations of hereditary type, we obtain the following characterizations for line graphs which have the set of all essential maximal cliques as a minimum edge clique cover (i.e. $cc(L(G)) = em(L(G))$), line graphs $L(G)$ which have the property that every maximal clique is essential (i.e. $cc(L(G)) = m(L(G))$), and line graphs with a unique edge maximal clique covering. Before we show the main theorems, notice that for line graphs there is a classical result by Whitney [17], which states the following:

Theorem 1. (H. Whitney, 1932) *For two graphs G and H which are not triangles, we have the following: G and H are isomorphic if and only if their line graphs are isomorphic.*

The figures of K_4, net, pendant-diamond, and octahedron please see Figure 3.

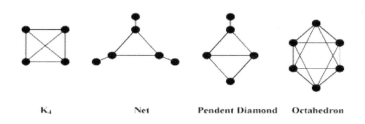

K_4 Net   Pendent Diamond   Octahedron

Fig. 3. K_4, net, pendant-diamond and octahedron

Then we may have the following characterizations of proper classes of line graphs with unique maximal edge clique covers in [15]:

Theorem 2. *Let $L(G)$ be the line graph of a graph G, then*

1. *$cc(L(G)) = em(L(G))$ if and only if G has no simplicial vertices of degree more than two.*
2. *$cc(L(G)) = m(L(G))$ if and only if G is K_4-free, net-free and pendant-diamond-free.*

Finally, we have the characterization of the line graphs with unique set of maximal cliques covering all the edges:

Theorem 3. *Let $L(G)$ be the line graph of a graph G and G is connected, then $L(G)$ has a unique minimum edge maximal clique covering except $L(G) \cong$ octahedron.*

Proof. Note that if G is connected, then clearly $L(G)$ is connected. Let Q_v be the maximal clique in $L(G)$ induced by the maximal star centered at v, for a vertex v of degree at least three in G.

Assume $L(G)$ is not isomorphic to the octahedron. Let v be a vertex in G of degree at least three. We claim that every minimum edge maximal clique covering of $L(G)$ contains the maximal clique Q_v induced by the maximal star at v in G. Suppose not, by the Whitney Theorem, then the minimum edge maximal clique covering must contain $\frac{d(d-1)}{2}$ triangles, where d is the degree of v in G. Now then the vertex v must be simplicial because otherwise Q_v is essential and hence in the minimum edge maximal clique covering. Let the neighbors of v be $v_1, ..., v_d$. Note that these $\frac{d(d-1)}{2}$ maximal cliques can be replaced with $d+1$ maximal cliques $Q_{v_1}, ..., Q_{v_d}$ and Q_v. We would obtain a smaller size of EMCC unless $d = 3$ and $G = K_4$, a contradiction. Hence the claim follows.

Now it is clear that $L(G)$ has a unique minimum edge maximal clique covering by including all possible edges in $L(G)$ induced by vertices of degree no more than two in G.

Q.E.D.

Therefore in particular we have shown that all line graphs have a unique, up to isomorphisms, edge maximal clique covering, since the graph octahedron has two but isomorphic edge maximal clique covers.

4 Maximal Clique Irreducibility versus Clique-Helly Properties

A *Clique-Helly* graph is a graph G for which the set of all maximal cliques **Q** obeys the *Helly property*. G is *hereditary clique-Helly* if every induced subgraph of G is clique-Helly. The smallest graph that is not clique-Helly is the *Hajós graph*, that is the first graph F_1 in Figure 1.

It is well-known that clique-Helly graphs are easy to be constructed. For any graph G, the graph $G \vee K_1$ (formed by adding a vertex v that is adjacent to all vertices of $V(G)$) is clique-Helly as every clique contain v. So every graph is an induced subgraph of a clique-Helly graph with just one more vertex. In particular, the $k-$partite complete graph $K_{3,...,3}$ with all parts of size 3 has $3k$ vertices but 3^k cliques, and is not clique-Helly when $k \geq 3$; on the other hand $K_{3,...,3} \vee K_1$ is clique-Helly with $3k+1$ vertices and 3^k cliques. We have just seen that there are both clique-Helly graphs and non-clique-Helly graphs with exponentially many cliques. In [11], E. Prisner regarded this observation as the reason why there is no polynomial recognition algorithm known for clique-Helly graphs up to now, and thus he considered the hereditary clique-Helly graphs and proved the following theorem, where ocular graphs are the four graphs F_1, F_2, F_3, F_4 in Figure 1.

Theorem 4. (E. Prisner, 1993) *A graph is hereditary clique-Helly if and only if it contains no ocular graph(see Figure 1) as induced subgraph.*

A *hypergraph* consists of a collection of vertices and a collection of edges; if the vertex set is V, then the edges are subsets of V. Any hypergraph on n vertices and m edges yields an $n \times m$ $(0,1)$-matrix $A = [a_{ij}]$, where $a_{ij} = 1$ if and only if vertex j is in edge i. The matrix A is a (edge-vertex) incidence matrix of the hypergraph. We avoid dealing with row and column permutations of the matrix here. Let a *chordless chain* of length k be a chain $x_1 E_1 x_2 E_2 ... x_k E_k x_{k+1}$ of distinct vertices x_i and distinct edges E_j with

$$E_i \cap \{x_1, x_2, ..., x_{k+1}\} = \{x_i, x_{i+1}\} \text{ for } i = 1, 2, ..., k.$$

A *chordless cycle* of length k is defined to be a chordless chain of length k except that $x_1 = x_{k+1}$. A $(0,1)$-matrix is *totally balanced* if it doesn't contain an incidence matrix of any graph cycle, of length at least 3, as a submatrix, where *graph cycles* means cycles in graphs rather than cycles in hypergraphs.

In [2], Anstee and Farber observed the following fact:

Theorem 5. (Anstee and Farber, 1984) *The incidence matrix of a hypergraph is totally balanced if and only if the hypergraph contains no chordless cycle of size greater than 2.*

The clique matrix of a graph G on the vertices $v_1, ..., v_n$ with maximal cliques $C_1, ..., C_k$ is the k by n matrix $C(G)$ whose (i, j) entry is 1 if v_j is in C_i and is 0 otherwise.

In [7], M. Farber proved the following theorem.

Theorem 6. (Farber, 1983) *The graph G is strongly chordal if and only if $C(G)$ is totally balanced.*

A *partial hypergraph* of a hypergraph $H = (V, \{S_1, ..., S_t\})$ is any hypergraph we can obtain by deleting hyperedges and vertices, that is, any hyperedge $P = (W, \{W \cap S_j : j \in J\})$, where $W \subseteq V$ and $J \subseteq \{1, ..., t\}$. The *underlying graph* $U(H)$ of the hypergraph has the same vertex set as H, and two distinct vertices are adjacent in $U(H)$ if they lie in some common hyperedge. A hypergraph H is *conformal*, if the set of its hyperedges are exactly the set of maximal cliques of $U(H)$.

In [11], E. Prisner proved the following theorem.

Theorem 7. (E. Prisner, 1993) *Let Θ denote the class of all conformal hypergraphs without graph cycle of length 3 as partial hypergraph. Then the underlying graphs of the members of Θ are exactly the hereditary clique-Helly graphs.*

Note that it is straightforward to see that for any conformal hypergraph H, the edge-vertex incidence matrix of H are exactly the clique matrix of $U(H)$. And note that the incidence matrix of a partial hypergraph $P = (W, \{W \cap S_j : j \in J\})$ of H is exactly the submatrix of the incidence matrix of H with certain rows and columns corresponding to W, $\{S_j : j \in J\}$. Hence the class Θ of hypergraphs in the above theorem exactly consists of all hypergraphs H such that the clique matrix of $U(H)$ contain no incidence matrix of any graph cycle, of length at least 3, as a submatrix, that is, $C(U(H))$ is a totally balanced matrix. Thus by Theorem 6 we know that the underlying graph of the members of Θ are strongly chordal. But Theorem 7 states that the underlying graphs of the members of Θ are exactly the hereditary clique-Helly graphs. Thus we know that all hereditary clique-Helly graphs are strongly chordal. Hence the underlying graphs of the members of the class Θ of hypergraphs in the above theorem have their clique matrices containing no incidence matrix of one graph cycle of length 3 as a sub-matrix.

Theorem 8. (E. Prisner, 1993) *All hereditary clique-Helly graphs have their clique matrices containing no incidence matrix of one graph cycle of length 3 as a submatrix.*

Now let G be strongly chordal, then by Theorem 6 we know that $C(G)$ contain no incidence matrix of any graph cycle of length at least 3 as a submatrix. Now we construct a hypergraph H with its vertex set $V(H)$ being exactly $V(G)$ and with its edge set being exactly the set of all maximal cliques of G. Then H is conformal and have its incidence matrix identical with $C(G)$ which contain no incidence matrix of any graph cycle of length at least 3 as a submatrix. Thus $H \in Q$. Thus by Theorem 7, $U(H) = G$ is hereditary clique-Helly.

Theorem 9. (E. Prisner, 1993) *Every strongly chordal graph is hereditary clique-Helly.*

As a generalization of Theorem 4, J. L. Szwarcfiter[12] characterized the clique-Helly graphs using the terminology of extended triangle. Let G be a graph and T a triangle of it. The *extended triangle* of G, relative to T, is the subgraph of G induced by the vertices which form a triangle with at least one edge of T. Let H be a subgraph of G. A vertex $v \in V(H)$ is *universal* in H whenever v is adjacent to every other vertex of H. In terms of extended triangle, Szwarcfiter characterized clique-Helly graphs in the following theorem.

Theorem 10. (J. L. Szwarcfiter, 1997) *G is clique-Helly graphs if and only if every extended triangle of G contains an universal vertex.*

This leads to a polynomial time algorithm for recognizing clique-Helly graphs.
A graph is called maximal clique irreducible if each maximal clique of G contains an edge which is not contained in any other maximal clique of G. Otherwise G is called reducible. Wallis and Zhang[13] characterized maximal clique irreducible graphs in the following theorem.

Theorem 11. (Wallis and Zhang, 1990) *A graph G is reducible if and only if there exists a set of maximal cliques*

$$\mathcal{F} = \{M_1, ..., M_t\}$$

such that the set of vertices contained in at least two maximal cliques in $\mathcal{F}$ form a maximal clique different from those in $\mathcal{F}$.

In [11], E. Prisner characterized hereditary irreducible graph as well as hereditary clique-Helly graph in the following theorem.

Theorem 12. (E. Prisner, 1993) *A graph G is hereditary clique-Helly if and only if G is hereditary irreducible.*

By the above theorem, we know that clique-Helly and irreducible are equivalent under hereditary operations. However, clique-Helly and irreducible graphs are incompatible. For example, the first graph in Figure 4 is clique-Helly but not irreducible, while the second and third are irreducible but not clique-Helly.

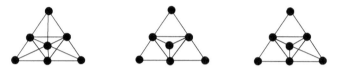

Fig. 4. Examples of Reducible Graphs and Non-Clique-Helly Graphs

In fact, in Figure 4 the first graph represents the class of graphs which are clique-Helly but not hereditary clique-Helly, and the second and the third graphs represent the class of graphs which are irreducible but not hereditary irreducible.

Theorem 13. *Let G be clique-Helly but not hereditary clique-Helly. Then G contain an induced ocular graph H together with another v adjacent to all vertices of H. On the other hand, let G be irreducible but not hereditary irreducible. Then G contain an induced ocular graph H together with another vertex v adjacent to all vertices of the middle triangle of H and adjacent to at most one of the other vertices of H.*

Proof. Let G be clique-Helly but not hereditary clique-Helly. Then by Theorem 4, G contains an induced ocular graph, say H, and by Theorem 10 the extended triangle of the middle triangle, say T, of this ocular graph must have universal vertex. But none of the 6 vertices of this ocular graph can be universal vertex of this extended triangle. Thus there must be another vertex in G adjacent to the 6 vertices of this ocular graph as the first graph in Figure 4.

Let G be irreducible but not hereditary irreducible. Then by Theorem 4 and Theorem 12, G contain an induced ocular graph, say H. Now assume that the middle triangle $\{a, b, c\}$ of H is a maximal clique, refer to the first graph in figure 17, then three different maximal cliques containing the three triangles $\{a, c, e\}, \{a, b, d\}, \{b, c, f\}$, respectively together with the maximal clique $\{a, b, c\}$ form a contradiction with Theorem 11. Thus there must be another vertex, say v_1, in G adjacent to a, b, c, since $\{a, b, c\}$ is not a maximal clique. Now if v_1 is adjacent to at least two of the three vertices d, e, f, say d, e, and $\{v_1, a, b, c\}$ is a maximal clique, then the three maximal cliques $\{v_1, a, c, e\}, \{v_1, a, b, d\}, \{b, c, f\}$ together with the middle maximal clique $\{v_1, a, b, c\}$ again form a contradiction with Theorem 11. Thus there exists another vertex, say v_2, in G adjacent to v_1, a, b, c, since $\{v_1, a, b, c\}$ is not a maximal clique. Now if v_2 is still adjacent to at least two of the three vertices d, e, f and $\{v_1, v_2, a, b, c\}$ is a maximal clique, then similarly we again arrive at a contradiction with Theorem 11.

Eventually we will find a vertex $v' \notin \{a, b, c, d, e, f\}$ adjacent to at most one of the three vertices d, e, f as the second or third graph in Figure 4.

Q.E.D.

References

1. R. Alter and C. C. Wang, *Uniquely intersectable graphs*, Discrete Mathematics, 18(1977), pp.217-226.
2. R. Anstee, Martin Farber *Characterizations of totally balanced matrices*, J. Algorithms, 5(1984), pp. 215V230.
3. P. Erdős, A. Goodman, and L. Posa, *The representation of a graph by set intersections*, Canadian Journal of Mathematics, 18(1966), pp.106-112.
4. F. Harary, *Graph Theory*, Addison-Wesley Publishing Company, Reading, MA, 1969.
5. N. V. R. Mahadev and T.-M. Wang, *A characterization of hereditary UIM graphs*, Congressus Numerantium, 126(1997), pp. 183-191.
6. N. V. R. Mahadev and T.-M. Wang, *On uniquely intersectable graphs*, Discrete Mathematics, 207(1999), pp. 149-159.
7. M. Farber, Characterizations of strongly chordal graphs, Discrete Math. 43 (1983) 173-189.
8. N. V. R. Mahadev and T.-M. Wang, *A characterization of hereditary UIM graphs*, Congressus Numerantium, 126(1997), pp. 183-191.
9. N. V. R. Mahadev and T.-M. Wang, *On uniquely intersectable graphs*, Discrete Mathematics, 207(1999), pp. 149-159.
10. R. J. Opsut and F. S. Roberts, *On the fleet maintenance, mobile radio frequency, task assignment, and traffic phasing problems*, G. Chartrand, Y. Alavi, D. L. Goldsmith, L. Lesniak-Foster, and D. R. Lick, eds., The Theory and Applications of Graphs, Wiley, NY, 1981, pp. 479-492.
11. E. Prisner, *Hereditary clique-Helly graphs*, Journal of Combinatorial Mathematics and Combinatorial Computing (JCMCC), 14 (1993) pp. 216-220.
12. J.L. Szwarcfiter, *Recognizing clique-Helly graphs*, Ars Combinatoria 45 (1997) 29-32.
13. W. D. Wallis and G.-H. Zhang, *On maximal clique irreducible graphs*, Journal of Combinatorial Mathematics and Combinatorial Computing (JCMCC), 8 (1990), pp. 187-193.
14. T.-M. Wang, *On characterizing weakly maximal clique irreducible graphs*, Congressus Numerantium 163 (2003), 177–188.
15. T.-M. Wang, *On line graphs which are weakly maximal clique irreducible*, Ars Combinatoria, 76 (2005), pp. 233–238.
16. D. West, *Introduction to graph theory*, Pretice Hall Inc., Upper Saddle River, NJ, 1996.

17. H. Whitney, *Congruent graphs and the connectivity of graphs*, Amer. J. Math., 54 (1932), pp. 150-168

On partitional labelings of graphs [*]

Rikio Ichishima and Akito Oshima

College of Humanities and Sciences, Nihon University,
3-25-40 Sakurajyousui Setagaya-ku, Tokyo 156-8550, Japan
e-mail address: `ichishim@chs.nihon-u.ac.jp`

Department of Mathematical Information Science, Tokyo University of Science,
1-3 Shinjuku-ku, Tokyo 162-8601, Japan
e-mail address: `akito_o@rs.kagu.tus.ac.jp`

Abstract. The notion of partitional labelings of graphs is introduced as a particular type of sequential labelings, and the cartesian product of a partitional graph and K_2 is shown to be partitional. Every sequential labeling is harmonious and felicitous. In this paper, the partitional property of some bipartite graphs including the n-dimensional cube Q_n is studied, and thus extends what was known about the sequentialness, harmoniousness and felicitousness of such graphs.

Key words: partitional labeling, sequential labeling, harmonious labeling, felicitous labeling, n-cube, graph labeling.

1   Introduction

Throughout this paper, we consider a graph G to be finite and simple, that is, there are no loops and multiple edges. Let $V(G)$ and $E(G)$ denote the vertex and edge sets of G, respectively. For any undefined graph theory terminology and notation, the authors refer the reader to Chartrand and Lesniak [1]. For the sake of brevity, we will denote the set of integers $\{m, m+1, \ldots, n\}$ by simply writing $[m, n]$.

Harmonious graphs naturally arose in the study by Graham and Sloane [7] of modular versions of additive bases problems stemming from error-correcting codes. They defined a graph G of order p and size q with $q \geq p$ to be *harmonious* if there exists an injective function $f : V(G) \to \mathbb{Z}_q$ such that when each edge $uv \in E(G)$ is labeled $f(u) + f(v) \pmod{q}$, the resulting edge labels are distinct. Such a function is called a *harmonious labeling*. If G is tree so that $q = p - 1$, exactly two vertices are labeled the same; otherwise, the definition is the same.

In [6], Grace defined a *sequential labeling* of a graph G of size q as an injective function $f : V(G) \to [0, q-1]$ such that when each edge $uv \in E(G)$ is labeled $f(u) + f(v)$, the resulting edge labels are $[m, m+q-1]$ for some positive integer m. A graph is called *sequential* if it admits a sequential labeling. Every sequential labeling induces a harmonious labeling; however, it is an open question whether or not every harmonious graph admits a sequential labeling.

---

[*] Dedicated to our friend and colleague Yasuhiro Fukuchi

Harmonious and sequential labelings have been the object of study for many papers. For recent contributions to these subjects and other types of labelings, the authors refer the reader to an excellent survey paper by Gallian [5].

To present our results, we define the cartesian product of two graphs. The *cartesian product* $G \cong G_1 \times G_2$ has $V(G) = V(G_1) \times V(G_2)$, and two vertices (u_1, u_2) and (v_1, v_2) of G are adjacent if and only if either

$$u_1 = v_1 \text{ and } u_2 v_2 \in E(G_2)$$

or

$$u_2 = v_2 \text{ and } u_1 v_1 \in E(G_1)$$

Hence, the n-dimensional cube Q_n can be defined inductively as $Q_0 \cong K_1$ and $Q_n \cong Q_{n-1} \times K_2$ for any positive integer n.

In Section 2, we show that Q_n is sequential for every integer $n \geq 4$. In Section 3, we introduce the notion of a partitional labeling, which is a sequential labeling with additional properties and, generalizing the argument in Section 2, prove that the cartesian product of a partitional graph and K_2 is partitional (see Theorem 2). It is clear that every partitional labeling is sequential and harmonious. It is also true that every partitional labeling is 'felicitous' (see [5] for the definition of a felicitous labeling). Thus, if a graph G is shown to be partitional, then by virtue of Theorem 2, it follows that for each nonnegative integer n, $G \times Q_n$ is sequential, harmonious and felicitous (see Theorem 4, and Corollaries 2 and 3 in Section 3).

2   Result on the n-dimensional Cube

The n-dimensional cube Q_n (also called hypercube) serves as useful models for a broad range of applications such as circuit design, communication network addressing, parallel computation and computer architecture; hence, we concern in this section the sequential labeling of Q_n.

The following proof is inspired by the work that Kotzig [8] carried out when showing that there exists an α-valuation of Q_n for every positive integer n. Utilizing the construction in his proof, Figueroa-Centeno and Ichishima [3] have recently verified that every n-dimensional cube Q_n is felicitous.

Theorem 1. *Let n be an integer with $n \geq 2$. Then the n-dimensional cube Q_n is sequential if and only if $n \neq 2, 3$.*

Proof. The n-dimensional cube Q_n has already been shown to be not harmonious by Graham and Sloane [7] when $n = 2, 3$. Thus, they are not sequential either.

For the converse, assume that n is an integer with $n \geq 4$, and proceed by induction on n. First, let Q_2 be the graph with $V(Q_2) = \{v_1, v_2, v_3, v_4\}$ and $E(Q_2) = \{v_1 v_3, v_1 v_4, v_2 v_3, v_2 v_4\}$. For every integer $n \geq 2$, construct the graph Q_{n+1} by using the decomposition

$$Q_{n+1} \cong A_{n+1} \oplus B_{n+1} \oplus C_{n+1}$$

with

$$V(A_{n+1}) = \{v_i | i \in 1, 2^{n-1}\} \cup \{v_i | i \in 2^n + 1, 2^n + 2^{n-1}\},$$
$$V(B_{n+1}) = \{v_i | i \in 1, 2^{n+1}\},$$
$$V(C_{n+1}) = \{v_i | i \in 2^{n-1} + 1, 2^n\} \cup \{v_i | i \in 2^n + 2^{n-1} + 1, 2^{n+1}\},$$
$$E(A_{n+1}) = \{v_i v_{2^n + j} | v_i v_{2^{n-1} + j} \in E(Q_n) \text{ and } i, j \in 1, 2^{n-1}\},$$
$$E(B_{n+1}) = \{v_i v_{2^{n+1} + 1 - i} | i \in [1, 2^n]\},$$
$$E(C_{n+1}) = \{v_{2^{n-1} + i} v_{2^n + 2^{n-1} + j} | v_i v_{2^{n-1} + j} \in E(Q_n) \text{ and } i, j \in 1, 2^{n-1}\}.$$

Notice then that in this construction, A_{n+1} and C_{n+1} are isomorphic to Q_n, and also that Q_{n+1} is represented as a bipartite graph with two partite sets $\{v_i | i \in [1, 2^n]\}$ and $\{v_i | i \in 2^n + 1, 2^{n+1}\}$ of the same cardinality 2^n.

With the aid of the preceding construction, we will prove that there exists a sequential labeling f_n of Q_n such that

$$f_n(v_i) \in 0, (n-1)2^{n-2} - 3 \qquad \text{if } i \in 1, 2^{n-1} - 2,$$
$$f_n(v_i) \in (n-1)2^{n-2} + 3, n2^{n-1} - 5 \quad \text{if } i \in 2^{n-1} + 3, 2^n,$$
$$f_n(v_{2^{n-1}-1}) = (n-1)2^{n-2}, \qquad f_n(v_{2^{n-1}}) = (n-1)2^{n-2} + 2,$$
$$f_n(v_{2^{n-1}+1}) = (n-1)2^{n-2} - 2, \quad f_n(v_{2^{n-1}+2}) = (n-1)2^{n-2} - 1,$$

and

$$\{f_n(u) + f_n(v) | uv \in E(A_n)\} = (n-1)2^{n-2} - 2, (n-1)2^{n-1} - 3,$$
$$\{f_n(u) + f_n(v) | uv \in E(B_n)\} = (n-1)2^{n-1} - 2, n2^{n-1} - 3,$$
$$\{f_n(u) + f_n(v) | uv \in E(C_n)\} = n2^{n-1} - 2, (3n-1)2^{n-2} - 3.$$

For $n = 4$, the above conditions are certainly satisfied by letting

$$(f_4(v_i))_{i=1}^{16} = (0, 2, 3, 5, 9, 7, 12, 14, 10, 11, 15, 16, 23, 24, 27, 26).$$

A sequential labeling of Q_4 is given in Fig. 1.

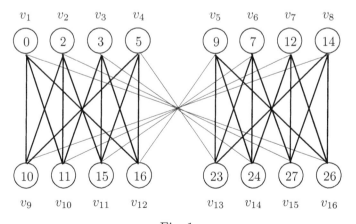

Fig. 1

For every integer $n \geq 4$, assume that there exists a sequential labeling f_n of Q_n with the aforementioned properties. To do this, define the vertex labeling

$$f_{n+1} : V(Q_{n+1}) \rightarrow [0, (n+1)2^n - 1]$$

such that

$$
\begin{aligned}
f_{n+1}(v_i) &= f_n(v_i), \\
f_{n+1}(v_{2^{n-1}+i}) &= f_n(v_i) + (n+1)2^{n-2}, \\
f_{n+1}(v_{2^n+i}) &= f_n(v_{2^{n-1}+i}) + (n+1)2^{n-2}, \\
f_{n+1}(v_{2^n+2^{n-1}+i}) &= f_n(v_{2^{n-1}+i}) + (n+2)2^{n-1}
\end{aligned}
$$

for each $i \in \overline{1, 2^{n-1}}$.

To verify that f_{n+1} is indeed a sequential labeling of Q_{n+1}, we compute the induced edge labels. For the edges in A_{n+1} and C_{n+1}, we have that

$$f_{n+1}(v_i) + f_{n+1}(v_{2^n+j}) = f_n(v_i) + f_n\ v_{2^{n-1}+j} + (n+1)2^{n-2}$$

and

$$f_{n+1}(v_{2^{n-1}+i}) + f_{n+1}\ v_{2^n+2^{n-1}+j} = f_n(v_i) + f_n\ v_{2^{n-1}+j} + (3n+5)2^{n-2},$$

respectively. This implies that

$$
\begin{aligned}
&\{f_{n+1}(u) + f_{n+1}(v) \,|\, uv \in E(A_{n+1})\} \\
&= \overline{n2^{n-1} - 2, n2^n - 3}
\end{aligned}
$$

and

$$
\begin{aligned}
&\{f_{n+1}(u) + f_{n+1}(v) \,|\, uv \in E(C_{n+1})\} \\
&= \overline{(n+1)2^n - 2, (3n+2)2^{n-1} - 3} .
\end{aligned}
$$

Next, observe that the edges in B_{n+1} are of two types, namely, they are either of the form $v_i v_{2^{n+1}+1-i}$ or $v_{2^{n-1}+i} v_{3 \cdot 2^{n-1}+1-i}$ $i \in \overline{1, 2^{n-1}}$. Hence, we have that

$$f_{n+1}(v_i) + f_{n+1}(v_{2^{n+1}+1-i}) = f_n(v_i) + f_n(v_{2^n+1-i}) + (n+2)2^{n-1}$$

and

$$f_{n+1}(v_{2^{n-1}+i}) + f_{n+1}(v_{3 \cdot 2^{n-1}+1-i}) = f_n(v_i) + f_n(v_{2^n+1-i}) + (n+1)2^{n-1},$$

respectively. This implies that

$$
\begin{aligned}
&\{f_{n+1}(v_i) + f_{n+1}(v_{2^{n+1}+1-i}) \,|\, i \in \overline{1, 2^{n-1}}\} \\
&= \overline{(2n+1)2^{n-1} - 2, (n+1)2^n - 3}
\end{aligned}
$$

and

$$
\begin{aligned}
&\{f_{n+1}(v_{2^{n-1}+i}) + f_{n+1}(v_{3 \cdot 2^{n-1}+1-i}) \,|\, i \in \overline{1, 2^{n-1}}\} \\
&= \overline{n2^n - 2, (2n+1)2^{n-1} - 3} ,
\end{aligned}
$$

which in turn implies that

$$\{f_{n+1}(u) + f_{n+1}(v) \,|\, uv \in E(C_{n+1})\}$$
$$= [n2^n - 2, (n+1)2^n - 3].$$

Finally, the inductive hypothesis together with the definition of f_{n+1} yields the conditions for the vertex labelings, which shows that f_{n+1} is an injective function.

Therefore, by induction, Q_n is sequential for $n \neq 2, 3$.

3   Results on Partitional Graphs

In this section, to generalize the idea of the construction technique used in the proof of Theorem 1, we introduce a new type of sequential labeling.

A sequential labeling f of a graph G of size $2t + s$ is called a *partitional labeling* if G is bipartite with two partite sets X and Y of the same cardinality s such that $f(x) \leq t+s-1$ for all $x \in X$ and $f(y) \geq t-s$ for all $y \in Y$, and there is a positive integer m such that the induced edge labels are partitioned into three sets $[m, m+t-1]$, $[m+t, m+t+s-1]$ and $[m+t+s, m+2t+s-1]$ with the properties that there is an involution π which is an automorphism of G such that π exchanges X and Y, $x\pi(x) \in E(G)$ for all $x \in X$, and $\{f(x) + f(\pi(x)) \,|\, x \in X\} = [m+t, m+t+s-1]$. A graph is called *partitional* if it admits a partitional labeling (see Fig. 1 shows an example).

With this definition in hand, we are now able to present the following result.

Theorem 2. *If G is partitional, then $G \times K_2$ is partitional.*

Proof. First, assume that G is a partitional graph of size $2t+s$ with a partitional labeling f. Then G is bipartite with two partite sets X and Y of the same cardinality s such that $f(x) \leq t+s-1$ for all $x \in X$ and $f(y) \geq t-s$ for all $y \in Y$, and there is a positive integer m such that the induced edge labels are partitioned into three sets $[m, m+t-1]$, $[m+t, m+t+s-1]$ and $[m+t+s, m+2t+s-1]$ with the properties that there is an involution π which is an automorphism of G such that π exchanges X and Y, $x\pi(x) \in E(G)$ for all $x \in X$, and $\{f(x) + f(\pi(x)) \,|\, x \in X\} = [m+t, m+t+s-1]$.

Next, let $H \cong G \times K_2$, and construct the graph H by using the decomposition

$$H \cong H_1 \oplus H_2 \oplus H_3$$

with

$$
\begin{aligned}
V(H_1) &= \{v_i | i \in [1,s]\} \cup \{v_i | i \in [2s+1, 3s]\}, \\
V(H_2) &= \{v_i | i \in [1, 4s]\}, \\
V(H_3) &= \{v_i | i \in [s+1, 2s]\} \cup \{v_i | i \in [3s+1, 4s]\}, \\
E(H_1) &= \{v_i v_{2s+j} | v_i v_{s+j} \in E(G), \, i \in [1,s] \text{ and } j \in [1,s]\}, \\
E(H_2) &= \{v_i v_{4s+1-i} | i \in [1, 2s]\}, \\
E(H_3) &= \{v_{s+i} v_{3s+j} | v_i v_{s+j} \in E(G), \, i \in [1,s] \text{ and } j \in [1,s]\}.
\end{aligned}
$$

Notice then that in this construction, H_1 and H_3 are isomorphic to G, H_2 is isomorphic to $2sK_2$, and H is a bipartite graph of size $4t + 4s$ with two partite sets $X = \{w_i | i \in [1, 2s]\}$ and $Y = \{w_i | i \in [2s+1, 4s]\}$ of the same cardinality $2s$ so that $\pi(w_i) = w_{2s+1-i}$ for each $i \in [1, 2s]$. Note that if we define a permutation π' of $V(H)$ by

$$\pi'(v_i) = v_{4s+1-i} \ (i \in [1, 4s]),$$

then π' is an involutive automorphism of H, and we have that

$$E(H_2) = \{v_i \pi'(v_i) | i \in [1, 2s]\}.$$

Now, define the vertex labeling $g : V(H) \to [0, 4t + 4s - 1]$ such that

$$
\begin{aligned}
g(v_i) &= f(v_i), \\
g(v_{s+i}) &= f(v_i) + t + s, \\
g(v_{2s+i}) &= f(v_{s+i}) + t + s, \\
g(v_{3s+i}) &= f(v_{s+i}) + 2t + 3s
\end{aligned}
$$

for each $i \in [1, s]$.

To verify that g is indeed a partitional labeling of H, with the induced edge labels of f in hand, we compute the induced edge labels of g. For the edges in H_1 and H_3, we have that

$$g(v_i) + g(v_{2s+j}) = f(v_i) + f(v_{s+j}) + t + s$$

and

$$g(v_{s+i}) + g(v_{3s+j}) = f(v_i) + f(v_{s+j}) + 3t + 4s,$$

respectively. This implies that

$$\{g(u) + g(v) | uv \in E(H_1)\} = [m + t + s, m + 3t + 2s - 1]$$

and

$$\{g(u) + g(v) | uv \in E(H_3)\} = [m + 3t + 4s, m + 5t + 5s - 1].$$

Next, observe that the edges in H_2 are of two types, namely, either of the form $v_i v_{4s+1-i}$ or $v_{s+i} v_{3s+1-i}$ $(i \in [1, s])$. Thus, we have that

$$g(v_i) + g(v_{4s+1-i}) = f(v_i) + f(v_{2s+1-i}) + 2t + 3s$$

and

$$g(v_{s+i}) + g(v_{3s+1-i}) = f(v_i) + f(v_{2s+i-1}) + 2t + 2s,$$

which in turn implies that

$$\{g(v_i) + g(v_{4s+1-i}) | i \in [1, s]\} = [m + 3t + 3s, m + 3t + 4s - 1]$$

and

$$\{g(v_{s+i}) + g(v_{3s+1-i}) | i \in [1, s]\} = [m + 3t + 2s, m + 3t + 3s - 1].$$

Consequently, we have that

$$\{g(u) + g(v) \,|\, uv \in E(H_2)\} = [m + 3t + 2s, m + 3t + 4s - 1].$$

It is now immediate that the induced edge labels of H are $4t + 4s$ consecutive integers. Finally, notice that g is an injective function, and

$$g(v_i) \leq (t + s - 1) + (t + s) < t' + s' - 1 \text{ for all } i \in [1, 2s]$$

and

$$g(v_{s+i}) \geq (t - s) + (t + s) > t' - s' \text{ for all } i \in [1, 2s],$$

where $t' = 2t + s$ and $s' = 2s$.

Therefore, g is a partitional labeling of H.

If we apply Theorem 2 repeatedly, then we obtain the following result.

Theorem 3. *If G is partitional, then $G \times Q_n$ is partitional for every nonnegative integer n.*

In light of Theorem 3 and the fact that every partitional labeling is sequential, harmonious and felicitous, we have the following result.

Corollary 1. *If G is partitional, then $G \times Q_n$ is sequential, harmonious and felicitous for every nonnegative integer n.*

Applying Theorem 3 and Corollary 1 with $G \cong Q_4$, we obtain the following result, which is a refinement of the 'if' part of Theorem 1.

Theorem 4. *Let n be an integer with $n \geq 4$. Then the n-dimensional cube Q_n is partitional, sequential, harmonious and felicitous.*

The next class of graphs we are concerned with is books. The book is the graph $S_m \times Q_1$, where S_m is the star with $m + 1$ vertices. In [4], Gallian and Jungreis have shown that the book $S_{2m} \times Q_1$ is sequential for every positive integer m. In fact, their sequential labeling of $S_{2m} \times Q_1$ satisfies the conditions to be partitional. Thus, we are able to state the following result.

Theorem 5. *For every positive integer m, the book $S_{2m} \times Q_1$ is partitional.*

The next class of graphs we are concerned with is the ladders. The ladder is the graph $P_m \times Q_1$, where P_m is the path with m vertices. In [2], the ladder $P_{2m+1} \times Q_1$ has been shown to be super edge-magic for every positive integer m, and it induces the partitional labeling of $P_{2m+1} \times Q_1$ by subtracting 1 from each vertex label; hence, we have the following result.

Theorem 6. *For every positive integer m, the ladder $P_{2m+1} \times Q_1$ is partitional.*

We are ending this paper with the following two results, which are consequence of Theorem 2.

Corollary 2. *For every two positive integers m and n, the generalized book $S_{2m} \times Q_n$ is partitional, sequential, harmonious and felicitous.*

Corollary 3. *For every two positive integers m and n, the generalized ladder $P_{2m+1} \times Q_n$ is partitional, sequential, harmonious and felicitous.*

Acknowledgments. The authors would like to express their gratitude to Professor Yoshimi Egawa for his continuous encouragement and precious assistance.

References

1. Chartrand, G., Lesniak, L.: Graphs and Digraphs, second edition. Wadsworth & Brooks/Cole Advanced Books and Software, Monterey (1986)
2. Figueroa-Centeno, R.M., Ichishima, R., Muntaner-Batle, F.A.: The place of super edge-magic labelings among other classes of labelings. Discrete Math. 231, 153–168 (2001)
3. Figueroa-Centeno, R.M., Ichishima, R.: The n-dimensional cube is felicitous. Bull. Inst. Combin. Appl. 41, 47–50 (2004)
4. Gallian, A.J., Jungreis, D.S.: Labeling books. Scientia. 1, 53–57 (1988)
5. Gallian, J.A.: A dynamic survey of graph labeling. Electron. J. Combin. 5, #DS6 (2008)
6. Grace, T.: On sequential labelings of graphs. J.Graph Theory. 7, 195–201 (1983)
7. Graham, R.L., Sloane, N.J.: On additive bases and harmonious graphs. SIAM J. Alg. Discrete Math. 1, 382–404 (1980)
8. Kotzig, A.: Decomposition of complete graphs into isomorphic cubes. J. Comb. Theory, Series B. 31, 292–296 (1981)

Bimagic Labelings

Alison Marr[1], N. C. K. Phillips[2], W. D. Wallis[2]

[1]Southwestern University, Georgetown, TX, USA 78626
[2]Southern Illinois University, Carbondale IL, USA 62901
marra&southwestern.edu, nckp&siu.edu,
wdwallis&math.siu.edu

Abstract. An *edge-magic* (total) *labeling* λ of a graph G is a one-to-one mapping from $V(G) \cup E(G)$ onto the set of integers $\{1, 2, \ldots n\}$ for which there exists a constant k such that $\lambda(x) + \lambda(xy) + \lambda(y) = k$ whenever x and y are adjacent vertices. In a *bimagic labeling*, there are two constants k_1 and k_2 such that all sums of the specified type equal one or other of those two sums. We discuss edge-bimagic labelings of graphs for which no edge-magic labeling exists. In particular, two cases are of special interest: when the number of edges with one sum is (approximately) the same as the number with the other; or when all edges but one have the common sum.

1 Definitions

A (total) *labeling* λ of a graph G is a one-to-one mapping from $V(G) \cup E(G)$ onto the set of integers $\{1, 2, \ldots n\}$. Various arithmetical properties of graph and digraph labelings have been studied (a good survey is [3]). In particular, *magic* denotes the requirement that all sums of labels of a certain kind have the same value. For example, a total labeling λ is *edge-magic* if there exists a constant k (the "magic constant") such that $\lambda(x) + \lambda(xy) + \lambda(y) = k$ whenever x and y are adjacent vertices.

Babujee [1,2] introduces the idea of a *bimagic labeling*, in which there are two constants k_1 and k_2 such that all sums of the specified type equal one or other of those two sums. For example, we would define an *edge-bimagic total labeling* of G to be a total labeling λ such that, when x and y are adjacent vertices of G, $\lambda(x) + \lambda(xy) + \lambda(y)$ equals either k_1 or k_2. (Babujee uses a more complicated terminology, but we have standardized to be consistent with [7].)

As one would expect, it is very easy to find bimagic labelings. However, it is of interest to consider bimagic labelings of graphs for which no magic labeling exists. In particular, two cases are of special interest. If the number of edges with one sum is the same as the number with the other, or (when $-E(G)-$ is odd) differs by 1, we refer to the labeling as *equitable*. When all edges but one have the common sum, the labeling is *almost magic*.

Our aim here is to give some elementary results, in the hope that further work will be forthcoming, particularly on the families we discuss.

2 Classes of Graphs

2.1 Wheels

The wheel W_n consists of an n-cycle C_n together with a central vertex c joined to the other n vertices. It is known (see [6]) that W_n has no edge-magic total labeling when $n \equiv 3 \pmod 4$. But an edge-bimagic total labeling is easy to construct. We know [4] that C_n has an edge-magic total labeling where the vertex labels are $\{1, 2, \ldots, n\}$, for any odd n. (The constant is $k = (5n+3)/2$.) Suppose λ is such a labeling, apply it to the vertices and edges of the cycle in a W_n. Set $\lambda(c) = 3n+1$; if x is any vertex of the cycle, set $\lambda(xc) = 3n+1-\lambda(x)$. This is a labeling of W_n with constants $(5n+3)/2$ and $6n+2$, so W_n is bimagic if n is odd. In particular, this works for $n \equiv 3 \pmod 4$. The labeling produced is equitable.

One particular case, the wheel W_3, is so small that labelings are easily produced by hand; an almost magic example was found in this way. An almost magic labeling of W_7 was constructed by computer. These two examples are shown below.

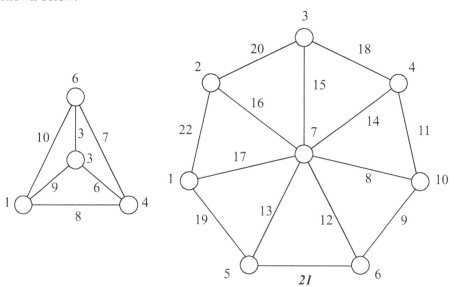

The existence of almost magic labelings of W_n for larger $n \equiv 3 \pmod 4$ is unresolved.

2.2 One-Factors

The one-factor nK_2, consisting of n independent edges, has an edge-magic total labeling if and only if n is odd. For n even, a bimagic total labeling is easily constructed: take an edge-magic total labeling of $(n-1)K_2$, and add a copy of K_2 with its three elements labeled $\{3n-2, 3n-1, 3n\}$.

2.3 Complete Graphs

The only complete graphs with edge-magic total labelings are K_2, K_3, K_5 and K_6. We know that K_4 has an edge-bimagic total labeling because $K_4 = W_3$. We now exhibit edge-bimagic total labelings of K_7 and K_8. The left-hand array below represents a suitable labeling of K_7 with magic constants 22 and 31; the diagonal entries are the vertex labels and the (i, j) entry is the label for the edge joining vertices i and j. Similarly, the right-hand array represents a labeling of K_8 with magic constants 27 and 41.

```
1 28 27 26 25 23 21        1 24 23 36 35 33 31 28
  2 17 16 24 22 20           2 22 21 34 32 30 27
    3 15 14 12 19              3 20 19 17 29 26
      4 13 11 18                 4 18 16 14 25
        5 10  8                    5 15 13 10
           7  6                       7 11  8
              9                          9  6
                                           12
```

We conjecture that there exists a constant N such that K_n has an EBTL if and only if $n \leq N$. If this is true, we now know $N \geq 8$. So far the search for an EBTL of K_9 has been unsuccessful.

2.4 Duplicated Graphs

Suppose G is an edge-magic graph with EMTL λ. Then it is easy to construct an EBTL of $2G$. One copy of G is labeled with λ. In the other copy, the label is increased by $|V(G)| + |E(G)|$. If k was the original magic constant, then the new labeling has constants k and $k + |V(G)| + |E(G)|$.

As an example, the graph $2K_3$, the union of two disjoint triangles, is not edge-magic. The problem is so small that an exhaustive computation is very short (and hand computation is feasible). Moreover a theoretical proof is now available [5]. But K_3 is edge-magic, so $2K_3$ is edge-bimagic. Similarly $2K_5$ and $2K_6$ are edge-bimagic. However, all these labelings are equitable; the more interesting question of almost magic labelings in much harder.

3 Strong Labelings

An edge-magic total labeling is called *strong* or *super-magic* if the smallest labels are affixed to the vertices. In other words, if there are v vertices, the vertex labels are $\{1, 2, \ldots, v\}$. Even cycles are known to have no strong edge-magic total labeling. But a strong edge-bimagic total labeling is available for every cycle. If there are $2t$ vertices, then the vertices are labeled $1, t+1, 2, t+2, \ldots,$ $t, 2t$; the edge from 1 to $t+1$ is labeled $4t$, and subsequent edges are $4t-1$, $4t-2, \ldots, 2t+1$.

4 Computations for Duplicated Complete Graphs

Our first computational results on $2K_n$ concerned the question: is $2K_n$ edge-magic?

In general, $2K_n$ has $2n$ vertices and $n(n-1)$ edges, a total of $n(n+1)$ elements. Suppose it has an edge-magic total labeling with constant k, and suppose the sum of the vertex labels is V and the sum of the edge-labels is E. $V + E$ equals the sum of all the labels:

$$V + E = 1 + 2 + \ldots + n(n+1) = n(n+1)(n^2 + n + 1)/2,$$

which is odd when $n \equiv 2 \pmod 4$. Each vertex lies in $n-1$ edges, so if we sum $\lambda(x) + \lambda(xy) + \lambda(y)$ over all edges xy, the result will be $(n-1)V + E = (n-2)V + (V+E)$. So we have

$$(n-2)V + (V+E) = k|E(G)| = kn(n-1).$$

When $n \equiv 2 \pmod 4$, the left hand term is odd, but the right-hand is even.

We have investigated $2K_n$ for the small values $n = 4$ and 5, and found that neither $2K_4$ nor $2K_5$ is edge magic. The computation for $2K_7$ is considerably larger.

When looking for almost magic labelings of $2K_n$, our technique was very simple. We first chose a value k and then attempted to apply the labels 1, 2, $\ldots, n(n+1)$ to the vertices and edges so that every edge-sum equalled k, using a backtrack search. However, we did not backtrack when this proved impossible. Rather, in that case, we recorded the fact that a problem had occurred, and went on to label the next element. We only backtracked when a second problem occurred. Surprisingly, this rather primitive approach seemed, in the case of $2K4$, to be just as efficient as any sophisticated algorithm we councocted.

We found that there are 140 almost magic labelings of $2K - 3$, 64 of $2K_6$ and 4 of $2K_5$. There are no examples for $2K_6$.

Examples for $2K_3$ (with $k = 18$, one edge sum is 10), $2K_4$ (with $k = 24$, one edge sum is 30) and $2K_5$ (with $k = 24$, one edge sum is 30) are shown below. In order to avoid cluttering the diagram, edge labels are omitted, except for the one edge label that gives rise to an anomalous sum; all other edge labels can be computed by subtracting the labels on the endpoints from k.

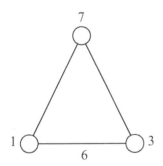
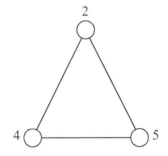

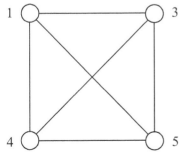
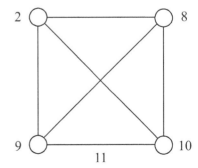

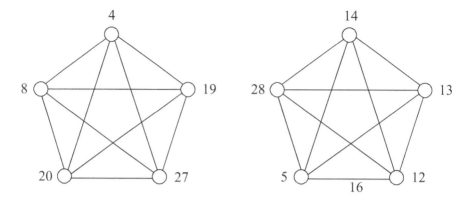

References

1. J. B. Babujee, Bimagic labeling in path graphs. *The Mathematics Education* **38**, 12–16 (2004).
2. J. B. Babujee, On edge bimagic labeling. *J. Combin., Inf., Syst. Sci.* **28**, 239–244 (2004).
3. J. A. Gallian, A dynamic survey of graph labeling. *Electronic J. Combin.*, *Dynamic Survey* #DS6, **5** (1998); Ninth Edition posted 1/20/2005.
4. R. D. Godbold and P. J. Slater, All cycles are edge-magic. *Bull. Inst. Combin. Appl.* **22**, 93–97 (1998).
5. D. MacQuillan and J. M. MacQuillan, Magic labeling of triangles. *Discrete Math.* (to appear).
6. G. Ringel and A. S. Llado, Another tree conjecture. *Bull. Inst. Combin. Appl.* **18**, 83–85 (1996).
7. W. D. Wallis *Magic Graphs*. Birkhauser, Boston (2001).

Improving Multikey Quicksort for Sorting Strings

Eunsang Kim and Kunsoo Park

School of Computer Science and Engineering, Seoul National University,
599 Gwanak-ro, Gwanak-gu, Seoul, 151-742, South Korea
{eskim, kpark}@theory.snu.ac.kr

Abstract. We present 'collect-center' partitioning to improve multikey Quicksort with 'split-end' partitioning due to Bentley and Sedgewick. It moves equal
elements to the middle like the 'Dutch National Flag Problem' partitioning approach and it uses two inner loops like Bentley and McIlroy's. Multikey Quicksort with 'collect-center' partitioning gives us 7~18% speed-up compared with
multikey Quicksort with 'split-end' partitioning.

Key words: Multikey Quicksort Speed-up, Split-end Partitioning, Dutch National Flag Problem

1 Introduction

The string sorting problem is to sort a set of strings in lexicographically nondecreasing
order. A well-known algorithm for this problem is multikey Quicksort due to Bentley
and Sedgewick [3]. Like regular Quicksort, it partitions input strings into a set smaller
than, a set equal to, and a set greater than a pivot; like radix sort, it moves onto the next
character once the given character of the current input is known to be equal to the pivot.

Hoare [7] sketched a basic multikey Quicksort in a section on "Multi-word Keys" :
"When it is known that a segment comprises all the items, and only those items, which
have key values identical to a given value over their first n words, then, in partitioning
this segment, comparison is made of the $(n+1)$th word of the keys." Bentley and McIlroy [2] implemented a ternary Quicksort that employs 'split-end' partitioning, which
moves equal elements to the ends and swaps them back to the middle. Bentley and
Sedgewick [3] presented multikey Quicksort for strings adopting these two approaches.

In this paper, we improve Bentley and Sedgewick's multikey Quicksort. First, we
improve the operation of swapping equal elements from the ends back to the middle
in 'split-end' partitioning. Then we improve multikey Quicksort by using another partitioning method called 'collect-center' partitioning, instead of 'split-end' partitioning.
This partitioning blends the 'Dutch National Flag Problem' (DNF) ternary partitioning
[1] and Bentley and McIlroy's. Like the DNF partitioning, it moves equal elements to
the middle directly; like Bentley and McIlroy's, it uses two inner loops, one of which

---

* This work was supported by FPR08-A1-021 of the 21C Frontier Functional Proteomics Project
from the Korean Ministry of Education, Science and Technology. The ICT at Seoul National
University provides research facilities for this study.

halts on a greater element and the other halts on a smaller element. Multikey Quick-sort with our partitioning method is about 7~18% faster than multikey Quicksort with 'split-end' partitioning.

Section 2 reviews multikey Quicksort, the DNF partitioning and 'split-end' partitioning. An improved 'split-end' partitioning and 'collect-center' partitioning are presented in Section 3 and 4 respectively. Section 5 shows experimental results with four partitioning methods.

2 Preliminaries

Multikey Quicksort We first consider multikey Quicksort. Its input is an array S of n pointers to strings and it sorts strings in lexicographically nondecreasing order. Like Quicksort, it partitions S to sub-arrays smaller than, equal to, and greater than a pivot; like radix sort, it moves onto the next character to partition the sub-array equal to a pivot. Then it sorts three sub-arrays recursively.

This recursive pseudo-code sorts the array S of n strings that are known to be identical in the first $d - 1$ characters. It is initially called as $sort(S, n, 1)$. $S_<$, $S_=$ and $S_>$ mean sub-arrays smaller than, equal to and greater than a pivot, respectively.

> sort(S, n, d)
>   if $n \leq 1$ return;
>   choose a pivot v;
>   partition S around v on d-th character to form sub-arrays $S_<, S_=, S_>$ of
>   sizes $n_<, n_=, n_>$;
>   sort($S_<, n_<, d$);
>   sort($S_=, n_=, d + 1$);
>   sort($S_>, n_>, d$);

Choosing a pivot and partitioning S around the pivot can be implemented in many ways. In order to improve multikey Quicksort, we focus on the partitioning method of multikey Quicksort in this paper.

Bentley and McIlroy's 'Split-end' Partitioning Bentley and Sedgewick adopted 'split-end' partitioning as the partitioning method of multikey Quicksort. 'Split-end' partitioning uses the following invariant.

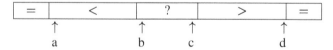

Indices b and c point to the left and right ends of the array of unchecked elements, respectively. An index a points to the beginning of the array of smaller elements, and d points to the end of the array of greater elements.

As index b moves to the right, it scans over smaller elements, swaps equal elements to the element pointed by a and stops at a greater element. Symmetrically it moves index c to the left, scans over greater elements, swaps equal elements to the element pointed by d and stops at a smaller element. That is, it swaps elements every time b and c meet equal elements. And then, it swaps elements pointed by b and c, increase and decrease these pointers and continues until b and c cross. At the end, it reaches the following state.

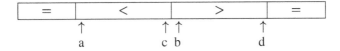

It swaps equal elements from the ends back to the middle.

This partitioning is usually efficient. In case of many equal elements, however, it wastes time swapping them to the end and back to the middle.

'Dutch National Flag Problem' Partitioning On the other hand, the DNF partitioning uses the following invariant.

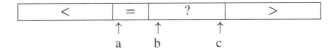

Indices b and c point to the left and right ends of the array of unchecked elements, respectively. An index a points to the beginning of the array of equal elements.

As index b moves to the right, it scans over equal elements, swaps smaller elements to the element pointed by a and greater elements to the element pointed by c. In case of greater elements, it doesn't move index b because the swapped element is not checked. That is, it swaps elements every time b meets a smaller or greater element.

If the input such as integers has a few equal elements, the DNF partitioning can be slower than 'split-end' partitioning. In case of many equal elements, however, it just scans over equal elements and can be very fast.

3   An Improved 'Split-end' Partitioning

In this section, we consider the operation of swapping equal elements from the ends back to the middle in 'split-end' partitioning. Let A be an array of equal elements at the left or right end and B be an array of smaller or greater elements in the middle. The operation of swapping equal elements from the ends back to the middle is the problem of swapping k elements in array A and k elements in array B, where the order of swapped elements does not need to maintain.

For swapping equal elements from the ends back to the middle, Bentley and Sedgewick [3] use the usual swap operation k times as follows. $A[i]$ means the ith element in an array A.

$$
\begin{aligned}
&\textbf{for } i \leftarrow 1 \text{ to } k \textbf{ do} \\
&\quad t \leftarrow A[i] \\
&\quad A[i] \leftarrow B[i] \\
&\quad B[i] \leftarrow t
\end{aligned}
$$

Now we define **movement** as an assignment operation and count movements of a swap operation. In the pseudo-code above, one swap operation consists of three movements. First, it copies an element in array A to a temporary space, t. Then, it copies an element in array B to array A. Finally, it copies the value in temporary space t to array B. So, the swap operation of k elements consists of $3k$ movements.

However, we can decrease the number of movements as follows. We call this method **batch-swap**.

```
1        t ← A[1]
2        for i ← 1 to k − 1 do
3            A[i] ← B[i]
4            B[i] ← A[i + 1]
5        od
6        A[k] ← B[k]
7        B[k] ← t
```

It moves the first element in array A to the temporary space t, and it gets an empty space in array A (line 1). Next, it moves the first element in array B to that empty space, and it gets an empty space in array B again and repeats this step. After the *for* loop, we move the last element in array B to an empty space in array A, and move the element in t to an empty space in array B. Each step is shown in Figure 1.

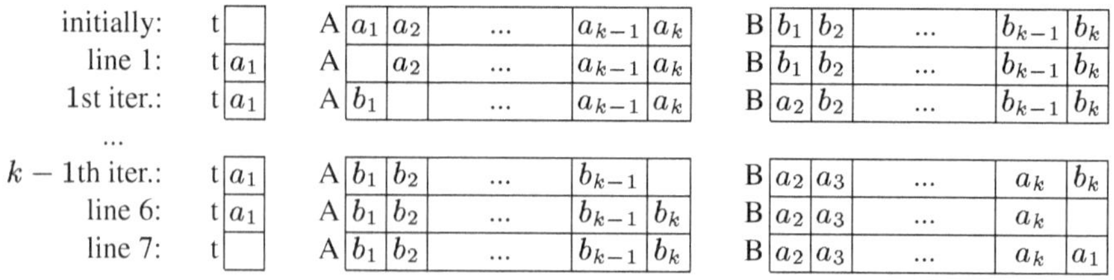

	t	A					B				
initially:		a_1	a_2	...	a_{k-1}	a_k	b_1	b_2	...	b_{k-1}	b_k
line 1:	a_1		a_2	...	a_{k-1}	a_k	b_1	b_2	...	b_{k-1}	b_k
1st iter.:	a_1	b_1		...	a_{k-1}	a_k	a_2	b_2	...	b_{k-1}	b_k
...											
$k − 1$th iter.:	a_1	b_1	b_2	...	b_{k-1}		a_2	a_3	...	a_k	b_k
line 6:	a_1	b_1	b_2	...	b_{k-1}	b_k	a_2	a_3	...	a_k	
line 7:		b_1	b_2	...	b_{k-1}	b_k	a_2	a_3	...	a_k	a_1

Fig. 1. Steps of batch-swap

Now we count the number of movements of the batch-swap. Because there are $k-1$ iterations, we need $2(k − 1)$ movements in the *for* loop. Since we need 3 movements outside the *for* loop, the batch-swap needs $2k + 1$ movements for swapping k elements.

4 'Collect-center' Partitioning

In this section, we propose another partitioning method for multikey Quicksort. We also present the C program of multikey Quicksort adopting this method.

4.1 Details of 'Collect-center' Partitioning

The 'split-end' partitioning is not efficient in case of many equal elements. If we can partition equal elements directly to the middle instead of swapping them to the end and back to the middle, this inefficiency can be removed. We call this method 'collect-center' partitioning. This partitioning blends the DNF partitioning [1] and Bentley and McIlroy's [2]. Like DNF partitioning's approach, it moves equal elements to the middle directly. Like Bentley and McIlroy's approach, it uses a loop invariant that a partitioning is processed at both ends of an array. That is, 'collect-center' partitioning uses the following invariant.

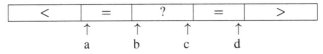

Indices b and c point to the left and right ends of the array of unchecked elements, respectively. An index a points to the beginning of the left array of equal elements, and d points to the end of the right array of equal elements symmetrically.

As b moves to the right, it scans over equal elements, swaps smaller elements to the element pointed by a and stops at a greater element. Symmetrically it moves index c to the left, scans over equal elements, swaps greater elements to the element pointed by d and stops at a smaller element. And then, it swaps elements pointed by b and c. But it doesn't increase and decrease these pointers because the element pointed by b is a smaller element and the element pointed by c is a greater element. So, it needs to swap the element pointed by b for the element pointed by a and swap the element pointed by c for the element pointed by d. After that, it moves those pointers and continues until b and c cross.

But it has an inefficiency that it needs to swap three times when pointers b and c are stopped. It first swaps elements pointed by b and c, next swaps elements pointed by a and b and elements pointed by c and d. For removing this inefficiency, we apply the batch-swap approach to this case. In order to get the same result of 3 swap operations, a smaller element pointed by c should be moved to the place pointed by a and a greater element pointed by b to the place pointed by d, and equal elements pointed by a and d should be moved to the places pointed by b and c, respectively. We can do movements as follows to satisfy these conditions.

1. $t \leftarrow$ element pointed by b
2.   move element pointed by a to b
3.   move element pointed by c to a
4.   move element pointed by d to c
5.   move t to d

The procedure above obtains the same result and it decreases 9 movements (3 swaps) to 5 movements. The 5 movements are shown in Figure 2. $a_=$ means the equal element pointed by a and $b_>$ the greater element pointed by b. The reverse order which starts with moving element pointed by c to t obtains the same result, too.

The order of movements is important. If the order of movements is changed, the result can be wrong when pointers a and b point to the same position or pointers c and d do. For example, we assume that pointers a and b point to the same position.

init.: t	<	$a_=$	=	$b_>$	?	$c_<$	=	$d_=$	>
line 1: t $b_>$	<	$a_=$	=		?	$c_<$	=	$d_=$	>
line 2: t $b_>$	<		=	$a_=$	?	$c_<$	=	$d_=$	>
line 3: t $b_>$	<	$c_<$	=	$a_=$	?		=	$d_=$	>
line 4: t $b_>$	<	$c_<$	=	$a_=$	?	$d_=$	=		>
line 5: t	<	$c_<$	=	$a_=$	?	$d_=$	=	$b_>$	>

Fig. 2. Steps of 5 movements instead of 3 swaps

1. $t \leftarrow$ element pointed by a
2. move element pointed by c to a
3. move element pointed by d to c
4. move element pointed by b to d
5. move t to b

If we change the order of movements as above, it overwrites a greater element pointed by b ($b_>$) with a smaller element pointed by c ($c_<$) (line 2) and it copies an overwritten smaller element pointed by b ($b_<$) to the position pointed by d in the end (line 4). The movements are shown in Figure 3. This result is wrong.

init.: t	<	$a_>, b_>$	?	$c_<$	=	$d_=$	>
line 1: t $a_>$	<	$b_>$	?	$c_<$	=	$d_=$	>
line 2: t $a_>$	<	$b_<, c_<$	?		=	$d_=$	>
line 3: t $a_>$	<	$b_<, c_<$	?	$d_=$	=		>
line 4: t $a_>$	<	$c_<$	?	$d_=$	=	$b_<$	>
line 5: t	<	$a_>$	?	$d_=$	=	$b_<$	>

Fig. 3. The wrong case of movements for swapping elements pointed by b and c

4.2   C Program of Multikey Quicksort with 'Collect-center' Partitioning

We assume that readers are familiar with the C programming language and explain the details of collect-center partitioning with C program. The C program of multikey Quicksort with 'collect-center' partitioning is shown in Figure 4.

The sort function is similar to the code of Bentley and Sedgewick [3], i.e., multikey Quicksort with 'split-end' partitioning. Its input is the array a of n pointers to strings. It sorts strings in lexicographically nondecreasing order with the $depth$-th characters of strings. In the code above, the pivot selection phase is omitted, which is the same with the code of Bentley and Sedgewick [3] choosing the median of three elements (and on larger arrays, the median of three medians of three) for the pivot v described by Sedgewick [4]. $p2c(i)$ macro (for 'pointer to character') accesses the $depth$-th character of string $a[i]$. Pointers pa, pb, pc and pd means pointers a, b, c and d, respectively. After

```c
void ccsort(char **a, int n, int depth)
{
    int r, v;
    char **pa, **pb, **pc, **pd, **pn, *t;
    ...    // pivot selection phase
    for (;;) {
        while (pb <= pc && (r = p2c(pb)-v) <= 0) {
            if (r < 0)   { swap(pa, pb); pa++; }
            pb++;
        }
        while (pb <= pc && (r = p2c(pc)-v) >= 0) {
            if (r > 0)   { swap(pc, pd); pd--; }
            pc--;
        }
        if (pb > pc) break;
        t = *pb;
        *pb = *pa;
        *pa = *pc;
        *pc = *pd;
        *pd = t;
        pb++;
        pc--;
        pa++;
        pd--;
    }
    pn = a + n;
    if ((r = pa-a) > 1)
        ccsort(a, r, depth);
    if (ptr2char(a + r) != 0)
        ccsort(a + r, pd+1-pa, depth+1);
    if ((r = pn-pd-1) > 1)
        ccsort(pn - r, r, depth);
}
```

Fig. 4. C program using our 'collect-center' partitioning for multikey Quicksort

partitioning, it sorts strings of three parts recursively. The most important difference between the two is just which elements pointers b and c scan over. In the code of Bentley and Sedgewick [3], pointers b and c scan over smaller and greater elements, respectively. But in this code, pointers b and c scan over equal elements. And the code of Bentley and Sedgewick [3] just swaps elements pointed by b and c when pointers b and c stop, but our code does slightly complex assignment operations adopting the batch-swap approach as we described in the preceding subsection. Finally, the code of Bentley and Sedgewick swaps equal elements from the ends back to the middle, but our code does not need this step because equal elements are already in the middle.

5 Experimental Results

We had experiments with four methods for multikey Quicksort, which are the DNF partitioning [1], Bentley and Sedgewick's 'split-end' partitioning [3], the improved 'split-end' partitioning using the batch-swap and 'collect-center' partitioning, . We downloaded the code of Bentley and Sedgewick at http://www.cs.princeton.ed u/~rs/strings. All the methods had been implemented in C, compiled with gcc 3.2, with option $-O3$ for maximum optimization. We have the experiments in 2.4GHz Xeon with 2 GB RAM, running GNU/Linux 2.4.20-28.

We measured the run-time of multikey Quicksort with various inputs. For each input, we experimented 1000 times repeatedly with shuffling pointers to strings in the array at each time, because multikey Quicksort can have different performances according to the order of input strings. The reported average times are measured with *gettimeofday()* function. As inputs, we use a set of large files, listed in Table 1. We use various inputs to compare the partitionings for different kinds of ratio of equal elements.

file	description	size(byte)
ran	A collection of strings of 1,000 random characters	50,049,998
kordic	A collection of about 630,000 Korean dictionary words encoded by EUC-KR	5,837,700
words	A set of words in /usr/share/dict/words file of Linux	4,951,020
fasta	*Mus Musculus* fragments set of DNA sequences from NCBI	224,803,104
bible	A set of verses of 'New International Version' bible	4,137,735
circ	Data set of library call numbers used in the DIMACS Implementation Challenge from http://theory.stan ford.edu/~csilvers/libdata/. We extract the set of unique keys from the file.	1,931,910
ran_ab	A collection of strings repeated the letters 'a' or 'b' 1,000 times randomly	50,049,998
html_cnn	A collection of about 1,000 html files from www.cnn.com	60,544,921
html_kr	A collection of about 1,000 html files from servers in Korea	47,690,332
aaa500	All the suffix of the string repeated the letter 'a' 500 times	125,750

Table 1. Input data set used for comparison

Table 2 shows sorting time of multikey Quicksort with four partitionings in milliseconds, listed in order of the ratio of equal elements. The ratio of equal elements is the average of the number of equal elements divided by n in $sort(a, n, depth)$ functions of $depth < 10$. Figure 5 shows the time ratio of 'collect-center' partitioning, the batch-swap and the DNF partitioning to 'split-end' partitioning.

The result shows that 'collect-center' partitioning is faster than 'split-end' partitioning in all the cases. 'Collect-center' partitioning gives us 2~18% speed-up. In addition, ratios of most of non-random strings such as dictionary words, web documents and DNA sequences are 0.4~0.8. 'Collect-center' partitioning gives us 7~18% speed-up in these cases.

file	ratio of equal elements	split-end	batch-swap	collect-center	DNF
ran	0.108737	42.4696	**41.3044**	41.4939	42.0272
kordic	0.400524	358.6713	346.5001	**332.0699**	339.5534
words	0.513000	291.3794	281.1821	**267.2211**	269.1517
fasta	0.633087	436.0785	428.8563	**399.6577**	408.069
bible	0.677879	28.3272	27.4417	**24.4105**	25.1406
circ	0.684815	47.283	45.8968	**38.2317**	39.0791
ran_ab	0.753825	58.2717	56.6157	**50.7555**	52.9012
html_cnn	0.798380	2,666.2008	2,631.8045	**2,262.9385**	2,317.721
html_kr	0.857262	1,869.7797	1,844.7099	**1,543.419**	1,583.3273
aaa500	0.998185	1.2745	1.28	**0.4658**	0.4767

Table 2. Sorting times in milliseconds. The 'ratio of equal elements' is the average of the number of equal elements divided by n in *sort(a, n, depth)* functions of $depth < 10$.

'Split-end' partitioning has a lot of movements when there are many equal elements and 'collect-center' partitioning does when there are a few equal elements. So, it can be supposed that 'split-end' partitioning is faster than 'collect-center' partitioning in case of a few equal elements. But the result in Figure 5 shows that 'collect-center' partitioning is faster than 'split-end' partitioning in all the cases. Figure 6 shows the time ratio without compiler's optimization, which shows the supposed result. Figure 7 shows the ratio of multikey Quicksort's total movements of 'collect-center' partitioning, the batch-swap and the DNF partitioning to 'split-end' partitioning without compiler's optimization.

References

1. Dijkstra, E.W. *A Discipline of Programming.* Prentice-Hall, Englewood Cliffs, NJ, 1976.
2. Bentley, J.L. and McIlroy, M.D. *Engineering A Sort Function.* Software-Practice and Experience 23, 1 (1993), 1249-1265.
3. Bentley, J.L. and Sedgewick, R. *Fast Algorithms for Sorting and Searching Strings.* Proceedings of the eighth annual ACM-SIAM symposium on Discrete algorithms, New Orleans, January, 1997, 360-369.
4. Sedgewick, R. *Implementing Quicksort Programs.* Communications of the ACM 21, 10 (October 1978), 847-857.
5. Larsson, N.J. and Sadakane, K. *Faster suffix sorting.* Technical Report LU-CS-TR:99-214, LUNDFD6/(NFCS-3140), 1999, 1-20.
6. McIlroy, P.M., Bostic, K. and McIlroy, M.D. *Engineering radix sort.* Computing Systems, vol.6, no.1, 1993, 5-27.
7. Hoare, C.A.R. *Quicksort.* Computer Journal 5, 1 (April 1962), 10-15.
8. Kernighan, B.W. and McIlroy, M.D. *UNIX Programmer's Manual, 7th Edition.* Bell Telephone Laboratories, Murray Hill, NJ (1979).
9. Sedgiwick, R. *Quicksort with equal keys.* SIAM J. Comp, 6, 240-267 (1977).

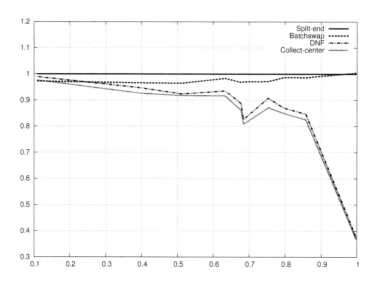

Fig. 5. The time ratio of 'collect-center' partitioning, the batch-swap and the DNF partitioning to 'split-end' partitioning

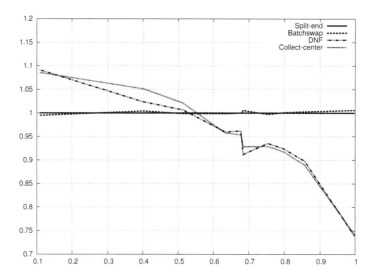

Fig. 6. The time ratio of 'collect-center' partitioning, the batch-swap and the DNF partitioning to 'split-end' partitioning with -O0 option

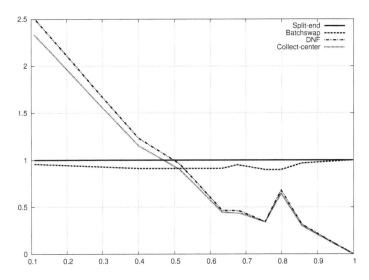

Fig. 7. The movement ratio of 'collect-center' partitioning, the batch-swap and the DNF partitioning to 'split-end' partitioning with -O0 option

Approximate Word Sequence Matching on an Inverted File Index*

Taehyung Lee[1], Sung-Ryul Kim[2]**, and Kunsoo Park[1]

[1] School of Computer Science and Engineering, Seoul National University,
{thlee,kpark}@theory.snu.ac.kr
[2] Division of Internet & Media, Konkuk University,
kimsr@konkuk.ac.kr

Abstract. We consider approximate word sequence matching on an inverted file. We propose an algorithm framework in which we can efficiently find all approximate occurrences of word sequences, without modifying the existing inverted file structure. We also present algorithmic details for approximate word sequence matching.

Key words: Approximate matching, Inverted file, Information retrieval, Search engine

1 Introduction

Given an *inverted file* $\mathcal{I}$ of a collection of d documents $\mathcal{D} = \{T^1, T^2, ..., T^d\}$, a pattern P of m words, and an integer k, we are to find all approximate occurrences of the pattern P in T^h, $1 \leq h \leq d$, where a *word sequence* edit distance between P and a subsequence of T^h is less than k.

The word sequence edit distance [16] between two sequences of words is given by the minimum number of edit operations needed to transform one sequence into the other, where an operation is an insertion, deletion, or substitution of a single *word*, not a character.

This problem arises from the Information Retrieval domain, especially from the Internet search engines. Virtually all current commercial search engines such as Google, Yahoo, and MSN Search are using the inverted file as their primitive text retrieval databases. The inverted file consists of a set of words, where each word is associated with the list of documents containing it in a given document collection $\mathcal{D}$. It has long been used as the underlying data structure of choice for the pattern matching problem, where P is a unit of text, such as a word or phrase, and $T \in \mathcal{D}$ is a natural language [24]. However, due to the limited facility of inverted files, most search engines provide restricted search options such as bag-of-word search, phrase search, or proximity search.

* This work was supported by FPR08-A1-021 of the 21C Frontier Functional Proteomics Project from the Korean Ministry of Education, Science and Technology. The ICT at Seoul National University provides research facilities for this study.
** Supported by Seoul R&BD Program(10581)

There have been many studies on character-based approximate matching over inverted files, but they have not been widely adapted to search engines in reality for efficiency reasons. In addition, a character-based edit distance metric itself sometimes does not fulfil exactly what search engine users expect. If a pattern is given as "approximate string matching", for example, character-based approximate matching with $k = 2$ will find a match in *"aproximate* string matching" with 1 deletion of the character 'p'. This is quite reasonable, and in practice it works fairly well with error sources such as human typing errors or optical character recognition errors, where a relatively small number of errors is presented in the unit of a character. However this is not the case when we consider the query given by a search engine user. This time the source of error is human memory, where error occurs more often in the form of insertion, deletion, or substitution of a single *word*. For example, a user might give a query string such as "approximate matching", or "string matching with errors" while he (or she) intended to find "approximate string matching". In this case, the character-based edit distance usually becomes much larger than the fixed k, thus we may fail to find a proper match. On the other hand, the word sequence edit distance will be more appropriate for this problem. This can happen in other areas such as the affiliation matching problem [5].

Therefore, if we solve the approximate word sequence problem on the inverted file index, we can provide search engines with a more flexible and readily available search option without any modification of the existing data structure. Since it is not necessary to perform a sequential access on the vocabulary, our algorithm seems to be as efficient as current search facilities. Thus we can increase the effectiveness of the search engine without losing the efficiency.

Approximate string matching has been one of the most active area of research in stringology for decades, since the simple dynamic programming solutions [17, 21] were proposed. There have been abundant works, which fall into four main streams of research, namely, dynamic programming algorithms, automata, bit-parallelism, and filtering algorithms [13]. We can classify them into *online* search since they do not preprocess the text. For more details, we refer to the excellent survey by Navarro [13].

In recent years, research focus has developed into *offline* search, as the volume of text data to be processed grows prominently. Suffix trees are the main index structures of choice for *full-text* indexing [19, 20]. However, for natural languages such as English text or HTML documents, the inverted file is the most efficient and practical index for matching.

There have been many studies into approximate matching on the inverted file: Glimpse [23], Agrep [22], Igrep [1], Signature hashing [3], etc. They find approximate matches starting at word boundaries only, but they use character-based distance metrics. Moreover, most of them inherently require sequential approximate searches over the whole vocabulary.

Risvik [16] introduced a notion of approximate word sequence matching. He used a word-based sparse suffix tree to solve the problem. However it seems not to be the proper data structure of choice for huge document collections due to the large space requirement of suffix tree itself.

In this paper, we solve the approximate word sequence matching problem over the inverted file. We propose an algorithm framework in which we can efficiently find all approximate occurrences of word sequences, without modifying the existing inverted file structure.

2 Preliminaries

We assume that the text can be divided naturally into *words*. We will use the notion of *word sequence*.

Definition 1 (Word sequence). *A word sequence is a sequence of separated and consecutive words. A word sequence $S[1..n] = s_1 s_2 ... s_n$ consists of n single words s_1, s_2, ..., s_n, which are separated by delimiters, such as white spaces, commas, periods, or other punctuations. We also define $S[i..j]$ as the subsequence of words $s_i s_{i+1} ... s_j$ of S and $S[i] = s_i$ as the i-th word of S.*

Let $T[1..n] = t_1 t_2 ... t_n$ be the text (or a document) of n words and $P[1..m] = p_1 p_2 ... p_m$ be the pattern of m words. We are given a collection of d documents $\mathcal{D} = \{T^1, T^2, ..., T^d\}$, where n_h denotes the number of words in the h-th document T^h. An occurrence of a word w in $\mathcal{D}$ is given by a pair $\langle h, j \rangle_w$ of a document index h and an in-document *word* position j. In other words, the pair $\langle h, j \rangle_w$ denotes that w occurs at the j-th word position in the h-th document $T^h \in \mathcal{D}$.

We now define an inverted file over $\mathcal{D}$. The inverted file $\mathcal{I}(\mathcal{V}, \mathcal{P})$ is composed of a vocabulary $\mathcal{V}$ and postings (or inverted lists) $\mathcal{P}$. A vocabulary $\mathcal{V}$ is a collection of all words in $\mathcal{D}$ and it supports an efficient look-up operation for a given query word.[3] Generally, all the words in vocabulary $\mathcal{V}$ are stored in lexicographical order so as to support efficient lookup operations. Postings $\mathcal{P}$ is a set of occurrence lists L_w, where L_w is a sorted list of indices of documents having a given word w. For each index $h \in L_w$, we also maintain an ordered list $l_{w,h}$ of in-document positions of w in T^h. Thus, if a word w occurs at the j-th word position in $T^h \in \mathcal{D}$, there must be $w \in \mathcal{V}$, $h \in L_w$ and $j \in l_{w,h}$ in $\mathcal{I}(\mathcal{V}, \mathcal{P})$. In Fig. 1, for example, the occurrence $\langle h, 2 \rangle_{\mathbf{boy}}$ of the word "boy" is represented by "**boy**" $\in \mathcal{V}$, $h \in L_{\mathbf{boy}}$ and $2 \in l_{\mathbf{boy},h}$. For notational convenience, given a pattern word p_i we will use L_{p_i} and L_i interchangeably if there is no confusion, and the same holds for $l_{p_i,h}$ and $l_{i,h}$.

Once we have a collection $\mathcal{D}$ and a pattern P, we can formally define the problem of finding an approximate occurrence of P as follows. An approximate matching is the problem of finding the occurrences of a pattern in a text where the pattern and the occurrence have a limited number of differences. We generally model this problem using a distance function that tells how similar two strings are. One of the best studied distance functions is *Levenshtein distance* or *edit*

---

[3] In general, inverted files are case-insensitive. Thus, we assume that all words are turned into lowercases in vocabulary and patterns.

Documents $\mathcal{D}$

$T^h =$ **The boy** is **the boy** left to her $\cdots$	
Word position $\underline{1}$ 2 3 $\underline{4}$ $\underline{5}$ 6 7 8 $\cdots$	

$T^{h'} =$ He is **the boy** who met a **boy** $\cdots$	
Word position 1 2 $\underline{3}$ $\underline{4}$ 5 6 7 $\underline{8}$ $\cdots$	

Vocabulary $\mathcal{V}$ Postings $\mathcal{P}$

Word	List
a	
boy	$L_{\mathbf{boy}} \quad \rightarrow \quad L_{\mathbf{boy}} = \cdots \boxed{h} \cdots \boxed{h'} \cdots$
he	
her	$l_{\mathbf{boy},h} = \boxed{2\,5} \cdots \quad l_{\mathbf{boy},h'} = \boxed{4\,8} \cdots$
is	
left	$\cdots \quad \cdots \quad \cdots \quad \cdots \quad \cdots \quad \cdots \quad \cdots \quad \cdots \quad \cdots$
met	
the	$L_{\mathbf{the}} \quad \rightarrow \quad L_{\mathbf{the}} = \cdots \boxed{h} \cdots \boxed{h'} \cdots$
to	
who	$l_{\mathbf{the},h} = \boxed{1\,4} \cdots \quad l_{\mathbf{the},h'} = \boxed{3} \cdots$
$\cdots$	

Fig. 1. Example of an inverted file for $\mathcal{D} = \{T^h, T^{h'}\}$

distance [11]. We use the edit distance metric as a metric for approximate word sequence matching.

In order to define word sequence edit distance, we first define edit operations on word sequences as follows.

Definition 2 (Word sequence edit operation). *Given a sequence of words* $X = x_1 x_2 ... x_n$, *we define word sequence edit operations as follows*

- *Insertion* ($\epsilon \rightarrow \alpha$): *insertion of a word* α *into sequence* X
- *Deletion* ($\alpha \rightarrow \epsilon$): *deletion of a word* α *from sequence* X
- *Substitution* ($\alpha \rightarrow \beta$): *substitution of a word* α *with a word* β ($\alpha \neq \epsilon$ *and* $\beta \neq \epsilon$)

The cost function $C(x \rightarrow y)$ *is a constant defined as 1 for insertion, deletion, and substitution.*

The edit operations are identical to those used for the character-based edit distance. The only difference is that we treat words rather than characters as the atoms of edit operations. Using the edit operations defined above, the edit distance for word sequences is defined as follows.

Definition 3 (Unit cost word sequence edit distance). *Given two sequences of words* $X = x_1 x_2 ... x_n$ *and* $Y = y_1 y_2 ... y_m$, *we define* $ED(X, Y)$ *as the minimum number of edit operations required to convert* X *into* Y, *or vice versa. This equals the minimum sum of the costs* $C(x_i \rightarrow y_j)$ *for a sequence of edit operations transforming* X *into* Y.

Now we define the problem of finding an approximate word sequence match over $\mathcal{D}$.

Definition 4 (Approximate word sequence match). *We say that an approximate match ending at j occurs in document T^h if there exists a sequence of occurrences $\langle h, j_1 \rangle_{p_{i_1}}, \langle h, j_2 \rangle_{p_{i_2}}, \ldots, \langle h, j_r \rangle_{p_{i_r}}$ such that*

(i) $j_1 < j_2 < \ldots < j_{r-1} < j_r = j$ *and* $1 \leq i_1 < i_2 < \ldots < i_r \leq m$
(ii) $ED(T^h[j_1..j_r], P) \leq k$

where k is the maximum number of errors allowed.

Note that, compared to the standard notion of approximate matching, this definition has a slightly stronger condition in which we constrain the occurrence of a match to be confined within $[j_1, j_r]$, where $t_{j_1} = p_{i_1}$ and $t_{j_r} = p_{i_r}$. In this paper we will solve the above problem to find every approximate occurrence of P by utilizing the inverted file over $\mathcal{D}$.

Before we proceed further, we will give a few more definitions related to a pattern. For a given pattern of m words $P[1..m] = p_1 p_2 ... p_m$, we define $W = \{p_i | i \in [1..m]\}$ as the set of pattern words and $m' = |W|$. Note that $m' \leq m$ since a word can occur multiple times in P. We count the number of occurrences of each word $w \in W$ as $count[w]$. We also define Pos as the set of the first position of w in P for all $w \in W$. Let $count'[1..m]$ be defined as $count'[i] = count[p_i]$ for all $i \in Pos$. To chain those multiple occurrences of a word into a single linked list, we define $next[i]$, for $i \in [1..m]$, is the smallest $j > i$ such that $p_i = p_j$, or $\perp$ if such j does not exist. For example, the pattern $P = $ "to be or not to be" will result in $W = \{$ "to" (2), "be" (2), "or" (1), "not" $(1)\}^4$, $Pos = \{1, 2, 3, 4\}$, and

i	1	2	3	4	5	6
P	to	be	or	not	to	be
$next[i]$	5	6	$\perp$	$\perp$	$\perp$	$\perp$
$count'[i]$	2	2	1	1	$-$	$-$

where '$-$' denotes an undefined value.

3 Main algorithm

The main algorithm is composed of 3 steps:

1. Preprocess the pattern sequence
2. Look up pattern words over $\mathcal{V}$ and intersect the occurrence lists *at document level*
3. For each document T^h in the intersection, match pattern words with text words *at in-document position level* to find an approximate match.

---

[4] A number inside parentheses denotes corresponding $count[w]$

For step 3 we will describe three different matching algorithms in the next three sections, and here we describe step 1 and 2.

In step 1, we preprocess the pattern P to compute W, Pos, $next[1..m]$ and $count'[1..m]$. This is done by incrementally building a trie T for W. The trie T is an augmented trie, whose leaf nodes correspond to pattern words $w \in W$. For each leaf node q corresponding to a pattern word w, we store two variables $pos[q]$ and $cnt[q]$ at node q. At the end of step 1, $pos[q]$ and $cnt[q]$ will have the first position of pattern word w in P and the number of occurrences of w in P, respectively.

Now we briefly describe how the algorithm works. The algorithm proceeds by reading pattern words p_m to p_1. Note that during the iteration, we maintain an invariant for T, which says that after reading p_i, for each leaf node q corresponding to $w \in W$, $pos[q]$ contains the minimum $j \in [i..m]$ such that $p_j = w$ and $cnt[q]$ has the number of occurrences of w in $P[i..m]$. When we read p_i, we walk down T from the root in order to locate leaf node q corresponding to p_i. If it does not exist in T, we create q and its ancestors appropriately. Then we update $pos[q]$ and $cnt[q]$ as follows. If node q is just created, we set $cnt[q]$ as 1 and $pos[q]$ as i. Otherwise, we increase the number of p_i's occurrences by adding 1 to $cnt[q]$. For $pos[q]$, we first chain the current occurrence i of p_i with its previous occurrence $pos[q]$ by recording the previous occurrence $pos[q]$ into $next[i]$, so that we have $p_{next[i]} = p_i$. And then, we update $pos[q]$ as i. Once we complete building trie T, by visiting all leaf nodes of T, we collect all the first occurrences $pos[q]$'s of $w \in W$ into Pos and compute $count'$ as $count'[pos[q]] = cnt[q]$.

Lemma 1. *The algorithm* PREPROCESS *runs in* $O(|P|)$ *time, where* $|P|$ *denotes the length of the pattern* P *in terms of characters.*

In step 2, we first look up pattern words over the vocabulary $\mathcal{V}$. This can be done efficiently by applying a binary search in the sorted array of words, or B-trees. Hashing techniques can be used here to provide faster look-up operations. Once we find a pattern word $w \in \mathcal{V}$, then we retrieve the corresponding occurrence list L_w from the postings $\mathcal{P}$.

Lemma 2. *The algorithm* LOOKUP *takes* $O(m' \log |\mathcal{V}| + \sum_{w \in W} |L_w|)$ *time if we use a binary search in the sorted array.*

After retrieving L_w for all pattern words $w \in \mathcal{V}$, the algorithm performs intersection of these L_w's. However, in contrast to the exact matching problem, we do not require that a document has to contain all the pattern words in order to have approximate occurrences. In other words, there can be *missing* pattern words in approximate occurrences.

Lemma 3. *If we have an approximate match in document* $T^h \in \mathcal{D}$, *then* T^h *must have a set* $U \subseteq W$ *of pattern words, which satisfies* $\sum_{w \in U} count[w] \geq m-k$.

Proof. We can prove this by contradiction. □

Algorithm 1 MAIN

1: $(W, Pos, next, count') \leftarrow$ PREPROCESS(P)
2: **for all** $i \in Pos$ **do**
3:     **if** LOOKUP(p_i, $\mathcal{V}$) = **true then** // Look up p_i on $\mathcal{V}$
4:         Retrieve occurrence list L_i from $\mathcal{P}$
5:         $\mathcal{C} \leftarrow \mathcal{C} \cup L_i$
6:     **end if**
7: **end for**
8: $H \leftarrow$ INTERSECT($\mathcal{C}, count', m - k$) // $(m - k)$-Intersect occurrence lists L_i's
9: **for all** $h \in H$ **do**
10:     $\mathcal{L} \leftarrow$ Retrieve all $l_{i,h}$'s for $i \in Pos$ if $h \in L_i$
11:     MATCH($\mathcal{L}, Pos, next[1..m], h$)
12: **end for**

We now define an $(m - k)$-intersection by using Lemma 3. The $(m - k)$-intersection is to find the set of documents T^h, where $h \in \bigcap_{w \in U} L_w$ such that $U \subseteq W$ is a set of pattern words which satisfies $\sum_{w \in U} count[w] \geq m - k$. Although we might not find an approximate occurrence in some T^h that satisfies the above condition, it is a necessary condition for having an approximate occurrence in T^h. The $(m - k)$-intersection can be achieved as follows. We attach the pattern index i with every document index $h \in L_{p_i}$ to make a list of pairs $\langle h, i \rangle$'s. Then we merge these lists of pairs into a single sorted list L in increasing order of h. For this purpose, we maintain a minimum heap, which contains the head of each sorted list L_{p_i}. We extract the minimum element from the heap. If the extracted element is from L_{p_i}, then we put the next element of L_{p_i} to the heap, if it exists. While constructing the merged list L, whenever we meet a run R of pairs with the same h's in L, we can check $\sum_{i, \langle h, i \rangle \in R} count'[i] \geq m - k$. If this holds, we put h into the resulting intersection H.

Lemma 4. *The $(m - k)$-intersection algorithm runs in $O(\log m' \cdot \sum_{i \in Pos} |L_{p_i}|)$.*

Once we have found the $(m - k)$-intersection H, we now apply a matching algorithm at in-document position level for each T^h in the intersection. For this purpose, we have to fetch all in-document position lists $l_{w,h}$ from Postings $\mathcal{P}$ if T^h is in the intersection. In step 3, given these in-document position lists of T^h, we can use one of three different matching algorithms. The following three sections describe these algorithms one by one.

4   Matcher 1

The algorithm MATCHER1 computes the dynamic programming table $D[0..m, 0..n]$ column by column, where the value of $D[i, j]$ denotes the minimum edit distance between $P[1..i]$ and a suffix of $T[1..j]$. However we can skip computations over consecutive columns which have no match between the pattern and the text and compute only the columns having one or more matches. This is done by using a Multiset data structure [12].

Before we get into the detail, we need to define some notation. Let $D[i,j]$ be the (i,j)-th entry of the dynamic programming table D. We call a column $D[0..m,j]$ as the matching column if $p_i = t_j$ for some $i \in [1..m]$. In the algorithm MATCHER2, we only maintain the current matching column C and the last matching column C'. A gap is the interval of consecutive columns without a match. Let δ denote the length of the gap. As the algorithm proceeds, whenever we encounter a gap, we compute the matching column C next to the gap using the Multiset data structure and the last matching column C'.

The Multiset is a structure which supports add, remove, maximum and minimum operations in worst-case $O(1)$ time over C'. It exploits the fact that consecutive entries in C' differ by at most one [18]. The details of the Multiset operations are described in [12]. We note that for more generalized distance metrics[5], we can use an MQUEUE [10] data structure, or a deque with heap order [6] instead, in order to get worst-case constant time processing.

The algorithm MATCHER1 proceeds as follows.

First we merge occurrence lists of pattern words into a single occurrence list L. Before doing this, we attach the first index i to every in-document position $j \in l_{i,h}$, in order to use this as an identifier of that pattern word. Then we sort the list L using a linear time sorter (e.g. radix sort) in the increasing order of all in-document positions j.

We now pop the occurrence j of p_i from L in the increasing order and compute each column j containing one or more matches from left to right. There arise three possible cases:

Case 1. (The last matching column is $j - 1$)
    We use the standard recurrence

$$D[i,j] = \min(D[i-1,j]+1, D[i,j-1]+1, D[i-1,j-1]+s(p_i,t_j)) \quad (1)$$

where $s(p_i, t_j) = 0$ if $p_i = t_j$ (this means $j \in L_{p_i}$) or 1 otherwise.
Case 2. (The last matching column column is $j - \delta - 1$, where $\delta \leq k$)
    We use the extended recurrence

$$D[i,j] = \min \begin{bmatrix} D[i-1,j]+1, \\ D[0,j-1]+(i-1)+s(p_i,t_j), \\ \min(D[i-\delta-1..i-1,j-\delta-1])+\delta+s(p_i,t_j), \\ D[i,j-\delta-1]+(\delta+1) \end{bmatrix}.$$
$$(2)$$

Note that $D[0,j-1]+(i-1)+s(p_i,t_j) = (i-1)+s(p_i,t_j)$ since $D[0,*] = 0$. When $s(p_i,t_j) = 0$, we can further simplify the above recurrence (2) as $D[i,j] = \min(i-1, \min(D[i-\delta-1..i-1,j-\delta-1])+\delta)$. Observe that $\min(D[i-\delta-1..i-1,j-\delta-1])+\delta$ can be computed in $O(1)$ time using Multiset.
Case 3. (The last matching column is $j - \delta - 1$, where $\delta > k$)

---

[5] For example, one can assign an arbitrary cost to each edit operation, or consider more sophisticated distance metrics such as the affine gap penalty model [10].

We initialize the preceding column, say the column $j - 1$, and use the standard recurrence (1).

Once we have computed all the entries of the current matching column, we check if the value in the last row of the column is less than k, If this is true, declare "Match" with a cost of $D[m, j]$. Otherwise, we continue the process for the next matching column.

Lemma 5. *The algorithm* MATCHER1 *correctly computes the entry* $D[0..m, j]$ *for every* $j \in L$ *if the entry value is less than* $k + 1$.

Proof. We prove the correctness by induction;

1. For $j = 0$, the algorithm correctly computes $D[i, 0] = i$
2. For all $j' \in L$ such that $j' < j$, if we assume $D[*, j']$ be correctly computed, then we can prove that $D[*, j]$ be correctly computed by
   - Case 1: Within intervals of consecutive columns containing matches, we calculate the standard recurrence relation (1).
   - Case 2: The main computation of recurrence relation (2) is $\min(D[i - \delta - 1..i - 1, j - \delta - 1])$. Using Multiset, we can store values of the last matching column, $D[i - \delta - 1..i - 1, j - \delta - 1]$, and extract the minimum element in constant time [12]. This guarantees correct calculation of recurrence relation (2) in the case that a gap occurs.
   - Case 3: Once $D[i, j] > k$, the exact value of $D[i, j]$ does not influence the existence of a match. If $\delta > k$, the last column entries $D[*, j']$ plus δ can never be less than k. Thus we safely ignore them for computation of $D[*, j]$.

□

Lemma 6. *The algorithm* MATCHER1 *runs in* $O(mn')$ *time, where* n' *is the length of* L, *i.e.* $n' = \sum_{i \in Pos} |l_{h,p_i}| \leq n$.

Proof. We can analyze the time complexity of MATCHER1 as follows

- SORT(L) takes $O(n')$ by using a linear sorter, for example, radix sort.
- The main loop iterates n' times
  - In case 1, 3, a column can be computed in $O(m)$
  - In case 2, Multiset guarantees worst-case $O(1)$ per operation, thus the calculation can be done in $O(m)$ per column.

□

5 Matcher 2

The second algorithm MATCHER2 utilizes sparse dynamic programming [4]. For a sparse set of matches, we only compute entries in the table D that correspond to matches. Each matching entry value has triangle-like regions that influence

Algorithm 2 MATCHER1($\mathcal{L}$, Pos, $next[1..m]$, h)

1: Initially we set $\mathsf{C}'[i] \leftarrow i$ for $\forall i \in [0, m]$
2: L: $\{(i, j) | j \in l_{i,h}$ if $l_{i,h} \in \mathcal{L}\}$, merged list of $l_{i,h}$ with the first index i attached
3: SORT(L) // Sort $(i, j) \in L$ by increasing order of j
4: **while** L is not empty **do**
5:     Pop (i, j) from L and $\delta \leftarrow j - j'$
6:     INITIALIZE(Multiset, ϕ)
7:     **for** $r \leftarrow 0$ to m **do**
8:       **if** $r = i$ **then**
9:         $s \leftarrow 0$ and $i \leftarrow next[i]$
10:       **else**
11:         $s \leftarrow 1$
12:       **end if**
13:       **if** $\delta = 1$ **then** // Case 1
14:         $\mathsf{C}[r] \leftarrow \min(\mathsf{C}[r-1] + 1, \mathsf{C}'[r] + 1, \mathsf{C}'[r-1] + s)$
15:       **else if** $1 < \delta \leq k$ **then** // Case 2
16:         ADD(Multiset, $\mathsf{C}'[r]$)
17:         $\mu \leftarrow$ MINIMUM(Multiset)
18:         $\mathsf{C}[r] \leftarrow \min(\mathsf{C}[r-1] + 1,\ r + s - 1,\ \mathsf{C}'[r] + \delta + 1,\ \mu + \delta + s)$
19:         REMOVE(Multiset, $C'[r - \delta - 1]$) if $r > \delta$
20:       **else** // Case 3
21:         $\mathsf{C}'[r] \leftarrow r$ and $\mathsf{C}[r] \leftarrow \min(\mathsf{C}[r-1] + 1, \mathsf{C}'[r] + 1, \mathsf{C}'[r-1] + s)$
22:       **end if**
23:     **end for**
24:     Declare "Match at j in T^h with cost $\mathsf{C}[m]$", if $\mathsf{C}[m] \leq k$
25:     $\mathsf{C}' \leftarrow \mathsf{C}$ and $j' \leftarrow j$
26: **end while**

---

following entries. We maintain column bounds and diagonal bounds of that regions using priority queue structures, so that we efficiently retrieve the most influencing previous value and compute the next entry value on demand.

Before we describe the algorithm, we give some definitions as follows. Let (i, j) be a point of a match, such that $p_i = t_j$, and S be the *sparse* set of points of all those matches in T^h. M denotes $|S|$, where $M \leq mn$. We can easily get S again by merging occurrence lists of pattern words as the algorithm MATCHER1 does. $(i', j') \prec (i, j)$ means (i', j') precedes (i, j), that is both $i' \leq i$ and $j' \leq j$ but not $(i', j') = (i, j)$. For a point $(i, j) \in S$, $(i, j_1) \in S$ (or $(i_1, j) \in S$) is the nearest preceding point of a match in the same row (or the same column, respectively). $UR(i, j)$ and $LR(i, j)$ are the upper and lower triangular region of influence of $(i, j) \in S$.

We briefly describe how the algorithm works. The recurrence of unit cost edit distance in S is

$$D[i, j] = \min \begin{bmatrix} D[i_1, j] + (i - i_1), \\ D[i, j_1] + (j - j_1), \\ \min\{D[i', j'] + \max(i - i' - 1, j - j' - 1)\} \end{bmatrix} \quad (3)$$

Algorithm 3 MATCHER2($\mathcal{L}, Pos, next[1..m], h$)

1: L: $\{(i,j)|j \in l_{i,h}$ if $l_{i,h} \in \mathcal{L}\}$, merged list of $l_{i,h}$ with the first index i attached
2: SORT(L) // Sort $(i,j) \in L$ by increasing order of j
3: **for** $i \leftarrow 0$ to m **do**
4:    $J[i] \leftarrow 0$
5: **end for**
6: **while** L is not empty **do**
7:    Pop (i,j) from L
8:    $i_1 \leftarrow 0$
9:    **while** $i \neq \perp$ **do**
10:      $j_1 \leftarrow J[i]$ // The maximum $j_1 < j$ such that $(i,j_1),(i,j) \in S$
11:      $J[i] \leftarrow j$
12:      Look up CBOUND and DBOUND to get (i',j_r) whose UR contains (i,j)
13:      Look up DBOUND$'$ to get $(i'',j_{r'})$ whose LR contains (i,j)
14:      $E[i,j] \leftarrow \min(E[i',j_r] + f(i',j_r,i,j), E[i'',j_{r'}] + f(i'',j_{r'},i,j))$
15:      $D[i,j] \leftarrow \min(D[i_1,j] + (i - i_1), D[i,j_1] + (j - j_1), i - 1, E[i,j])$
16:      Update CBOUND, DBOUND or DBOUND$'$ with (i,j)
17:      $i_1 \leftarrow i$ // The maximum $i_1 < i$ such that $(i_1,j),(i,j) \in S$
18:      $i \leftarrow next[i]$
19:    **end while**
20:    **if** $i_1 \neq m$ **then**
21:      Compute $D[m,j]$ by doing the same calculations inside the above **while** statement with a exception that we have no match in (m,j).
22:    **end if**
23:    **if** $D[m,j] \leq k$ **then** // Check if a match occurs
24:      Declare "Match" at j with cost $D[m,j]$
25:    **end if**
26: **end while**

---

for $(i,j),(i',j') \in S$ and $(i',j') \prec (i,j)$. The main computation in (3) can be represented by

$$E[i,j] = \min\{D[i',j'] + f(i',j',i,j)\} \qquad (4)$$

where f is a linear function[6] in each of $UR(i',j')$ and $LR(i',j')$.

This case will be called the triangle case and it can be computed using the algorithm which runs in $O(n + M \log\log\min(M, n^2/M))$ time [4] with slight modification. Note that we have to take care of 0's of the row $D[0,*]$. It can be done by simply taking $D[0,*] + i - 1 = i - 1$ into consideration when we minimize (3). Instead of using the computation by rows like the original paper, using the column-wise computation will give the worst-case $O(n + M \log\log m)$ time by using Johnson's data structure [9].

Lemma 7. *The algorithm* MATCHER2 *runs in* $O(n + M \log\log m)$ *time, where* $M = |S|$.

---

[6] $f = \max(i - i' - 1, j - j' - 1)$ is $i - i' - 1$ if $(i,j) \in LR(i',j')$; otherwise, it is $j - j' - 1$. Therefore, f is linear in both cases.

6 Matcher 3

The third algorithm applies the bit-parallelism of the NFA simulation, which is known as BPR approach [22]. The NFA for approximate string matching is well defined by $m \times (k+1)$ states, where every row denotes the number of errors seen and every column represents matching a pattern prefix [22]. The algorithm BPR packs each row i of the NFA in a different machine word R_i, with each state represented by a bit. For each new text character, all the transitions of the automaton are simulated using bit operations among the $k+1$ bit masks. For more details, refer to the original work [22].

We adapt this algorithm to our algorithm MATCHER3 as follows. Basically we treat each pattern word as a single character. Therefore reading a new text word can be simulated as having the next occurrence j of a word p_i from the merged occurrence list L. Note that we must take care of the previous consecutive occurrence of unmatched text words, before feeding a new word to the automaton.

Algorithm 4 shows the pseudocode for the algorithm MATCHER3. In the preprocessing phase, we build a table B, which stores a bit mask $b_m...b_1$ for each pattern word, such that the mask in $B[c]$ has the i-th bit set if $p_i = c$. This work should be done in step 1 of the main algorithm. In searching phase, like BPR, whenever we read the new text word t_j, we update the rows of NFA using the bit mask $B[t_j]$. In our algorithm, this is equivalent to saying that whenever we have the next occurrence j of p_i, we update the rows of NFA using the bit mask $B[p_i]$, while we must take care of the previous occurrences of unmatched text words $t_{j-\delta}...t_{j-1}$ by updating rows δ times with the empty bit mask 0^m in advance.

The time complexity of the above algorithm is straightforward [22].

Lemma 8. *The algorithm* MATCHER3 *runs in* $O(k\lceil m/\omega\rceil n)$ *in the worst case, where ω is the size of the machine word in bits.*

Notice that for short pattern, for example $m \leq \omega$, this can be $O(kn)$, which is very competitive and practically fast among online search techniques [14].

7 Discussions

MATCHER1 is as flexible as simple dynamic programming methods. We may apply more generalized metrics, such as general cost edit distance model or linear gap penalty model, if we replace Multiset by MQUEUE or deque with heap order. Moreover we can take additional edit operations into consideration: "Transposition" is one of those, which frequently occurs in the form of $(\alpha\beta \to \beta\alpha)$ in practice.

On the other hand, from the practical point of view, MATCHER3 is the solution of choice. Navarro et. al. [14] shows that if $m \leq \omega$ and k is small enough (e.g. 1 or 2), BPR runs very fast and outperforms other algorithms. It can be easily inferred that the same holds in our algorithm.

Algorithm 4 MATCHER3($\mathcal{L}, Pos, next[1..m], h$)

Preprocessing phase

1: **for** $i \leftarrow 1$ to m **do**
2:    $B[i] \leftarrow 0^m$
3: **end for**
4: **for all** $c \in Pos$ **do**
5:    $i \leftarrow c$
6:    **while** $i \neq \perp$ **do**
7:       $B[c] \leftarrow B[c] \mid 0^{m-i}10^{i-1}$
8:       $i \leftarrow next[i]$
9:    **end while**
10: **end for**

Searching phase

1: **for** $i \leftarrow 0$ to k **do**
2:    $R_i \leftarrow 0^{m-i}1^i$
3: **end for**
4: $j' \leftarrow 0$
5: L: $\{(i,j)|j \in l_{i,h}$ if $l_{i,h} \in \mathcal{L}\}$, merged list of $l_{i,h}$ with the first index i attached
6: SORT(L) // Sort $(i,j) \in L$ by increasing order of j
7: **while** L is not empty **do**
8:    Pop (i,j) from L
9:    $\delta \leftarrow j - j'$
10:    **while** $\delta > 1$ **do**
11:       Update R_i's with the bit mask 0^m
12:       $\delta \leftarrow \delta - 1$
13:    **end while**
14:    Update R_i's with the bit mask $B[i]$
15:    **if** $R_k \& 10^{m-1} \neq 0^m$ **then** // Check if the last bit of R_k set
16:       Declare "Match" at j
17:    **end if**
18:    $j' \leftarrow j$
19: **end while**

We can easily convert the above matchers to the longest common subsequence (LCS) problem with slight modifications. For MATCHER2, for instance, the linear function $f(i', j', i, j) = 1$ and we are looking for the maximum rather than the minimum. We can adapt the algorithm of Hunt and Szymanski [8] to our algorithmic framework.

We now describe possible extensions of our work. A simple modification of the dynamic programming algorithm, which retains all its flexibility, improves the time complexity to $O(kn)$ [18]. We might apply this to MATCHER1 to achieve $O(kn')$ time.

In [16], Risvik suggests an enhanced edit distance metric by allowing the use of similarity at the character level when the substitution operation is used. Its cost function is defined as $C(\alpha, \beta) = ed(\alpha, \beta)/\min(|\alpha|, |\beta|)$, where $ed(\alpha, \beta)$ is a character-based edit distance between words α and β. For computing this, we might scan the whole vocabulary to find neighboring words of given pattern words, and modify merge and intersection algorithm with newly introduced neighboring word lists. This is another challenging topic.

Compressed indexing and block indexing [2] are next challenging topics of research. It becomes more and more important to maintain smaller index size without losing search efficiency, as the data size grows. Compressed full-text self-indexing [15] is an attractive alternative, but the existing compressed indexes for secondary memory are usually slower than their uncompressed counterparts. However, the recent work [7] gave a possibility toward a practical disk-based compressed text index, which provides good I/O times for searching and, at the same time, takes much less space than the existing indexes. Adapting block indexing to our algorithm or developing a new compressed index which supports approximate matching might enable us to have efficient and succinct index with flexible search operations.

References

1. Araujo, M.D., Navarro, G., Ziviani, N.: Large Text Searching Allowing Errors. In: WSP, pp. 2–20 (1997)
2. Baeza-Yates, R., Navarro, G.: Block Addressing Indices for Approximate Text Retrieval. J. Am. Soc. Inf. Sci. (JASIS) 51, pp. 69–82 (2000)
3. Boitsov, L.M.: Using Signature Hashing for Approximate String Matching. Computational Mathematics and Modeling. 13(3), pp. 314–326 (2002)
4. Eppstein, D., Galil, Z., Giancarlo, R., Italiano, G.F.: Sparse Dynamic Programming I: Linear Cost Functions. J. ACM. 39(3), pp. 519–545 (1992)
5. French, C.J., Powell, A.L., Shulman, E.: Applications of Approximate Word Matching in Information Retrieval. Technical report, Univ. of Virginia (1997)
6. Gajeuska, H., Tarjan, R.E.: Deque with Heap Order. IPL. 22, pp.197–200 (1986)
7. González, R., Navarro, G.: A Compressed Text Index on Secondary Memory. In: IWOCA, pp. 80–91 (2007)
8. Hunt, J.W., Szymanski, T.G.: A fast algorithm for computing longest common subsequeces. Communication of the ACM. 20, pp. 350–353 (1977)
9. Johnson, D.B.: A priority queue in which initialization and queue operation take $O(\log\log D)$ time. Math. Systems Theory. 15, pp. 295–309 (1982)
10. Kim, J.W., Park. K.: An Efficient Alignment Algorithm for Masked Sequences. Theoretical Computer Science. 370, pp. 19–33 (2007)
11. Levenshtein, V.I.: Binary codes capable of correcting spurious insertions and deletions of ones. Problems of Information Transmission. 1, pp. 8–17 (1965)
12. Mäkinen, V., Navarro, G., Ukkonen, E.: Approximate Matching of Run-Length Compressed Strings. Algorithmica. 35, pp. 347–369 (2003)
13. Navarro, G.: A Guided Tour to Approximate String Matching. ACM Computing Surveys. 33(1), pp. 31–88 (2001)
14. Navarro. G., Raffinot, M.: Flexible Pattern Matching in Strings: Practical on-line algorithms for texts and biological sequences. Cambridge University Press (2002)
15. Navarro, G., Mäkinen, V.: Compressed full-text indexes. ACM Computing Survey. 39(1), article 2 (2007)
16. Risvik. K.M.: Approximate Word Sequence Matching over Sparse Suffix Tree. In: 9th Annual Symposium on Combinatorial Pattern Matching, pp. 65–79 (1998)

17. Sellers, P.: The theory and computation of evolutionary distances: pattern recognition. J. Algorithms. 1, pp. 359–373 (1980)
18. Ukkonen, E.: Finding approximate patterns in strings. J. Algorithms. 6, pp. 132–137 (1985)
19. Ukkonen, E.: Approximate string matching with q-grams and maximal matches. Theoretical Computer Science. 92, pp. 191–211 (1992)
20. Ukkonen, E.: Approximate string matching over suffix-trees. In: 4th Annual Symposium on Combinatorial Pattern Matching, pp. 229–242 (1993)
21. Wagner, R.A., Fisher, M.J.: The string-to-string correction problem. J. ACM. 21, pp. 168–173 (1974)
22. Wu, S., Manber, U.: Fast Text Searching Allowing Errors. Communication of the ACM. 35(10), pp. 83–91 (1992)
23. Wu, S., Manber, U.: GLIMPSE: A tool to search through entire file systems. Technical report, Dept. of CS, Univ. of Arizona (1993)
24. Zobel, J., Moffat, A.: Inverted Files for Text Search Engines. ACM Computing Surveys. 38(2) (2006)

Cop-Win Graphs with Maximal Capture-Time (extended abstract)

Tomáš Gavenčiak

Department of Applied Mathematics, Charles University,
Malostranské Náměstí 25, Prague, Czech Republic
http://kam.mff.cuni.cz/~gavento
gavento@kam.mff.cuni.cz

Abstract. A cop and robber game is a two-player vertex-pursuit combinatorial game where the players stand on the vertices of a graph and alternate in moving to adjacent vertices. The cop wins when he captures the robber, robber's goal is to avoid capture. The game has been studied in various modifications, many of which have an interesting relationship to certain well-known graph parameters, such as tree-width.

In this paper we present the matching upper bound $n - 4$ for the maximum length of a cop and robber game (the capture-time) on a cop-win graph of order n. We also analyze the structure of the class of all graphs attaining this maximum and describe an inductive construction of the entire class. We show that this class is rather rich, namely, that it contains exponentially many non-isomorphic graphs.

Key words: capture-time, cop and robber games, game theory

1 Preliminaries

In this article we consider only undirected simple loopless finite graphs. A graph is a tuple (V, E) of a set of vertices and a set of edges, respectively. We use $N_G(v)$ to denote the *neighborhood* of a vertex v in G, that is the set of all the vertices sharing an edge with v. We use $N_G[v] = N_G(v) \cup \{v\}$ for the *closed neighborhood* of v. The *degree* of v denoted by $\deg_G(v)$ is the number of edges of G incident with v and $\Delta(G)$ is the maximum degree of vertices of G.

A *path* on n vertices, denoted P_n, is a graph on n vertices labeled $v_1, \ldots v_n$ with edges $v_1 v_2, v_2 v_3, \ldots v_{n-1} v_n$. A *cycle* on n vertices, or C_n, is a graph P_n with an additional edge $v_n v_1$ forming a cycle. *Distance* of u and v, or $\text{dist}_G(u, v)$ is the length of a shortest path connecting u and v. $G - e$ and $G - v$ denote the graph G without the edge e or without the vertex v and all incident edges, respectively.

Graph G' is a *subgraph* of (or *contained* in) G, if $V_G' \subseteq V_G$ and $E_G' \subseteq E_G$. Graph G' is an *induced subgraph* of G if G' is a subgraph of G and contains all the possible edges. A *component* C of a graph G is a inclusion-maximal subgraph such that for every two vertices u and v in C, C contains a path from u to v. Graph G is connected if it contains only one component. A *bridge* is an edge such that its removal increases the number of components of the graph.

We say that a vertex v of G is *dominated* by a vertex u if $N_G[v] \subseteq N_G[u]$, in other words u and v are adjacent and every neighbor of v is also a neighbor of u. The set of all the vertices dominating v in G (except v itself) is denoted $\text{dom}_G(v)$. Any $u \in \text{dom}_G(v)$ is called a *dominator* of v. Where no confusion can arise, we drop the subscripts and write simply V, E, $N(v)$, ...

2 The game

Given a graph G the *cop and robber game* on G is a game for two players – the cop and the robber. First, the cop selects his starting vertex, then the robber selects hers, and then they alternate in moving across the edges of G. In their turn, both players can either move to some neighborhood vertex or stay at their current vertex. They both see each other and have complete information about the game. The cop wins if at some time he shares a vertex with the robber, the robber wins if she is never captured.

Given a graph G, the described combinatorial game on G has a winning strategy for one of the players. If that player is the cop, we call the graph *cop-win*, otherwise we call it *robber-win*. For detailed information about the existence of winning strategies in the various combinatorial games with complete information we recommend the general paper from B. Banaschewski and A. Pultr [9]. Note that every disconnected graph is clearly robber-win and therefore we consider only connected graphs from now on.

The *capture-time*[1] $\text{ct}(G)$ is the smallest number of cop's moves (excluding the initial placement) always sufficient to capture robber (regardless of her strategy).

---

[1] The capture-time is sometimes called the *search-time* or the *cop-time*.

The maximum possible capture-time among all cop-win graphs on n vertices is denoted as $\mathtt{ct_{max}}(n)$.

This game was first proposed and analyzed independently by Nowakowski and Winkler [6] and Quilliot [10][11]. They characterized the cop-win graphs and Nowakowski and Winkler proposed a polynomial algorithm determining whether a given graph is cop-win and also its capture-time.

Later, many other similar games were proposed, adding more cops, allowing the cops to fly, making robber invisible, faster or lazy, allowing to stay at edges, and many more. Some of these games have very nice properties and some relate to interesting graph parameters such as *tree-width*. See [1] for a good survey.

3   Main results

In their paper A. Bonato et al. [2] consider the capture-time of finite and infinite graphs and also prove some upper bounds for the capture-time both in general and for various graph classes. They proved that on every cop-win graph on n vertices, the cop wins in at most $n-3$ steps, and constructed an example showing that $n-4$ steps are needed. Here we close the gap by improving the upper bound on $\mathtt{ct_{max}}(n)$ to $|V_G| - 4$ for $|V_G| \geq 7$.

Theorem 1. *For all $n \geq 7$ we have $\mathtt{ct_{max}}(n) = n - 4$, for $n \leq 7$ we have* $\mathtt{ct_{max}}(n) = \lfloor \frac{n}{2} \rfloor$.

After the proof of the main theorem we describe the structure of the graphs with the maximal capture-time. In the search for all the graphs with maximal capture-time we found these to have a nice inductive structure.

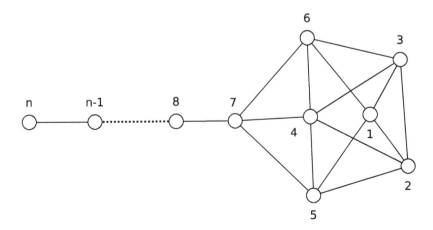

Fig. 1. Graph H_n

Here we introduce the class $\mathcal{M}$ of all the graphs with the maximal capture-time among all the cop-win graphs of the same size. We define the graph H_n to be a graph on n vertices containing a base graph H_7 with an attached path on $n - 7$ vertices (see Figure 1).

The following theorems describe the structure of all the graphs in $\mathcal{M}$ on at least 8 vertices. These conditions are not only necessary but also sufficient and they lead to an efficient inductive construction of the entire class $\mathcal{M}$.

Theorem 2. *For every graph $G \in \mathcal{M}$ on $n \geq 8$ vertices we have:*

- *The graph G has the graph H_7 as an induced subgraph.*
- *The graph G has the graph H_n as a subgraph. The subgraph is uniquely determined up to the symmetry of H_7.*
- *The graph G has exactly one dominated vertex v.*
- *The graph $G - v \in \mathcal{M}$ and $G - v$ has exactly one dominated vertex u.*
- *Vertex v is adjacent to u but to none of $\mathrm{dom}_{G-v}(u)$.*

Theorem 3. *Every graph in $\mathcal{M}$ on at least 8 vertices can be generated by the following algorithm:*

- *Start with $G_0 = H_7$ and $u_0 = 7$. Vertex u_i will always be dominated in G_i.*
- *Repeat for $i = 1, 2, \ldots$*
  - *Let G_{i+1} be G_i with a new vertex u_{i+1}.*
  - *Select a neighborhood N_{i+1} of u_{i+1} in any way so that u_{i+1} is dominated in G_{i+1}, $u_i \in N_{i+1}$ and $\mathrm{dom}_{G_i}(u_i) \cap N_{i+1} = \emptyset$.*
  - *Connect u_{i+1} to all vertices N_{i+1}.*

Each graph of $\mathcal{M}$ is generated in at most two ways by this algorithm.

We conclude the results with a simple lower bound on the size of $\mathcal{M}$.

Theorem 4. *For every $n \geq 8$ the class $\mathcal{M}$ contains at least 2^{n-8} graphs on n vertices.*

A double exponential upper bound follows from the number of all graphs on n vertices. The exact size of $\mathcal{M}$ is open and not of much interest.

4   Cop-win graphs and capture-time

In this section we prove Theorem 1. We start with a simple characterization of cop-win graphs.

4.1   Cop-win graphs

We start with a characterization of the cop-win graphs.

Lemma 1 ([6], [2, **Theorem 1**]). *Graph G is cop-win if and only if it is either a single vertex or it is obtained from a cop-win graph G' by attaching one new vertex v such that v is dominated in G by some vertex from G'. In that case, $\mathrm{ct}(G') \leq \mathrm{ct}(G) + 1$.*

Note that Lemma 1 and the fact that $\mathtt{ct}(1) = 0$ immediately give the upper bound $\mathtt{ct_{max}}(n) \leq n - 1$. A similar observation was used in the paper by Bonato et al. [2].

As a technical tool to be used later in case analysis, we prove a local property of all the cop-win graphs.

Lemma 2 (Triangle lemma). *If $G = (V, E)$ is a cop-win graph, then each edge uv of G is either a bridge or there is a vertex $w \in V$ adjacent to both u and v. Such w is said to form a triangle over uv.*

Proof. Suppose that G has a non-bridge edge uv such that no C_3 in G contains uv. Then there is a shortest path $P = v_0 v_1 \ldots v_{k-1}$ with $u = v_0$ and $v = v_{k-1}$ in $G - uv$. The edge set $P \cup uv$ forms a cycle C in G of the length $k \geq 4$.

We construct a winning strategy for the robber:

Let $R(w) = (\mathrm{dist}_{G-uv}(w, v_0) + 2) \bmod k$. Note that the distance is measured in the graph without uv, so $R(v_j) = (j + 2) \bmod k$. Whenever the cop is on a vertex c, the robber should move to $r = v_{R(c)}$ on C, where she cannot be attacked, because r is in a distance at least 2 from c. When the cop moves to c', his distance from v_0 changes at most by 1, so the robber can always move to an appropriate vertex on C. When the cop moves from the vertex c with $R(c) = k - 2$ to c' with $R(c') = k - 1$, the robber should move across uv from v_{k-1} to v_0. When the cop decides to start at c_0, the robber starts at $v_{R(c_0)}$. □

4.2 Lower bound

The lower bound follows from the following lemmas and the observation in Lemma 1.

Lemma 3. *We have $\mathtt{ct_{max}}(7) = 3$.*

Proof. This lemma could be proved by an analysis of all the cop-win graphs on 7 vertices. Instead, below in Lemma 4 we analyze all the cop-win graphs on 6 vertices using Lemma 2 and show that the only cop-win graph on 6 vertices with $\mathtt{ct}(G) \geq 3$ is P_6.

Therefore, by Lemma 2 and Lemma 1 every graph on 7 vertices with $\mathtt{ct}(G) \geq 4$ must be a P_6 extended by attaching a new dominated vertex v to some interval of P_6. If $\deg(v) \geq 2$ or $\deg(v) = 1$ and v is adjacent to some other vertex than an endpoint of P_6, then the cop should start at one of the middle vertices of the P_6 and capture the robber in at most 3 moves. If $\deg(v) = 1$ and it is adjacent to an endpoint of P_6, then the new graph is a P_7 and $\mathtt{ct}(P_7) = 3$. This shows that all these graphs have $\mathtt{ct} \leq 3$ so $\mathtt{ct_{max}}(7) \leq 3$.

On the other hand, a simple robber's strategy gives $\mathtt{ct}(P_7) = 3$. □

Lemma 4. *The only cop-win graph on 6 vertices with $\mathtt{ct} \geq 3$ is P_6.*

The proof of this lemma is of a technical nature and can be found in Appendix A. The proof goes by a case analysis of all possible cop-win graphs on 6 vertices by their maximum degree.

4.3 Upper bound

In their paper [2], A. Bonato, P. Golovach, G. Hahn, and J. Kratochvíl show an explicit construction giving $\mathtt{ct_{max}}(n) \geq n - 4$ for $n \geq 7$. Here we show a simpler construction with the minimum number of edges. We use the graphs H_n defined in Section 3 on Figure 1.

Lemma 5. $\mathtt{ct}(H_n) = n - 4$ *for all* $n \geq 7$.

Proof. We show both cop's and robber's time-optimal strategies. The cop's optimal strategy is to start at vertex 2 (or at 3, symmetrically). In this situation the robber may start only at vertices $6, 7, \ldots n$ to avoid immediate capture. If she starts at 6, the cop moves to 3, then she can move only to 7. The cop then moves to 4 and the robber must flee to the tail. In this case, the cop will win in at most $n - 4$ moves. If the robber decides to start at vertex $7, 8, \ldots n$, the cop moves to 4 and then pursues the robber into the tail and captures her in at most $n - 3$ moves.

The only place, where cop can corner and capture a clever robber is the end of the tail, as there is no other dominated vertex. From every other position robber can escape an immediate capture. Therefore it suffices to show that the robber can safely play at least 1 move in H_7 before entering the vertex 7 and then the tail. From that moment it takes the cop $n - 5$ moves to capture her, because at the moment she enters the vertex 7, the cop is in a nonadjacent vertex. If the cop starts somewhere in H_7, the robber always has an option to start safely at some vertex other than 7. In case that the cop moves to the tail before her, she stays in H_7.

These two strategies show that $\mathtt{ct}(H_n) = n - 4$. $\qquad\square$

Now we are ready to prove Theorem 1 about the exact value of $\mathtt{ct_{max}}$.

Proof of Theorem 1. The proof consists of the analysis of all the cop-win graphs on at most 6 vertices. First we note, that $\mathtt{ct_{max}}(1) = 0$, $\mathtt{ct_{max}}(2) = \mathtt{ct_{max}}(3) = 1$ and $\mathtt{ct_{max}}(4) = \mathtt{ct_{max}}(5) = 2$ by a simple analysis of all small cop-win graphs. The cases for 1 and 2 vertices are trivial and there are only two cop-win graphs of order 3 to check – P_3 and K_3.

The cop-win graphs on 4 vertices either have $\Delta(G) = 3$ and therefore $\mathtt{ct}(G) = 1$ (the cop can start in any vertex of degree 3), or $\Delta(G) = 2$ and the graph is a P_4 with $\mathtt{ct}(P_4) = 2$.

The situation with cop-win graphs on 5 requires a short analysis:
If $\Delta(G) = 4$, then $\mathtt{ct}(G) = 1$. If $\Delta(G) = 2$, then $G = P_5$. The only interesting case is $\Delta(G) = 3$. In that case the cop starts at any v with $\deg(v) = 3$. The robber must start at the only safe vertex u such that $u \notin N[v]$. If the vertex u is adjacent to one vertex of $N[v]$, the cop moves there and captures the robber with his next move. If u is adjacent to two vertices of $N[v]$, then these two must be adjacent (otherwise the graph is robber-win according to the Triangle Lemma) and the cop can move to any of these two vertices capturing the robber in his

second move. When u is adjacent to all three neighbors of v, these three must be connected by at least two edges (again by the Triangle Lemma) and therefore one has $\deg(w) = 4$, a contradiction with $\Delta(G) = 3$.

By Lemma 3, Lemma 4 $\mathtt{ct}_{\max}(6) = \mathtt{ct}_{\max}(7) = 3$

From Lemma 1 follows that for all $n \geq 7$ $\mathtt{ct}_{\max}(n) \leq n - 4$.

Graphs P_n and H_n show that the bound is tight. Note that $\mathtt{ct}(P_n) = \lfloor \frac{n}{2} \rfloor$ and $\mathtt{ct}(H_n) = n - 4$ by Lemma 5. $\square$

5 Graphs with the maximal $\mathtt{ct}$

In this section we prove the remaining theorems and closely examine the results about the structure of the class $\mathcal{M}$.

5.1 The structure of $\mathcal{M}$

We start this section with some auxiliary observations.

Lemma 6. *For a dominated vertex v, $\mathrm{dom}_G(v)$ is a complete subgraph of G. For any $u \in N[v]$, $\mathrm{dom}_G(v) \cup \{u\}$ is also a complete subgraph.*

Proof. The observation follows from the definition of domination. Every neighbor of v must be adjacent to all the vertices dominating v. $\square$

Lemma 7. *If the cop has a winning strategy of length at most k on graph G, he can choose one dominator d_i for every dominated vertex u_i in G and then choose his strategy so that he either captures the robber after k moves in some $N[u_i]$ by moving from d_i or captures her sooner.*

Proof. The cop plays using an optimal winning strategy on $G' = G \backslash \{u_1, u_2, \dots\}$. If the robber is at u_i, he plays as if she was at d_i. If the robber is still free after $k - 1$ cop's moves, she must stand at some u_i with the cop at d_i (otherwise she would have a strategy for surviving k cop's moves, a contradiction). Then she can not move outside $N[d_i]$ and the cop wins in k moves. $\square$

The theorem itself is a combination of the following lemmas.

Lemma 8. *The only cop-win graph on 7 vertices extensible (by adding a dominated vertex) to a graph $G \in \mathcal{M}$ on 8 vertices with $\mathtt{ct}(G) \geq 4$ is H_7 with $\mathtt{ct}(H_7) = 3$.*

Proof. This has been proved only by a computer-based examination of all the non-isomorphic cop-win graphs on at most 8 vertices. We used the Nauty software [5] to generate all the non-isomorphic graphs on $n \leq 8$ vertices. We implemented an algorithm similar to the one proposed in the above-mentioned paper "The search-time of a graph" [2] to recognize the cop-win graphs and to calculate their capture-time.

The details of the algorithm and an implementation in Python are omitted. We originally used the same program to verify Lemma 4 and to analyze the structure of the cop-win graphs with the maximum capture-time.

This lemma could be proved using a method similar to the one used in the proof of Lemma 4 (a case-analysis by $\Delta(G)$ and bridges, using Lemma 2 on all non-bridge edges), but that would be inconveniently long and too technical. □

The structural results about the unique dominated vertex are proved by induction from the lemma above. The inductive step itself does not depend on the exhaustive search, however the base does.

Lemma 9. *All the graphs $G \in \mathcal{M}$ on at least 8 vertices have exactly one dominated vertex v, in all v is adjacent to the unique vertex u dominated in $G' = G - v$ and v is not adjacent to any $\mathrm{dom}_{G'}(u)$.*

Proof. The graph H_7 has only one dominated vertex 7, but $H_7 - 7$ has more dominated vertices. The proof is by induction on $|V_G|$, the verification of the theorem for all graphs of order 8 is a byproduct of the computer-based examination of all cop-win graphs on at most 8 vertices described in the proof of Lemma 8.

The verification can be also done by hand, as there are only 10 such graphs of order 8. All these can be constructed from G_7 by a method described in Theorem 3 below: Choose a dominating set of the new vertex as a subset of $\{5, 6, 7\}$ and choose the non-dominating neighbors so that the chosen set is really dominating (note that the case $\{5, 6, 7\}$ is impossible). The new vertex must be adjacent to the vertex 7 and must not be adjacent to the vertex 4. Then verify the conditions for all these graphs.

Every cop-win graph contains at least one dominated vertex. If G contains only one dominated vertex v_1, we call the dominated vertex in the graph $G' = G - v_1$ as v_2. In this case, the vertex v_2 is well-defined, as G' has exactly one dominated vertex by the induction hypothesis.

If some cop-win $G \in \mathcal{M}$ with $|V_G| = n$ contains two dominated vertices v_1 and v_2, at least one of the graphs $G - v_1$ and $G - v_2$ has to be in $\mathcal{M}$ (otherwise G could not have $\mathrm{ct}(G) = n - 4$ according to Lemma 2), without loss of generality suppose that this is $G' = G - v_1$. By the induction hypothesis $\mathrm{ct}(G') = n - 5$ and G' has exactly one dominated vertex v_2.

We use the notation v_1, v_2 and G' for both the cases, the only difference being that v_2 may or may be not dominated in G. Note that v_2 is always dominated in G' and $\mathrm{dom}_{G'}(v_2) \subseteq \mathrm{dom}_G(v_2) \cup \{v_1\}$. The graph $G \in \mathcal{M}$ clearly can not have more than two dominated vertices. We consider these two cases:

If v_1 and v_2 are nonadjacent or v_1 is adjacent to some $d_2 \in \mathrm{dom}_{G'}(v_2)$, the cop can use an optimal strategy for G' preferring to capture the robber at v_2 from d_2 and in situation with the robber on v_1 play as if she were on some (fixed) vertex d_1 dominating v_1 in G. In $n - 5$ moves cop either captures the robber or his strategy makes him move to d_1 (where he supposed to capture the robber, who is at v) or he moves to d_2 (with the robber at v_2). But from both situations the robber cannot escape from cop's neighborhood and the cop will capture her in his next move. The latter situation is illustrated in Figure 2. Note that d_1 can be equal to d_2.

Both these situations imply that $\text{ct}(G) = \text{ct}(G')$, a contradiction.

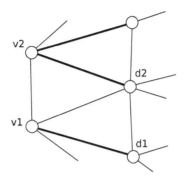

Fig. 2. v_1 is adjacent to $d_2 \in \text{dom}_{G'}(v_2)$

In the case when u and v are adjacent and v_1 is not adjacent to $\text{dom}_{G'}(v_2)$ the vertex v_2 is not dominated in G by any vertex, as it has a neighbor v_1 not adjacent to $\text{dom}_{G'}v_2$ and v_1 also does not dominate v_2. □

Lemma 10. *All $G \in \mathcal{M}$ on at least 8 vertices contain H_7 as an induced subgraph and H_n as a subgraph. The embedding of H_n is uniquely determined (up to the symmetry of H_7).*

Proof. According to Lemma 8 and Lemma 9 the only graphs on 8 vertices with $\text{ct}(G) \geq 4$ are those obtained from G_7 by an addition of a dominated vertex v to H_7 such that v is adjacent to vertex 7. All the other graphs in the class are constructed from these by adding vertices and edges from the new vertices, so they also contain H_7 as an induced subgraph and all the new vertices form a (generally non-induced) path from vertex 7.

The numbering of vertices by a repeated removal of a dominated vertex with numbers $n \dots 7$ is unique (as there is always only a single dominated vertex), the remaining 7 vertices are isomorphic to a H_7 in one of two possible ways. This determines at most two embeddings of H_n in G. □

Proof of Theorem 2. The above lemmas show all the properties of graphs in $\mathcal{M}$ on at least 8 vertices. Note that these properties do not hold for smaller graphs, these can have more dominated vertices and generally contain neither G_7 nor any of its subgraphs except the trivial cases. □

5.2   The construction of $\mathcal{M}$

In this section we show that the conditions stated in Theorem 2 are also sufficient for the new graph to belong to $\mathcal{M}$.

Lemma 11. *If we extend $G' \in \mathcal{M}$ with dominated vertex v_2 by a dominated vertex v_1 adjacent to v_2 and nonadjacent to $\mathrm{dom}_{G'}(v_2)$, the resulting G is also in $\mathcal{M}$.*

Proof. In G', the robber has an optimal strategy where cop captures her at v_2 after she passed her last move. She can use this strategy also in G: when the cop is at v_1, she plays as if he was at some d_1 dominating v_1. If the cop plays optimally, this results in the robber standing at v_2 and the cop at some $d_2 \in \mathrm{dom}_{G'}(v_2)$, but instead of passing her move, she moves to v_1, which is nonadjacent to d_2 and therefore prolongs the game by at least one move. $\square$

From this lemma follows the described inductive construction of $\mathcal{M}$.

Proof of Theorem 3. By generating the graphs as described in the algorithm we generate only graphs belonging to $\mathcal{M}$ according to Lemma 11.

For contradiction suppose that not all the graphs on at least 8 vertices in $\mathcal{M}$ are generated and the smallest graph omitted is $G \in \mathcal{M}$. According to Lemma 8 the graph G has at least 9 vertices. From Theorem 2 it follows that G has a single dominated vertex v. The graph $G - v$ is in $\mathcal{M}$ and has at least 8 vertices so we generate it. The contradiction follows from the fact that $N(v)$ satisfies the conditions of the algorithm as a neighborhood for v, so G is generated from $G - v$.

The construction is efficient as the choices of all the N_i are uniquely determined (up to one symmetry of G_7) in the resulting graph $G \in \mathcal{M}$ on at least 8 vertices by Lemma 10. $\square$

Note that whenever the dominated vertex v of $G \in \mathcal{M}$ has no neighbors except it's dominators, then every graph generated from it by the above construction is just a path connected to G. This follows from the fact that all the neighbors of appended vertices are from $N[v] \setminus \mathrm{dom}_G(v) = \{v\}$. This directly implies that whenever $G \in \mathcal{M}$ has a bridge, one side of the bridge is a path.

5.3   Exponential size of $\mathcal{M}$

From Lemma 10 we get a unique labeling (up to the symmetry of H_7) of vertices of every member $G \in \mathcal{M}$ on at least 8 vertices. This allows an explicit construction of exponentially many graphs (with respect to their size) of $\mathcal{M}$.

Proof of Theorem 4. We inductively construct sets $M_i \subseteq \mathcal{M}$ of graphs of size i such that the dominated vertex of every graph has only one dominating vertex in $\{1, \ldots 6\}$, has all possible neighbors in $\{1 \ldots 7\}$ and $|M_i| \geq 2^{i-7}$.

First, let M_7 contain only the graph H_7. From every graph $G \in M_i$ we generate at least two graphs $G' \in M_{i+1}$. Let v be the vertex dominated in G and let v' be it's single dominator. v has at least 2 neighbors in $\{1, \ldots 6\}$ (the previous dominator can not be a neighbor) and we can choose any of these as the dominator u' for the new dominated vertex u.

It suffices to check that by connecting the vertex u to v and all the vertices $(N[u'] \setminus \{v'\})/cap\{1, \ldots 7\}$, $\text{dom}(u)$ remains a single vertex. All dominators must be adjacent to both u' and v and therefore also to v'. But for every choice of v' and u' in $\{1, \ldots 6\}$ for every vertex $w \in N(u) \cap N(v)$ there is some neighbor of u' nonadjacent to w. This is proved by examining all of the cases.

This proves that all the generated graphs have the desired property. Every member of M_i is isomorphic to at most one other graph in M_i (according to Lemma 10), so $\mathcal{M}$ has the desired property. $\qquad\square$

A higher bound could be probably proved by allowing more neighbors and dominators outside $\{1, \ldots 7\}$ in the above construction and/or by allowing bigger dominator sets.

Acknowledgment

I would like to thank Jan Kratochvíl for introducing me to the problem and for all the provided advice.

References

1. Brian Alspach: *Searching and Sweeping Graphs: A Brief Survey*, Le Matematische, Vol. LIX, 2004
2. A. Bonato, P. Golovach, G. Hahn, and J. Kratochvíl, *The search-time of a graph*, submitted (2006)
3. Reinhard Diestel: *Graph Theory (Third Edition)*, Springer-Verlag, Heidelberg, 2006
4. Geňa Hahn, *Cops, robbers and graphs* Université de Montréal, submitted (2006)
5. Brendan D. McKay: *Practical Graph Isomorphism*, Congressus Numerantium, Vol 30, pp 45-87, 1981.
   Nauty web page: `http://cs.anu.edu.au/~bdm/nauty/`
6. R. Nowakowski and P. Winkler, *Vertex-to-vertex pursuit in a graph*, Discrete Math. 43 (1983), pp 235 - 239.
7. T. D. Parsons: *Pursuit-evasion in a graph*, Theory and Applications of Graphs, pp 426–441, Springer-Verlag, 1976
8. T. D. Parsons: *The search number of a connected graph*, Proc. 10th Southeastern Conf. Combinatorics, Graph Theory, and Computing, pp. 549–554, 1978
9. B. Banaschewski P. Pultr: *Tarski's fixpoint lemma and combinatorial games*, Order, Springer Netherlands
10. A. Quilliot, *Jeux et Points Fixes sur les graphes*, Ph.D Dissertation, Université de Paris VI, 1978
11. A. Quilliot, *Problèmes de jeux, de point fixe, de connectivité et de représentation sur des graphes, des ensembles ordonnés et des hyper-graphes*, Université de Paris VI, 1983

A   Proof of Lemma 4

Proof. We analyze all connected cop-win graphs on 6 vertices separately according to their maximum degree. We estimate ct of every case and also write some optimal starting vertices for the cop to prove the low capture-time of the case. Dashed edges in the illustration figures represent possible edges (sometimes not all of these edges may be present at once).

- A graph G with $\Delta(G) \in \{0,1\}$ is not connected and therefore not cop-win.
- A connected graph G with $\Delta(G) = 2$ is either a path P_6 or a cycle C_6. C_6 is not a cop-win graph and $\text{ct}(P_6) = 3$.
- If G with $\Delta(G) = 3$ has no bridge, then every edge has to be in some C_3 according to the Lemma 2. Select an arbitrary v with $\deg(v) = 3$. We denote the neighbors of v v_1, v_2 and v_3. $v_1 v$ must be in C_3, but v can have no more neighbors. Therefore v_1 is adjacent to, say, v_2. A similar argument shows that v_3 must be adjacent to v_1 or v_2, without loss of generality suppose it is adjacent to v_2. The remaining vertices p, q must be connected to v; suppose that $v_1 p$ is an edge (v_2 can have no more neighbors). There is no way to form a C_3 over $v_1 p$. The situation is drawn in Figure 3. No cop-win graph of order 6 with $\Delta = 3$ and without a bridge exists.

  If G with $\Delta(G) = 3$ has a bridge which divides the vertices in the ratio $3 : 3$, G clearly has $\text{ct} = 2$, the cop can start in any endpoint of the bridge. If G has no bridge with the ratio $3 : 3$, but has a bridge uv with the ratio $2 : 4$ with 2 vertices on the side of u, then v must have $\deg(v) = 3$, otherwise G would have a bridge with the ratio $3 : 3$. The situation is drawn in Figure 3. Whichever of the remaining possible edges exist, $\text{ct}(G) = 2$ with the cop starting at v.

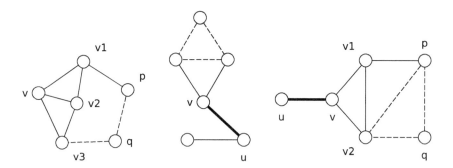

Fig. 3. The cases of $\Delta(G) = 3$ and no bridge, a bridge with the ratio $2 : 4$, and a bridge with the ratio $1 : 5$

  If G with $\Delta(G) = 3$ has a bridge uv with the ratio $1 : 5$ with a single vertex on the side of u, but has no bridges with the ratio $2 : 4$ or $3 : 3$ then $\deg(v) = 3$. Denote the neighbors of v, v_1 and v_2. The edges $v_1 v$ and $v_2 v$ can't be both bridges (they can't both have the ratio $1 : 5$ and the other possible ratios are forbidden), so one of them has to be in a C_3, but this can happen only if v_1 and v_2 are adjacent. The remaining vertex p is connected to v_1 or v_2, without loss of generality suppose that it is connected to v_1. The situation is drawn in Figure 3. If pv_1 is a bridge, remaining q cannot be connected to p (pv_1 would be bridge with ratio $2 : 4$), so q is adjacent to v_2. If pv_1 is not a bridge, it must be in C_3, but that is possible only if pv_2 is an edge. In that case q is adjacent only to p. In both cases $\text{ct}(G) = 2$ with the cop starting at v_1 or v_2.

- In G with $\Delta(G) = 4$ denote by v one vertex with $\deg(v) = 4$ and u the single vertex non-adjacent to v. If u has a single neighbor v_1, clearly $\text{ct}(G) = 2$ with the cop starting at v. If u has neighbors v_1 and v_2, these have to be adjacent (to form C_3 over uv_1) and again $\text{ct}(G) = 2$ with the cop starting at v. If u has three neighbors v_1, v_2 and v_3, there have to be at least two edges between v_1, v_2 and v_3 to form C_3s over uv_1, uv_2 and uv_3. Without loss of generality suppose that v_1v_2 and v_2v_3 are edges. Then the cop starting at v can capture the robber either immediately or after moving to v_2. This case is illustrated in Figure 4.

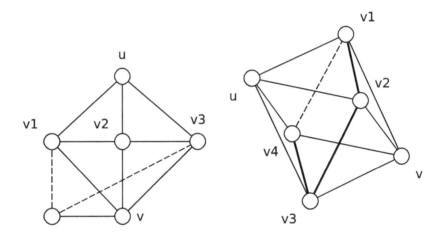

Fig. 4. Cases of $\Delta(G) = 4$ with $\deg(u) = 3$ and $\deg(u) = 4$

In the last case when u has four neighbors v_1, v_2, v_3 and v_4, there have to be at least two edges e_1 and e_2 between v_{1-4} to form C_3s over uv_1, uv_2, uv_3 and uv_4. No v_i can be connected to all the other v_i (it would have $\deg(v_i) = 5$). If e_1 and e_2 share an endpoint, there has to be an additional e_3 forming P_4 (together with e_1 and e_2) on the set $\{v_1, v_2, v_3, v_4\}$. If e_1 and e_2 do not share an endpoint, there must be another edge connecting the endpoints of e_1 and e_2, otherwise G would be robber-win. Both these situations necessarily lead to a graph isomorphic to that in Figure 4, where a concrete situation with the path $v_1v_2v_3v_4$ is drawn. If v_1v_4 is an edge, the graph is the net of a regular octahedron which is robber-win (the robber can always move to the vertex opposite of the cop). If v_1v_4 is not an edge, the cop wins in 2 moves by starting at v_2 or v_3 and therefore $\mathsf{ct}(G) = 2$.

- G with $\Delta(G) = 5$ has $\mathsf{ct}(G) = 1$ because the cop can start in a vertex with the maximum degree and capture the robber with his first move.

This shows that every cop-win graph G on 6 vertices except P_6 has $\mathsf{ct}(G) \leq 2$. $\mathsf{ct}(P_6) = 3$ and that finishes the proof. □

K-Phase Oscillator Synchronization
for Graph Coloring

Sof Anthony Lee and Raymond Lister

School of Software
The University of Technology Sydney
PO Box 123, Broadway, NSW 2007, Australia
{sofianto,raymond}@it.uts.edu.au
http://www.uts.edu.au

Abstract. This paper investigates ways of applying oscillator synchronization to graph coloring. A previous method based on the generalization of the Aihara model is sensitive to the varying degree of the vertices in the graph and there is a strong tendency for the network to form suboptimal limit cycles on regular graphs. Other models such as those by Wu and Nakaguchi, Jin'no & Tanaka do not generalize well into greater than 2-coloring. In this paper, we present ways to overcome these problems and describe the results of our experiments on graphs requiring more than 2 colors. Our k-phase model enhances the coloring performance over the previous similar models. We further attempt to formalize and analyze the categorical behavior of these systems and discuss connections to other optimization methods.

Key words: graph coloring, phase coupled oscillators, k-phase synchronization

1 Introduction

The synchronization phenomena of phase coupled oscillators both in natural and man made systems are well documented and have been investigated by many researchers, in particular the synchronous flashing of fireflies [1][2][3] and the mating calls of rain frogs [4]. A typical synchronization of such systems is an in-phase synchronization where each unit synchronizes into the same phase. While other systems may also exhibit synchronization into multiple phases, for example, the mating calls of rain frogs.

A particular interest from computational point of view is the possibility of exploring the viability of this mechanism in problem solving. For example, it is possible to adopt the in-phase synchronization for solving the time synchronization problem in ad hoc sensor networks. This is particularly important for tasks involving routing and detection of events in such a network. A network protocol based on this method has been developed [5][6][7].

Another form of computational use of the oscillator synchronization is the exploration of the multi phase synchrony for graph coloring. A typical strategy used for achieving this idea is to associate an oscillator to a vertex and the phase of the oscillator to a color. In this method, two neighboring oscillators will try to synchronize into different phases which represent different colors. The computational model that describes the interaction between the oscillators in these methods usually consists of one or more interacting differential equations.

The attraction of using coupled oscillators as a computational tool can be attributed to the parallel distributed nature of the system. A network of coupled oscillators in nature is very fast because each unit acts or changes its phase in parallel and interacts as part of the whole synchronization process. If we are to use sequential computing paradigms to implement the network of phase coupled oscillators, it means that we would need to compute the differential equation for each oscillator in the network. This would obviously be very time consuming. But with an appropriate model, a realization of the oscillator network can be achieved using a network of relatively simple electronic circuits.

1.1 Previous Work

Wu [8] investigated a model of coupled oscillators for performing graph coloring. His model uses an array of coupled RLC circuits, each RLC circuit is an oscillator associated with a vertex in a graph and a coupling resistor that connects two oscillators represents an edge between two vertices. The computation model consists of two differential equations representing an interweave wiring of the coupling resistor between each RLC oscillator, are shown below:

$$C\frac{dv_j}{dt} = -i - g(v_j) - \frac{1}{R_c}G(v)_j \tag{1}$$
$$C\frac{di_j}{dt} = v_j - \frac{i_j}{R}$$

The simulation is performed over a time t and is repeated if the coloring is unsatisfactory. The test instances consist of 2-and 3-colorable graphs with 4 to 16 vertices. Most of 2-colorable graphs in the test instances are properly colored and 80% of the 3-colorable graphs are correctly colored.

Nakaguchi, Jin'no & Tanaka [9] used a Hysteresis Neural Network (HNN) for solving the n-queens problem. The model can be seen as coupled hysteresis oscillators when an inhibitory feedback is used. The general equation of the HNN is

$$\lambda_i \frac{d}{dt}x_i = -x_i - \sum_{j=1}^{n}\omega_{ij}y_j + \mu$$
$$y_i = h(x_i), \quad (i = 1, ..., N) \tag{2}$$
$$= \begin{cases} 1 \ (x \geq 0) \\ 0 \ (x \leq 0) \end{cases}$$

The HNN can theoretically fall into a limit cycle state because the energy function will not necessarily decrease monotonically. In this case, the optimal solution may not always be found. Their experimental results indicate there can be many limit cycle states, but they introduced noise/perturbation to the system to avoid the problem. They subsequently demonstrated a hardware implementation of the system [10], but the model is limited to 2-colorable graphs. Despite this limitation, the speed of the network of hardware oscillators is independent of n, because the hardware oscillators change in parallel, while the computer simulation calculates the differential equation for each cell in the $n x n$ cells sequentially, thus the computation time increases rapidly as n increases.

Lee & Lister [11] developed a computer simulation of the Aihara anti phase model [4] for coloring 2-colorable graphs. They subsequently generalized the model to any k-colorable graphs. In their system, the modification of the oscillator phase θ is described by the following equation:

$$\frac{d\theta_i}{dt} = \omega_i + \frac{K}{N} \sum_{j=0}^{n} \sin(\theta_j - \theta_i - \beta), \quad (i = 1, ..., N) \tag{3}$$

However, the model tends to produce optimal solutions for complete graphs and complete k-partite graphs with equal or near equal partition size. Apart from these graphs, the method is suboptimal for arbitrary graphs including most trivial cases. The main problems faced by the generalized model include:

1. Graphs with varying degree of vertices,
2. Graphs with two or more sub graphs forming dissimilar phase separations,
3. Failure for the network to synchronize and
4. Formation of suboptimal limit cycles in the network.

A typical characteristic of synchronization is $d\theta \to 0$ as $t \to \infty$, whereas in a limit cycle state, the $d\theta$ does not approach zero. Even though the synchronization condition is achieved, the network may not necessarily form a minimum the number of phases/colors. This has been shown in [11] by using an example on the icosahedron.

2 Motivation for K-Phase Algorithms

While both Wu and Nakaguchi, Jin'no & Tanaka focused on the synchronization and results of the experiments, less emphasis was in place to investigate the cases where the network does not synchronize optimally. Lee & Lister however provide descriptive analysis for the suboptimal cases in their model. These models typically do not require the parameter for the number of colors/phase difference (non-parameterized). Usually, the phase difference among the oscillators will be automatically realized when the network converges into a stable state. Nevertheless, the mathematical proof of convergence has only been available for the anti phase model (2-colorable graphs), as demonstrated in [4][8][12]. The mathematical analysis of the three or more coloring appears to be complex in that the previous papers have resorted to providing empirical results. Another model where a fired oscillator causes the neighbor to change its phase to one of the discrete k phases is discussed in [13].

2.1 A General Framework

In order to avoid the problem with dissimilar phase separation among the oscillators, we developed an algorithmic model of the coupling between an oscillator and its neighbors in such a way that the oscillators will eventually move toward the multiple k phase separation, where k is the parameter given to the simulator. Given a set of oscillators in a 2π periodic system, with a parameter k, where k is the number of colors into which we expect the network of oscillators will converge. The phase separation, τ can be simply formulated as,

$$\tau = 2\pi/k \tag{4}$$

Fig. 1 illustrates how the oscillators will eventually move towards attractor points at $(v_0, ..., v_4)$ and thus form a k-phase synchronization for $k = 5$. The dials indicate the phase of the oscillators in a 2π periodic system (circle). The arrows indicate the intended directions into which the oscillators will converge. For the purpose of identification, let us also denote the area between the dials as a sector; therefore, there are five sectors in Fig. 1.

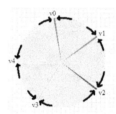

Fig. 1. Attractor points in a k-phase synchronization

2.2 A Basic Model

In [4][11], the update function follows $f(\theta) = \sin(\theta - \theta_j - \beta)$ where θ, θ_j and β are the phase of the current and neighboring oscillator, and the frustration parameter. In order to realize the general k-phase scheme above, we stipulated the following equation.

$$f(\theta) = \begin{cases} 0 & (\theta \approx j\tau), j = (0, 1, ..., k) \\ \theta - \sin(\tau) & (\theta \approx 2\pi) \\ \theta + \sin(\tau - \theta) & (\theta < \tau) \\ \theta - \sin(\theta - j\tau) & (j = max(j), where\ j\tau < \theta) \end{cases} \tag{5}$$

The main feature of this algorithm is the movement by $sin(x)$. This move is required to get close to the multi k-phase difference and once an oscillator is near the multi k-phase, the oscillator is attracted to these positions and the change becomes small or reduces to zero. Although this model can successfully color some of the simple graphs and avoid problem (1) and (2) as described earlier, in many cases the model produces suboptimal coloring and often fails to synchronize. Based on tests using the Peterson graph[14], the cause is related to the limit cycle problem (4) where the network cycles around a number of states indefinitely.

Other suboptimal behaviors include the oscillators inappropriately moving backward as dictated by condition 4 in the model leading to a limit cycle and oscillators moving to an attractor point that is occupied by its neighbors.

2.3 Revision of the Basic Model

We developed a revision of the basic model. In this revision, an oscillator will first try to move backward, i.e. move to the lower attractor point at j. If position j is already occupied by any of the neighbors, the oscillator will then try moving toward the attractor at $j + 1$. Failing these two options, the oscillator will move to another randomly selected sector.

$$
f(\theta) = \begin{cases}
0 & (\theta \approx j\tau), j = (0, 1, ..., k) \\
\theta - \sin(\tau) & (\theta \approx 2\pi) \\
\theta + \sin(\tau - \theta) & (\theta < \tau) \\
\theta - \sin(\theta - j\tau) & (j = max(j), j\tau < \theta \ \& \ j\tau \neq \theta_i, \forall i) \\
\theta - \sin(\theta - q\tau) & (q = max(j) + 1, q\tau < \theta \ \& \ q\tau \neq \theta_i, \forall i) \\
s\tau + (\theta - j\tau) & (s = rnd(k), if \ all \ above \ conditions \ failed)
\end{cases}
\tag{6}
$$

Our initial tests indicate this model performs better than those in [8][11] for the same test instances.

2.4 The Hyperbolic Algorithm

In this algorithm, the response of an oscillator to a stimulus follows a hyperbolic function similar to the function $f(x) = 1/x$. The idea behind this function is to increase the effect experienced by an oscillator when it is close to a firing oscillator. In comparison to the generalized Aihara algorithm, the frustration parameter β causes the oscillator to move away from the firing oscillator by $sin(\beta)$. However, in the model that uses the function $f(x) = 1/x$, the resulting calculation cannot be used directly, because the value can be very large when x is small.

We devised the following method to utilize such a function. Firstly, let x be the absolute phase difference between the firing oscillator and the oscillator receiving the stimulus (the neighbor), and τ is the required phase separation.

We stipulated a function that will reduce to zero when x is approaching τ. Such a function can be simply written as

$$f(x) = \frac{x}{\tau} - 1 \qquad (7)$$

Since $f(x)$ can be very large when two oscillators are close to each other, the oscillator would initially accumulate the sum of $f(x)$. When the oscillator's turn to fire arrives, it uses the sum to decide whether to move forward if the sum is positive or move backward if the sum is negative.

2.5 Computational Experience

Due to the intricacy in visualizing the DIMACS graph coloring instances [15], we tested our revised model and the hyperbolic algorithm on the graph instances used in the generalized Aihara model [11]. These instances are relatively simple, but they are useful in studying the behavior of models based on oscillator coupling. The instances, in Table 1 can be downloaded from http://www-staff.it.uts.edu.au/s̃ofianto/uts_instance_col.zip.

Results for K-Phase Algorithm. For each of the above instances, we performed 100 runs. The phase of the oscillators was randomly initialized. In order to automatically detect whether the network has converged to k phases, we implemented a modified k-means clustering algorithm. This algorithm is set to cluster the phases to within a narrow margin, for example, 20% of the phase separation from the centre of a cluster. At the end of each simulation, the k-means clustering routine is invoked to examine the number of phases/colors into which the graph has converged.

Table 2 shows the results of the simulation. The first column corresponds to the instance name, the second column is the chromatic number χ_G and the subsequent columns are the frequency of the number of color k was produced by the simulation. The network converges very fast, typically within 10 oscillator cycles.

The performance of this algorithm is generally better than the generalized Aihara model in [11]. At the very least, this algorithm is capable of coloring simple arbitrary graphs whereas it is not the case for the generalized Aihara model. By examining Table 2, we can see that the network converged at 100% rate for several instances, including the Ising32x8 graph, and for other instances, the convergence rate is above 90%. A lower rate of convergence is noticeable on a number of instances, particularly the hexagons, the dodecahedron and the Ising on torus.

Further examination of these cases indicates that the suboptimal limit cycle still persists but improved. For its simplicity we show a limit cycle on the 1x7Ring graph as illustrated in Fig. 2. The phase of the oscillators occupied half of the sphere continuously where oscillators reaching the firing state are located in alternate positions; and this causes to the network to traverse the same path

Table 1. Test instances

Instance name	Description
Ising32x8	This instance consists of vertices arranged in a 32x8 grid where each vertex is connected to the neighbors on the left, right, above and bottom.
Ising32x8 Torus	This graph is the same as the Ising32x8 and it is embedded in a torus.
2-Partite-6	This is a 2-partite graph with 6 vertices in each partition.
3-Partite-6	This is a 3-partite graph with 6 vertices in each partition.
7-Partite-2	A 7-partite graph with each partition contains 2 vertices.
3-PartiteS3	This is a 3-partite graph with 3 unequal partition sizes.
4-PartiteS4	This is a 4-partite graph with 4 unequal partition sizes.
2Subgraphs	This instance is used to illustrate two overlapping graphs.
Hexagon1	A tessellation of triangles forming a hexagon.
Hexagon2	Another tessellation of triangles forming 2 adjoining Hexagons.
Hexagon3	Another tessellation of triangles forming 3 adjoining Hexagons.
Hexagon6	Another tessellation of triangles forming 6 adjoining Hexagons.
Tetra	Tetrahedron, the first Platonic solid.
Cube	A cube, another platonic solid.
Octa	An octahedron.
Dodeca	A dodecahedron.
Icosa	An Icosahedron.
1x7Ring	A graph consists of 7 vertices forming a ring.
2Triangles	A simple graph consists of 2 triangles linked by an edge.
Peterson	The Peterson graph.

indefinitely. Further, we observed the phase change ($d\theta$) in the oscillators do not approach zero as time t increases.

The chances for this alternate firing to occur are obviously fairly low for the 1x7Ring graph. The alternate firing configuration found here is for $k = 6$. The results from Table 2 indicate the existence of other forms of limit cycles for $k = 4$ and $k = 5$.

Like other iterative improvement algorithms where local minima exist, the limit cycles in the k-phase algorithm have the same behaviors. One of the common schemes to avoid the limit cycle is to shake the state of the network/graph. To achieve this, we allowed a certain percentage of the vertices to randomly change its phase. If this is insufficient, the percentage is increased to a maximum 100 which means the whole network is perturbed. With similar test settings as before, most of the test instances are solved with a small amount of perturbations and the coloring performance is better than without perturbation. The result is shown in Table 3.

Several hard instances require more perturbation. In particular, the dodecahedron seems to resist synchronization even after the perturbation coverage reaches 100%. As indicated in Fig. 3, the amount of cycles required to completely solve all 100 tests run on the dodecahedron increases exponentially.

Table 2. Performance of the k-phase algorithm (* 15 tests produce k=8)

Instance name	χ_G	Avg # color	The # times k is achieved					
			k=2	k=3	k=4	k=5	k=6	k=7
Ising32x8	2	2	100	-	-	-	-	-
Ising32x8 torus	2	2.92	56	-	44	-	-	-
2-Partite-6	2	2.27	82	12	3	3	-	-
3-Partite-6	3	3.61	-	66	22	-	12	-
7-Partite-2	7	7.21	-	-	-	-	-	85*
3-PartiteS3	3	3.20	-	80	20	-	-	-
4-PartiteS4	4	4.23	-	-	77	23	-	-
2Subgraphs	4	4	-	-	100	-	-	-
Hexagon1	3	3.39	-	63	35	2	-	-
Hexagon2	3	3.74	-	37	52	11	-	-
Hexagon3	3	4.05	-	40	17	39	4	-
Hexagon6	3	4.62	-	28	8	41	22	1
Tetra	4	4	-	-	100	-	-	-
Cube	2	2	100	-	-	-	-	-
Octa	3	3.02	-	98	2	-	-	-
Dodeca	3	4.56	-	46	5	4	40	5
Icosa	4	4.55	-	-	51	42	7	-
1x7Ring	3	3.08	-	93	6	1	-	-
2Triangles	3	3	-	100	-	-	-	-
Peterson	3	3.70	-	58	16	20	6	-

Clearly some graphs are harder to solve than others. This relates to the structure of the graphs. For the case of the dodecahedron, it seems to be related to the triangle free nature of this graph, for which the algorithm seems to be not so effective. Further tests using the larger DIMACS instances indicate the model is less effective on large random graphs.

Discussion on the Hyperbolic Algorithm. By using the same testing procedure, the results (not shown here due to limited space) show that the convergence rate of the hyperbolic algorithm is relatively low in comparison to the k-phase algorithm. The main reasons are that the sum of $f(x)$ experienced by each oscillator may not necessarily be the same for each oscillator, this can be directly deduced by applying the algorithm to the hexagon instance.

The case where optimal and sub optimal colorings are achieved is when the oscillators synchronize into the pattern in Fig. 4. Clearly, the initial starting point influences the final outcome. The initial starting point of the chart on the right leads to a sub optimal solution, while the starting point in the chart on the left leads to optimal coloring. The 0 and 360 degrees in the chart on the left, map to the same color.

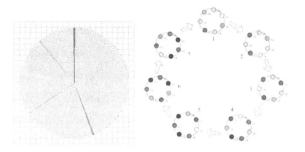

Fig. 2. Alternate firing of oscillators

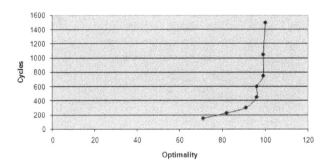

Fig. 3. Exponential trend to achieve optimality

2.6 Complexity

When analyzed from sequential computation point of view, the runtime complexity of the k-phase algorithms is $O(n^3)$ per cycle. However, it is difficult to mathematically quantify the number of oscillator cycles required for the network to reach a stable equilibrium point. Our experiments show an average of 10 cycles on the test instances if the network synchronizes. When the oscillators are implemented as hardware such as in [10], the complexity can be measured in the number of oscillator cycles. An order $O(1)$ cycles is indicated in [8] based on empirical results. This seems to agree with the result in [10] where the convergence time is constant for the n-queen problem. However, this is only the case if the network converges. If the network does not converge or resists convergence, our

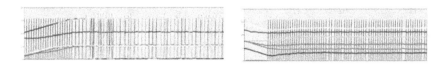

Fig. 4. Optimal and sub optimal state of the 1x7 ring. The (x, y) axis are the time and phase.

Table 3. Results using the k-phase model with perturbation

Instance name	χ_G	Avg # color	The # times k is achieved					
			k=2	k=3	k=4	k=5	k=6	k=7
Ising32x8	2	2	100	-	-	-	-	-
Ising32x8 torus	2	2	100	-	-	-	-	-
2-Partite-6	2	2	100	-	-	-	-	-
3-Partite-6	3	3	-	100	-	-	-	-
7-Partite-2	7	7	-	-	-	-	-	100
3-PartiteS3	3	3	-	100	-	-	-	-
4-PartiteS4	4	4.03	-	-	97	3	-	-
2Subgraphs	4	4	-	-	100	-	-	-
Hexagon1	3	3	-	100	-	-	-	-
Hexagon2	3	3	-	100	-	-	-	-
Hexagon3	3	3	-	100	-	-	-	-
Hexagon6	3	3.09	-	97	1	-	-	2
Tetra	4	4	-	-	100	-	-	-
Cube	2	2	100	-	-	-	-	-
Octa	3	3.02	-	98	2	-	-	-
Dodeca	3	3.8	-	71	4	5	14	6
Icosa	4	4	-	-	100	-	-	-
1x7Ring	3	3	-	100	-	-	-	-
2Triangles	3	3	-	100	-	-	-	-
Peterson	3	3	-	100	-	-	-	-

test (Fig. 3) on the dodecahedron shows an exponential time for computing the chromatic number at 100% rate. This result seems to be in line with $P \neq NP$.

3   Experiments with the Anti Voter Algorithm

A known method that is similar to the firing mechanism in the coupled oscillator method is the Anti Voter algorithm. The randomized Anti Voter algorithm for 3-colorable graphs was proposed in [18]. In this model, a vertex is associated with a clock. When a clock in a vertex rings, a neighbor is selected randomly; and if the color of the vertex and the neighboring vertex is the same, the vertex then randomly selects a color which is different from the neighbor's color; otherwise the vertex does not change its color.

Petford & Welsh [18] performed tests using 3-partite graphs of equal partition sizes. Each partition has between 4 and 20 vertices. Their test results indicate that the model appears to be optimal for the transition function $p_i(s_1, s_2, s_3) \sim 4^{-s_i}$. The transition function is a probability for a vertex to change to color i where i is the color range, that is, $1 \leq i \leq 3$, for 3 colors. In the conclusion, Petford & Welsh indicated that it will take a lot of experiments to find a good transition function; it could be that a good transition function must be empirically found. For the purpose of comparison, we implemented the Anti Voter algorithm and performed tests on the same instances used in the previous experiments. The results are not shown here due to space limitation. Generally, the Anti Voter does not perform as well as the k-phase algorithm, particularly on the $k > 3$ instances, the hexagons and platonic solids. A low success rate of coloring is seen across all the test instances.

4 Perspective Using Traditional Sequential Algorithms

Distributed graph coloring algorithms based on one virtual or physical computation element per vertex and communication via message passing have a typical coloring performance of $O(\Delta + 1)$ [19]. From this perspective, the k-phase algorithm falls into the same category; even though, the coloring performance is usually better than the theoretical bound. Obviously, most of the well known graph coloring algorithms will be able to optimally color the instances we used here. For our purpose, we implemented a simple greedy algorithm and a DSATUR algorithm.

The k-phase algorithm performs better than the simple greedy algorithm for the test cases used here. For larger arbitrary graphs with vertices up to 1000, the convergence rate of the k-phase decreases as the graph size increases. This is primarily due to the increase number of limit cycles in the larger random graphs. Nevertheless, for certain types of graphs, for example on the Ising32x8, hexagons, dodecahedron, icosahedron and Peterson graphs, the simple greedy method does not perform as well as the k-phase algorithm.

The DSATUR colors most of these graphs optimally, except for a few where the coloring is often suboptimal. When the vertex degree ordering and preliminary maximum clique are implemented on the DSATUR, all the instances are colored optimally, i.e. the chromatic number is found. Clearly the size of the instances used here is small and thus trivial for sequential exact algorithms.

5 Formalization and Descriptive Analysis

In this section, we analyze and formalize the behavior of the multiphase (k-phase) synchronization for graph coloring. We can deduce the general constraints and characteristics of the coupled oscillator methods for graph coloring as follows:

1. A network of oscillators represents a graph $G = (V, E)$
2. A vertex is typically associated with an oscillator; essentially an oscillator is a vertex with a number of properties
3. An edge is represented by a coupling link/connection between the oscillators
4. An oscillator has a phase which can represent a color of the vertex
5. The phase of an oscillator can have a value within $[0..2\pi]$
6. An oscillator only "knows" and interacts its neighbors via a coupling/update function, f
7. The coupling function updates the phase of an oscillator in such a way that the phase will eventually be different from the neighbors' phases
8. The synchronization process aims to ultimately bring the network to a stable equilibrium point in the phase space.

5.1   Descriptive analysis of limit cycles and stable equilibriums

A desired characteristic of the synchronization of a phase coupled oscillators model is that, the amount of phase change $\Delta\theta$ approaches zero as the simulation time is increased. Formally,

$$\Delta\theta_i \rightarrow 0 \ and \ t \rightarrow \infty, \quad \forall \ oscillator \ i \tag{8}$$

Whereas in a limit cycle state, $\Delta\theta$ fluctuates and does not approach zero. Even though the synchronization condition in Eq 8 is achieved, the network may not necessarily form a minimum number of phases/colors. This has been shown in [11] by using the icosahedron as a test case. Based on the results in [11] and our experiments with k-phase oscillator synchronization, we can describe the network states over a synchronization process into the categories as shown in Fig 5. In Fig 5, k is the number of distinct phases in the network/graph

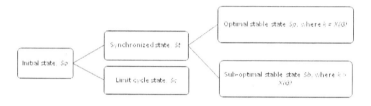

Fig. 5. States of a network over a synchronization process.

and $\chi(G)$ is the chromatic number of graph G. In the beginning, the network state is randomly initialized, this is the So state. During or after the simulation, the network can end in either a stable synchronized state (St) or a limit cycle state (Sc). In a limit cycle state (Sc), the phases of the oscillators may not represent a correct coloring of the graph because the amount of phase change $\Delta\theta$ fluctuates, thus the phase of the oscillator becomes arbitrary. Whereas in an St state, the coloring is always correct. However, the St state can be one of the multiple stable states, the optimal stable state (Sp) and the suboptimal stable state (Sb). Obviously, the desired state is the Sp state where the number of distinct phases is minimum and the other stable state is the Sb where the number of distinct phases is not minimum.

The reasons for the existence of the above states relate mainly to the initial state of the network and to an extent, the quality of the model used. Fig 4 plots the states of a network taken using the Poincar stroboscopic technique[16]. When an oscillator fires, its phase is reset to zero, at this time, we plot the phase of all other oscillators in the network. The y axis represents the phase of the oscillators and the x axis is time. The chart on the left shows an optimal stable state (Sp) of the network; the 0 and 2π in the chart are considered the same point due to the firing state of the oscillators. While the chart on the right with a different starting point, depicts the network in a suboptimal state (Sb).

Finding the initial state to begin the simulation is certainly difficult. A most common method is to perturb the state of the network so that the path to the optimal equilibrium Sp state is established. This however involves random trials which could be computationally expensive. In some hard test instances, the number of cycles required in the perturbation process increases exponentially.

It is certainly not possible to try every single update equations imaginable. In general, this leads to a question as whether there exists a model conforming to the general context set out earlier in this section, synchronizing into the optimal Sp state. Obviously, the NP harness of the graph coloring problem already provides a perspective to this question. In addition to this, we can formalize the general computation model for the phase coupled oscillation as follows:

Consider a network or a graph G with vertices $(v_1, v_2, , v_n)$ and the graph is associated with a network of oscillators satisfying the general context set out earlier in Section 5. Let the phases of all oscillator in the network be described by a set $(\theta_1, \theta_2, , \theta_n)$. We can use this set to describe the state of the network in a phase space for a given in time t as:

$$S_t = \{\theta_{i,t}\}, \quad \forall \text{ oscillator } i, \text{ and } t \text{ is the } t^{th} \text{ cycle} \tag{9}$$

To simplify the model, we measure the time t in unit of oscillator cycle. The transformation from an oscillator state $\theta_{i,t}$ to $\theta_{i,t+1}$ is defined by the update function $f(\theta)$. The total effect experienced by an oscillator in a cycle can be written as:

$$\Delta\theta_i = \sum_{k=0}^{n} f(\theta_{ik}), \quad \forall k \text{ are neighbors of } i \tag{10}$$

If we define a function $F(S)$ as the function that transforms S_t to $S(t+1)$, then the function $F(S)$ can be treated as a global transformer in a hyper dimension space. This could possibly be expressed in term of the Laplacian matrix because the matrix captures the structure of the graph[17]. Whereas the update function $f(\theta)$ only affects the individual oscillator. Typically, an analysis using a time series method is useful for this type of system, although the outcome might be specific to the specific phase coupled oscillator model being used.

Nevertheless, for the purpose of graph coloring, we know the objective of function $F(S)$, that is to map the state of the network such that the final state consists of a minimum number of phases. Formally, let $g(S)$ is a function that clusters all the phases in S into distinct clusters k and $g(S)$ returns k.

$$F : S_t \to S_{t+1} \text{ such that } g(S_{t+1}) \to \chi(G), \text{ as } t \to \infty \tag{11}$$

This formulation indicates an iterative global process that successively transforms a state S_t and uses the result of the transform as an input to compute the next state. Surprisingly, the local transform $f(\theta)$ at the oscillator level has had some success, ultimately this is limited because $f(\theta)$ does not "see" the global view as F does in Equation 11.

When a synchronized state St is achieved in a network of phase coupled oscillators, the result tends to produce an even phase separation among the oscillators. This provides an interesting twist due to the poly phase trigonometric property. Let the phase separation among the oscillators be $\tau = 2\pi/k$ where k is the number of distinct phases, then

$$\cos(\theta) + \cos(\theta + a) + \cos(\theta + 2a) + ... + \cos[\theta + (k-1)a] = 0 \qquad (12)$$

Where a is $1/n$ multiple of 2π. The same property holds for the sin function. If $h(S)$ returns a set containing the distinct cluster of the phase values, the synchronized state St must then satisfy,

$$\sum_{i=0}^{k} \cos[h(S)]_i = 0 \qquad (13)$$

It then becomes possible to redefine the transformer F in terms of other optimization techniques such as the linear programming with the constraints set out in the above equations.

6 Conclusion

Constructing a technique for the k-phase synchronization proves to be a challenging task. Not only from the point of view of establishing a mathematical proof of the analysis for the case of $k > 2$, but also from empirical approaches. The k-phase algorithms improve the coloring performance over the previous similar models. The issues include the restriction that the units only know their immediate neighbors and the suboptimal limit cycles.

We formalize the characteristics of the phase coupled oscillator methods in performing graph coloring and the reasons for the limitations of any such systems. For future work, we shall investigate ways of utilizing some of the known transformers such as the Laplace or the Fourier transforms and the viability of incorporating other optimization methods.

References

1. Attenborough, D.: The trials of life. William Collins Sons and Co and BBC books, London (1990)
2. Buck, E., Buck, J.: Synchronous fireflies. Scientific American (1976) 74-85
3. Mirollo, R.E., Strogatz, S.H.: Synchronization of Pulse-Coupled Biological Oscillators. SIAM Journals on Applied Mathematics 50 (1990) 1645-1662
4. Aihara, I., Kitahata, H., Aihara, K., Yoshikawa, K.: Periodic rhythms and anti-phase syn-chronization in calling behaviors of Japanese rain frogs. Vol. 2006. University of Tokyo, http://www.i.u-tokyo.ac.jp/mi/mi-e.htm (2006)
5. Hong, Y.W., Scaglione, A.: Time synchronization and reach-back communications with pulse-coupled oscillators for UWB wireless ad hoc networks. Proceedings of the IEEE con-ference on Ultra Wideband Systems and Technologies, Reston, VA (2003) 190-194

6. Lucarelli, D., Wang, I.-J.: Decentralized Synchronization Protocols with Nearest Neighbor Communication. SenSys'04. ACM, Baltimore, Maryland (2004)
7. Werner-Allen, G., Tewari, G., Patel, A., Welsh, M., Nagpal, R.: Firefly-Inspired Sensor Network Synchronicity with Realistic Radio Effects. SenSys'05. ACM, San Diego, California (2005)
8. Wu, C.W.: Synchronization in coupled chaotic circuits and systems. World Scientific Publishing Co, Singapore (2002)
9. Nakaguchi, T., Jin'no, K., Tanaka, M.: Theoretical Analysis of Hysteresis Neural Network Solving N-Queens Problems. Proc. of IEEE/ISCAS'99, Orlando, Florida (1999) 555-558
10. Nakaguchi, T., Jin'no, K., Tanaka, M.: Hardware combinatorial optimization problems solver by hysteresis neural networks. The 2001 IEEE International Symposium on Circuits and Systems, 2001 (ISCAS 2001), Vol. 3, Sydney, Australia (2001) 565-568
11. Lee, S., Lister, R.: Experiments in the Dynamics of Phase Coupled Oscillators When Applied to Graph Colouring. In: Dobbie, G., Mans, B. (eds.): The thirty-first Australasian Computer Science Conference (ACSC2008), Vol. 74. Conferences in Research and Practice in Information Technology (CRPIT), Wollongong (2008) 83-89
12. Jin'no, K., Taguchi, H., Yamamoto, T., Hirose, H.: Dynamical hysteresis neural networks for graph coloring problem. The 2003 IEEE International Symposium on Circuits and Systems, 2001 (ISCAS 2003), Vol. 5, Bangkok, Thailand (2003) 737-740
13. Lee, S.A.: Firefly Inspired Distributed Graph Coloring Algorithms. International Conference on Parallel and Distributed Processing Techniques and Applications (PDPTA08), Las Vegas (2008)
14. Holton, D.A., Sheehan, J.: The Petersen Graph, Vol. 7. Cambridge University Press (1993)
15. DIMACS: DIMACS implementation challenges: NP Hard Problems: Maximum Clique, Graph Coloring, and Satisfiability. Vol. 2007. `http://mat.gsia.cmu.edu/COLOR/instances.html` (1993)
16. Strogatz, S.H.: Nonlinear dynamics and chaos. Addison-Wesley, Reading, Massachusetts (1994)
17. Chung, F.R.K.: Spectral Graph Theory. American Mathematical Society, Providence, RI (1997)
18. Petford, A.D., Welsh, D.J.A.: A randomised 3-colouring algorithm. Discrete Mathematics 74 (1989) 253-261
19. Finocchi, I., Panconesi, A., Silvestri, R.: Experimental analysis of simple, distributed vertex coloring algorithms Proceedings of the thirteenth annual ACM-SIAM symposium on Discrete algorithms. Society for Industrial and Applied Mathematics, San Francisco, California (2002)

Self-similar planar graphs as models for complex networks [*]

Lichao Chen[1], Francesc Comellas[2] and Zhongzhi Zhang[1]

[1] Department of Computer Science and Engineering and
Shanghai Key Lab of Intelligent Information Processing,
Fudan University, Shanghai 200433, China
zhangzz@fudan.edu.cn
[2] Departament de Matemàtica Aplicada IV,
Universitat Politècnica de Catalunya
Avda. Canal Olímpic s/n, 08860 Castelldefels, Catalonia, Spain
comellas@ma4.upc.edu

Abstract. In this paper we introduce a family of planar, modular and self-similar graphs which have small-world and scale-free properties. The main parameters of this family are comparable to those of networks associated with complex systems, and therefore the graphs are of interest as mathematical models for these systems. As the clustering coefficient of the graphs is zero, this family is an explicit construction that does not match the usual characterization of hierarchical modular networks, namely that vertices have clustering values inversely proportional to their degrees.

1 Introduction

Research and studies performed in the last few years show that many networks associated with complex systems, like the World Wide Web, the Internet, telephone networks, transportation systems (including power and water distribution networks), social and biological networks, belong to a class of networks now known as small-world scale-free networks, see [1,11] and references therein. These networks exhibit a small average distance and diameter (compared to a random network with the same number of nodes and links) and, in many cases, a strong local clustering (nodes have many mutual neighbors). Another important common characteristic is that the number of links attached to the nodes usually obeys a power-law distribution (is scale-free). By introducing a new measuring technique, it has recently been discovered that many real networks are also self-similar, see [15,16]. Moreover, a degree hierarchy in these networks is sometimes related to the modularity of the system which they model.

---
[*] Research supported by the Ministerio de Educación y Ciencia, Spain, and the European Regional Development Fund under project TEC2005-03575 and by the Catalan Research Council under project 2005SGR00256. L. Chen and Z. Zhang are supported by the National Natural Science Foundation of China under Grant No. 60704044, the Postdoctoral Science Foundation of China under Grant No. 20060400162, and the Huawei Foundation of Science and Technology (YJCB2007031IN).

Most of the network models considered are probabilistic, however in recent years a deterministic approach has proven useful to complement and enhance the probabilistic and simulation techniques. Deterministic models have the strong advantage that it is often possible to compute analytically many network properties, which may be compared with experimental data from real and simulated networks. Some deterministic models have been proposed which are very often based on iterative constructions such that, at each step, one or more vertices are connected to certain subgraphs (for example, the so called k-trees [4]). Another technique produces graphs by duplication of certain substructures, see [5]. Here we propose a new family of graphs which generalize these methods by introducing at each iteration a more complex substructure than a single vertex. The result is a family of planar, modular, hierarchical and self-similar graphs, with small-world scale-free characteristics and with clustering coefficient zero. We note that some important real life networks, for example the networks associated to electronic circuits or Internet [11], have these characteristics as they are modular, almost planar and with a reduced clustering coefficient and have small-world scale-free properties. Thus, these networks can be modeled by our construction. A related family of graphs based on triangles, and which therefore has a high clustering coefficient was introduced in [18].

2   Hierarchical modular graphs

Several authors classify as hierarchical graphs, graphs with a modular structure and a strong connectedness hierarchy of the vertices which produces a power-law degree distribution. Moreover, they consider that the most important signature of hierarchical modularity is given by a clustering distribution with respect to the degree according to $\mathcal{C}(k) \propto 1/k$, see [2,8]. We recall that the clustering coefficient c_v of a vertex v is the ratio of the total number of existing connections between all δ_v its nearest neighbors and $\delta_v(\delta_v - 1)/2$, the number of all possible connections between them. The clustering of the graph is obtained averaging over all its vertices. In this section we define and analyze a family of hierarchical modular graphs, which are scale-free, planar and have clustering coefficient zero. They prove the existence of hierarchical graphs which do not have the above-mentioned relationship between the clustering coefficient and the degrees of the corresponding vertices.

Deterministic models for simple hierarchical networks have been published in [12,13]. These models consider the recursive union of several basic structures (in many cases, complete graphs) by adding edges connecting them to a selected root vertex. These and other hierarchical graphs have been considered when modeling metabolic networks in [10,14]. Hierarchical modularity also appears in some models based on k-trees or clique-trees, where the graph is constructed by adding at each step one or more vertices and each is connected independently to a certain subgraph [8,6,17]. The introduction of the so-called *hierarchical product of graphs* in [7] allows a generalization and a rigorous study of some of these models.

In [15,16], Song, Havlin and Makse relate the scale-free and the self-similarity properties as they verify that many self-similar graphs associated to real life complex systems have a fractal dimension and provide a connection between this dimension and the exponent of the degree power-law. However, a classical scale-free mode, the preferential attachment by Barbasi-Albert [1], which many authors consider a paradigm for these networks, has a null fractal dimension. This is not a paradox as the Barabási-Albert model lacks modularity because its generation process is based on the individual introduction of vertices.

In the next subsection we give details of our construction which is also based on an iterative process. However, the introduction at each step of a certain substructure allows the formation of modules and results in a final graph with a self-similar structure.

2.1 Iterative algorithm to generate the graph $H(t)$

The graph $H(t)$ is constructed as follows: For $t = 0$, $H(0)$ is C_4, a length four cycle. We define now as *generating cycle* a cycle C_4 whose vertices have not been introduced at the same iteration step and *passive cycle* a cycle C_4 which does not verify this property. For $t \geq 1$, $H(t)$ is obtained from $H(t-1)$ by considering all their generating cycles C_4 and connecting, vertex to vertex, to each of them a new cycle C_4. This operation is equivalent to adding to the graph a cube Q_3 by identifying vertex to vertex the generating cycle with one of the cycles of Q_3. The process is repeated until the desired graph order is reached.

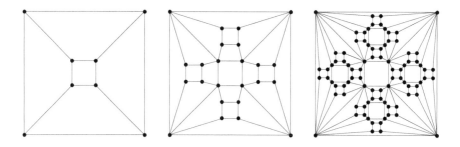

Fig. 1. Graphs $H(t)$ produced at iterations $t = 1, 2$ and 3.

2.2 Recursive modular construction

The graph $H(t)$ can be also defined as follows: For $t = 0$, $H(0)$ is the cycle C_4. For $t \geq 1$, $H(t)$ is produced from four copies of $H(t - 1)$ by identifying, vertex to vertex, the initial passive cycle of each $H(t-1)$ with each of four consecutive cycles of Q_3 (leaving two opposite cycles of Q_3 free), see Fig. 2.

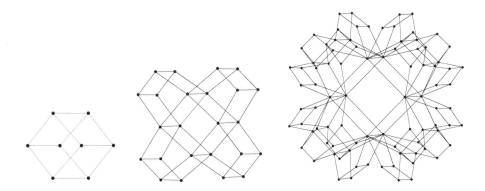

Fig. 2. Modular construction of $H(t)$ for $t = 1, 2$ y 3. At step t, we merge four copies of $H(t-1)$ to four cycles of the cube Q_3, leaving opposite cycles free. See the text for details.

2.3 Properties of $H(t)$

Order and size of $H(t)$.— We use the following notation: $\tilde{V}(t)$ and $\tilde{E}(t)$ denote, respectively, the set of vertices and edges introduced at the step t, while $V(t)$ and $E(t)$ denote the set of vertices and edges of the graph $H(t)$. $\tilde{C}(t)$ is the number of generating cycles C_4 at step t, which will be used to produce the graph $H(t+1)$.

Note that at each iteration, any generating cycle is replaced by four new generating cycles and one passive cycle. Therefore: $\tilde{C}(t+1) = 4 \cdot \tilde{C}(t), t \geq 1$ and $\tilde{C}(0) = 1$. Thus $\tilde{C}(t) = 4^t$. Moreover, each generating cycle introduces at the next iteration four new vertices and eight new edges. As a consequence, $\tilde{V}(t) = 4 \cdot \tilde{C}(t-1) = 4 \cdot 4^{t-1}$ and $\tilde{E}(t) = 8 \cdot \tilde{C}(t-1) = 8 \cdot 4^{t-1} = 2 \cdot 4^t$, thus:

$$|V(t)| = \sum_{i=0}^{t} \tilde{V}(t) = \frac{4^{t+1} + 8}{3} \qquad |E(t)| = \sum_{i=0}^{t} \tilde{E}(t) = \frac{2 \cdot 4^{t+1} + 4}{3} \qquad (1)$$

Degree distribution.— Intially, at $t = 0$, the graph is a single generating cycle C_4 and its four vertices have degree two.

When a new vertex i is added to the graph at iteration t_i $(t_i \geq 1)$, it has degree 3. We denote by $C(i, t)$ the number of generating cycles at iteration t which will produce new vertices that will connect to vertex i at step $t+1$. At iteration t_i, when vertex i is introduced, the value of $C(i, t_i)$ is 2. According to the construction process of the graph, at each iteration, each new neighbor

Step	Vertices	Edges	Number of active cycles
0	4	4	1
1	8	12	4
2	24	44	16
3	88	172	64
...	...	...	...
t	$\frac{4^{t+1}+8}{3}$	$\frac{2\cdot4^{t+1}+4}{3}$	4^t
...	...	...	...

Table 1. Number of vertices, edges and generating cycles of $H(t)$ at each step.

of i belongs to two generating cycles where i is also a vertex. If we denote as $k(i,t)$ the degree of vertex i at step t, then we have the following relationship: $C(i,t) = k(i,t) - 1$.

We now compute $C(i,t)$. As we have seen above, each generating cycle to which i belongs, produces two new generating cycles which also have i as a vertex. Thus $C(i,t) = 2 \cdot C(i, t-1)$. Using the initial condition $C(i,t_i) = 2$, we have $C(i,t) = 2^{t-t_i+1}$. Therefore the degree of vertex i at the step t is

$$k(i,t) = 2^{t-t_i+1} + 1. \qquad (2)$$

Note that the initial four vertices of step 0 follow a different process. In this case $C(i,0) = 2^t$ and $k(i,t) = 2^t + 1$. Thus, at step t the initial four vertices of the graph have the same degree than those introduced at step 1.

From equation (2) we verify that the graph has a discrete degree distribution and we use the technique described by Newman in [11] to find the cumulative degree distribution $P_{\text{cum}}(k)$ for a vertex with degree k: $P_{\text{cum}}(k) = \sum_{\tau \leq t_i} |V(\tau)|/|V(t_i)| = (4^{t_i+1} + 8)/(4^{t+1} + 8)$.

Replacing t_i, from equation (2), in the former equation $t_i = t+1-\ln(k-1)/\ln 2$ we obtain $P_{\text{cum}}(k) = (16 \cdot 4^t \cdot (k-1)^{-2} + 8)/(4^t + 8)$, which for large values of t, allows us to write $P_{\text{cum}}(k) \sim k^{1-\gamma_k} = k^{-2}$, and therefore the degree distribution, for large graphs, follows a power-law with exponent $\gamma_k = 3$. Research on networks associated to electronic circuits (these networks show planarity, modularity and a small clustering coefficient) gives similar values for their degree power-law distribution [9,11]. More precisely, the largest benchmark considered –a network with 24097 nodes, 53248 edges, average degree 4.34 and average distance 11.05– has a degree distribution which follows a power-law with exponent 3.0, precisely the same as in our model, and it has a small clustering coefficient $C = 0.01$.

Diameter.— At each step we introduce, for each generating cycle, four new vertices which will form a new cycle C_4 (and these vertices are among them at maximum distance 2). As all join the graph of the former step with one new edge, the diameter will increase by exactly 2 units. Therefore $D_t = D_{t-1} + 2$. $t \geq 2$. As $D_1 = 3$, we have that the diameter of $H(t)$ is $D_t = 2 \cdot t + 1$ if $t \geq 1$.

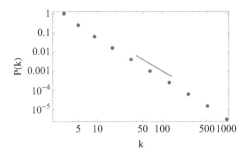

Fig. 3. Log-log representation of the cumulative degree distribution for $H(10)$ with $|V| = 1398104$ vertices. The reference line has slope -2.

Therefore, from Eq.1, and as for t large $t \sim \ln|V_t|$ we have in this limit that $D_t \propto \ln|V_t|$.

Average distance.— The average distance of $H(t)$ is defined as:

$$\bar{d}_t = \frac{1}{|V(t)|(|V(t)| - 1)/2} \sum_{i,j \in V(t)} d_{i,j}, \qquad (3)$$

where $d_{i,j}$ is the distance between vertices i and j. We will denote as S_t the sum $\sum_{i,j \in V(t)} d_{i,j}$.

The modular recursive construction of $H(t)$ allows us to calculate the exact value of $\bar{d}_t$. At step t, $H(t+1)$ is obtained from the juxtaposition of four copies of $H(t)$, which we label H_t^φ, $\varphi = 1, 2, 3, 4$, on top of the cube Q_3 (see Figs. 2 and 4). The copies are connected one to another at the vertices which we call *connecting vertices* and we label w, x, y, z, o, r, s, and a. The other vertices of $H(t+1)$ will be called *interior vertices*. Thus, the sum of distances distance S_{t+1} satisfies the following recursion:

$$S_{t+1} = 4\,S_t + \Delta_t - 4. \qquad (4)$$

where Δ_t is the sum over all shortest paths whose endvertices are not in the same $H(t)$ copy and the last term compensates for the overcounting of some paths between the connecting vertices –for example, $d(w, o)$ is included both in H_t^1 and H_t^2–. Note that the paths that contribute to Δ_t must all go through at least one of the eight connecting vertices. The analytical expression for Δ_t is not difficult to find. We denote as $\Delta_t^{\alpha,\beta}$ the sum of all shortest paths with endvertices in H_t^α and H_t^β. $\Delta_t^{\alpha,\beta}$ excludes the paths such that either endvertex is a connecting vertex. Then the total sum Δ_t is

$$\Delta_t = \Delta_t^{1,2} + \Delta_t^{1,3} + \Delta_t^{1,4} + \Delta_t^{2,3} + \Delta_t^{2,4} + \Delta_t^{3,4} + 20$$
$$+ \sum_{\substack{i \in H_t^3 \cup H_t^4, \\ i \notin x,r,s,y,a,z}} (d_{w,i} + d_{o,i}) + \sum_{\substack{i \in H_t^1 \cup H_t^4, \\ i \notin w,o,a,z,s,y}} (d_{x,i} + d_{r,i})$$

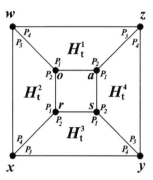

Fig. 4. Illustration of the classification of nodes in H_t^{φ}, $\varphi = 1, 2, 3, 4$.

$$+ \sum_{\substack{i \in H_t^1 \cup H_t^2, \\ i \notin x, r, o, w, a, z}} (d_{s,i} + d_{y,i}) + \sum_{\substack{i \in H_t^2 \cup H_t^3, \\ i \notin w, o, x, r, s, y}} (d_{a,i} + d_{z,i}), \tag{5}$$

where the term 20 comes from the sum of $d_{w,s}$, $d_{w,y}$, $d_{o,s}$, $d_{o,y}$, $d_{x,a}$, $d_{x,z}$, $d_{r,a}$, and $d_{r,z}$.

By symmetry, $\Delta_t^{1,2} = \Delta_t^{1,4} = \Delta_t^{2,3} = \Delta_t^{3,4}$, $\Delta_t^{1,3} = \Delta_t^{2,4}$, and $\sum_i d_{w,i} = \sum_i d_{o,i} = \sum_i d_{x,i} = \sum_i d_{r,i} = \sum_i d_{s,i} = \sum_i d_{y,i} = \sum_i d_{a,i} = \sum_i d_{z,i}$, and

$$\Delta_t = 4\Delta_t^{1,2} + 2\Delta_t^{1,3} + 8 \sum_{\substack{i \in H_t^3 \cup H_t^4, \\ i \notin x, r, s, y, a, z}} d_{w,i}. \tag{6}$$

To calculate Δ_t, we classify the interior vertices of $H(t+1)$ into four different classes according to their distances to each of the four vertices w, x, y, and z. Vertices w, x, y, and z are not classified into any of these classes which we represent as P_1, P_2, P_3, and P_4, respectively. This classification is represented in Fig. 4. By construction, for an arbitrary interior vertex v, there must exist one of the above mentioned vertices (say w) satisfying $d_{v,w} < d_{v,x}$, $d_{v,w} < d_{v,y}$, and $d_{v,w} < d_{v,z}$. All the interior vertices nearest to w (resp. x, y, and z) are assigned to class P_1 (resp. P_2, P_3, and P_4). The total number of vertices of H_t that belong to the class P_τ ($\tau = 1, 2, 3, 4$) is denoted by N_{t,P_τ}. Since the four vertices w, x, y, and z play a symmetrical role, classes P_1, P_2, P_3, and P_4 are equivalent. Thus, $N_{t,P_1} = N_{t,P_2} = N_{t,P_3} = N_{t,P_4}$ which will be abbreviated to N_t from now on. We have

$$N_t = \frac{|V_t| - 4}{4} = \frac{4^t - 1}{3}. \tag{7}$$

We denote by L_{t+1,P_1} ($L_{t+1,P_2}, L_{t+1,P_3}, L_{t+1,P_4}$) the sum of distances between vertices w (x, y, z) and all interior vertices $v \in P_1$ (P_2, P_3, P_4) of $H(t+1)$. Because of the symmetry, $L_{t+1,P_1} = L_{t+1,P_2} = L_{t+1,P_3} = L_{t+1,P_4}$ that will be written as L_{t+1} for short. Taking into account the second method of constructing $H(t)$, see Fig. 4, we can write the following recursive formula for L_{t+1}:

$$L_{t+1} = 4\,L_t + 2\,N_t + 1. \tag{8}$$

We can solve Eq. (8) inductively, with initial condition $L_1 = 1$, and we have

$$L_t = \frac{1}{18}\left(3t \cdot 4^t + 2 \cdot 4^t - 2\right). \tag{9}$$

We now return to compute Eq. (6), with $\Delta_t^{1,2}$ given by the sum

$$\Delta_t^{1,2} = \sum_{\substack{u \in H_t^1, u \notin \{w,o,a,z\}; \\ v \in H_t^2, v \notin \{w,x,r,o\}}} d_{u,v} = \sum_{i=1}^{4}\sum_{j=1}^{4} d_{P_i^{t,1},P_j^{t,2}}, \tag{10}$$

where $P_i^{t,1}$ and $P_j^{t,2}$ are the vertex classes P_i and P_j of H_t^1 and H_t^2, respectively, and $d_{P_i^{t,1},P_j^{t,2}}$ is the sum of distances $d_{u,v}$ for all vertices $u \in P_i \subset H_t^1$ and $v \in P_j \subset H_t^2$.

We have:

$$d_{P_1^{t,1},P_1^{t,2}} = \sum_{\substack{u \in P_1 \subset H_t^1, \\ v \in P_1 \subset H_t^2}} d_{u,v} = \sum_{\substack{u \in P_1 \subset H_t^1, \\ v \in P_1 \subset H_t^2}} (d_{u,o} + d_{o,r} + d_{r,v}) = 2N_t L_t + N_t^2. \tag{11}$$

Following the same process, we obtain $d_{P_i^{t,1},P_j^{t,2}}$ for the different values of i and j, which we use in Eq. (10) giving:

$$\Delta_t^{1,2} = 32\,N_t L_t + 24\,(N_t)^2. \tag{12}$$

Analogously, we can obtain

$$\Delta_t^{1,3} = 32\,N_t L_t + 32\,(N_t)^2. \tag{13}$$

Now, to find an expression for Δ_t, the only thing left is to evaluate the last term of Eq. (6), which can be obtained as above

$$8 \sum_{\substack{i \in H_t^3 \cup H_t^4, \\ i \notin x,r,s,y,a,z}} d_{w,i} = 64\,L_t + 128\,N_t. \tag{14}$$

Finally, and combining the former expressions, we write the exact result for the average distance of $H(t)$, $\bar{d}_t$, as

$$\bar{d}_t = \frac{4}{3} \cdot \frac{10 + 14 \cdot 4^t + 3(t+1)16^t}{10 + 13 \cdot 4^t + 4 \cdot 16^t}. \tag{15}$$

Notice that for a large order $(t \to \infty)$ $\bar{d}_t \simeq t + 1 \sim \ln |V_t|$, which means that the average distance shows a logarithmic scaling with the order of the graph, and has a similar behavior as the diameter (the graph is small-world).

Strength distribution.— The strength of a node in a network is associated to resources or properties allocated to it, as the total number of publication of an author, in the case of the network associated to the Erdős number; the total number of passengers in the world-wide airports network, etc.

In our case we associate to each vertex the area of the passive cycle, defined by the four vertices introduced at a given step. For this purpose we assume a uniform construction of the graph. At the initial step the area is $\mathcal{A}_0$ and we denote as $\mathcal{A}_t$ the area of the passive cycle introduced at step t. By convention, we establish that the area of this cycle is one fifth of the area of the cycle where it connects (as each introduction of a passive cycle is associated to the simultaneous introduction of four generating cycles). Therefore we have $\mathcal{A}_t = (\frac{1}{5})^t \mathcal{A}_0$. A vertex i introduced at t_i will have strength $s(i, t_i) = (\frac{1}{5})^{t_i} \mathcal{A}_0$ and it will keep it i n further steps $t > t_i$. As we want to find the strength distribution for all vertices of the graph at step t, we have that $s(i, t_i) = (\frac{1}{5})^{t_i - t} \cdot \mathcal{A}_t$.

Using equation (2) we obtain the following power-law for the correlation between the strength and the degree of a vertex:

$$s(i, t) = \frac{1}{5} \mathcal{A}_t (k(i, t) - 1)^{\ln 5 / \ln 3}, \tag{16}$$

which for large values of the degree k leads to $s(k) \sim k^{\ln 5 / \ln 3}$.

We should mention that similar exponents have been found for the relation between the strength and the degree of the node of real life networks like the airports network, Internet and the scientist collaboration graph [3].

After a similar analysis to the calculation of the degree distribution, we find that the strength distribution also follows a power law with exponent:

$$\gamma_s = 1 + 2 \frac{\ln 2}{\ln 5}. \tag{17}$$

It has been shown that if a weighted graph with a non-linear correlation between strength and degree $s(k) \sim k^\beta$ and the degree and strength distributions follows power laws, $P(k) \sim k^{-\gamma_k}$ and $P(s) \sim s^{-\gamma_s}$, then there exists a general relationship between γ_k and γ_s given by $\gamma_s = \frac{\gamma_k}{\beta} + \frac{\beta - 1}{\beta}$ [3].

From the former relationship, and as we have $\gamma_k = 3$ and $\beta = \ln 5 / \ln 3$, the exponent of the strength distribution is $\gamma_s = 3 \frac{\ln 2}{\ln 3} + \ln 2(\frac{\ln 5}{\ln 2} - 1)/\ln 5$, and we obtain the same value γ_s (17) which was computed directly.

3   Conclusion

The family of graphs introduced and studied here has as main characteristics planarity, modularity, degree hierarchy, and small-world and scale-free properties. At the same time the graphs have clustering zero. A combination of modularity and scale-free properties is present in many real networks like those associated to living organism (protein-protein interaction networks) and some social and technical networks [13,14]. The added property of a small clustering coefficient appears also in some technological networks (electronic circuits, Internet, P2P) and social networks [11]. Therefore our model, with a null clustering coefficient, could be considered to model these networks and also it can be used to study other properties without the influence of the clustering. The deterministic character of the family, as opposed to usual probabilistic models, should facilitate the exact computation of many network parameters.

On the other hand, simple variations of our model allow the introduction of clustering. As an example, by adding to each passive cycle an edge we can introduce two triangles for each cycle and therefore obtain a planar graph with non-zero clustering. Replacing in the construction each passive cycle by a complete graph K_4 will produce a family with a relatively large clustering coefficient. However the graph will no longer be planar.

References

1. R. Albert, A.-L. Barabási. Statistical mechanics of complex networks, *Rev. Mod. Phys.* 74:47–97, 2002.
2. A.-L. Barabási and Z.N. Oltvai, Network biology: Understanding the cell's functional organization, *Nature Rev Genetics* 5:101–113, 2004.
3. A. Barrat, M. Barthélemy, R. Pastor-Satorras, A. Vespignani. *Proc. Natl. Acad. Sci. U.S.A.* 101: 3747, 2004.
4. L.W. Beineke, R.E. Pippert. Properties and characterization of k-trees. *Mathematika* 18:141–151, 1971.
5. F. Chung, L. Lu, T.G. Dewey, D.J. Galas. Duplication models for biological networks. *J. Comput. Biol.* 10(5):677–87, 2003.
6. F. Comellas, G. Fertin, A. Raspaud. Recursive graphs with small-world scale-free properties, *Phys. Rev. E* 69: 037104, 2004.
7. L. Barriere, F. Comellas, C.Dalf, M.A. Fiol. The hierarchical product of graphs. *Discrete Appled Math.* En prensa, 2008.
8. S.N. Dorogovtsev, A.V. Goltsev, J.F.F. Mendes. Pseudofractal scale-free web. *Phys. Rev. E* 65: 066122, 2002.
9. R. Ferrer i Cancho, C. Janssen, R.V. Solé Topology of technology graphs: Small world patterns in electronic circuits. *Phys. Rev. E* 64:046119, 2001.
10. H. Jeong, B. Tombor, R. Albert, Z.N. Oltvai, A.-L. Barabási. The large-scale organization of metabolic networks. *Nature* 407: 651–654, 2000.
11. M.E.J. Newman, The structure and function of complex networks, *SIAM Review* ,45:167–256, 2003 .
12. J.D. Noh, Exact scaling properties of a hierarchical network model, *Phys. Rev. E* 67:045103, 2003 .
13. E. Ravasz, A.-L. Barabási. Hierarchical organization in complex networks. *Phys. Rev. E* 67:026112, 2003 .
14. E. Ravasz, A. L. Somera, D. A. Mongru, Z. N. Oltvai, A.-L. Barabási. Hierarchical organization of modularity in metabolic networks. *Science* 297: 1551–1555, 2002.
15. C. Song, S. Havlin, H. A. Makse. Self-similarity of complex networks. *Nature*, 433:392–395, 2005.

16. C. Song, S. Havlin, H. A. Makse. Origins of fractality in the growth of complex networks. *Nature Phys.*, 2:275–281, 2006.
17. Z.Z. Zhang, F. Comellas, G. Fertin, and L.L. Rong. High dimensional Apollonian networks. *J Phys A: Math Gen* 39:1811–1818, 2006.
18. Z.Z. Zhang, S. Zhou, L. Fang, J. Guan, Y. Zhang. Maximal planar scale-free Sierpinski networks with small-world effect and power law strength-degree correlation. *Europhys. Lett.* 79:38007, 2007.

On the pagenumber of the cube-connected cycles

Yuuki Tanaka[1] and Yukio Shibata[2]

[1] Information Science Center, Kyushu Institute of Technology, Japan
[2] Department of Computer Science, Graduate School of Engineering, Gunma University, Japan

Abstract. In this manuscript, we treat the book embedding of the cube-connected cycles. The book embedding of graphs is one of the graph layout problems and has been studied widely. We show that the pagenumber of $CCC(n)$, $n \geq 4$ is three and that of $CCC(3)$ is two. This result is optimal since $CCC(n)$ can not be embedded in two pages for $n \geq 4$.

1   Introduction

A *book* consists of a line, called the *spine*, and halfplanes called *pages* sharing the spine as a common boundary. A book embedding of a graph is defined by an assignment of vertices to distinct points on the spine and an assignment of edges to pages such that on each page there is no crossing of edges. The *pagenumber* of a graph G is the minimum number of pages in which G can be embedded. Book embeddings have been studied for many classes of graphs in [2], [3] etc. The book embedding problem has been motivated by several areas of computer science such as sorting with parallel stacks, single-row routing, and the design of fault-tolerant processor arrays [2]. The pagenumbers of several graph classes have been studied such as de Bruijn graphs [4], hypercubes [5], complete bipartite graphs [6], planar graphs [9] et al. The book embedding problem is one of the graph layout problems. As a most of the problems, it is hard to obtain an optiomal layout. For general graph G, this problem is NP-complete even if the assignment of vertices to the spine is given [2].

Preparata et al. [7] have proposed the cube-connected cycles as a versatile network for parallel computation and has been widely studied by many researcher. Konoe et al. showed that the cube-connected cycles $CCC(n)$ can be embedded in at most $n - 1$ pages in [5]. Authors previously showed that the pagenumber of the cube-connected cycles is at most five [8]. In this manuscript, we improve these results and obtain the pagenumber of the cube-connected cycles.

2   Preliminaries

In this section, we introduce some notations and definitions. We use a bold letter $\mathbf{x}$ as a binary sequence $x_0 x_1 \cdots x_{n-1}$ of length n and $\mathbf{x}^i$ as a sequence $x_0 x_1 \cdots x_{i-1} \overline{x_i} x_{i+1} \cdots x_{n-1}$ where $\overline{x_i}$ represents the inverse of x_i.

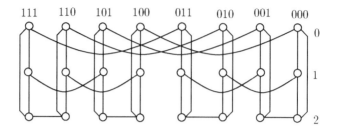

Fig. 1. A cube-connected cycles $CCC(3)$.

Let $n \geq 3$ be a positive integer. The n-dimensional cube-connected cycles $CCC(n)$ has the vertex set $V(CCC(n)) = \{(k, \mathbf{v}) | 0 \leq k \leq n-1, v_i \in \{0,1\}\}$ and each vertex $v = (k, \mathbf{v})$ in $V(CCC(n))$ is adjacent to vertices $u = (k+1 \bmod n, \mathbf{v})$ and $x = (k-1 \bmod n, \mathbf{v})$ (these edges are called C-edges) and $y = (k, \mathbf{v}^k)$ (called S-edges). For a vertex $v = (k, \mathbf{v})$ in $CCC(n)$, the value k is called the *level* of v. The sequence $v_0 v_1 \cdots v_{n-1}$ is called the *sequence* of v.

The 3-dimensional cube-connected cycles $CCC(3)$ is illustrated in Figure 1. In Figure 1, C-edges are drown as vertical lines and S-edges are drown as horizontal arcs.

On other terminology and notation, we refer to [1].

3 An optimal book embedding of the cube-connected cycles.

In this section, we show that the cube-connected cycles can be embedded in at most three pages. Since $CCC(3)$ is not an outerplanar graph but a hamiltonian planar graph, the following result is easy to obtain.

Proposition 1. *The pagenumber of $CCC(3)$ is two.*

Figure 2 shows a 2-page book embedding of $CCC(3)$. Before turning to the main results, we define a function B as follows:

$$B(\mathbf{x}) = \sum_{i=0}^{n-1} 2^{n-1-i} \left(\bigoplus_{j=0}^{i} x_j \right). \tag{1}$$

For the function B, we prove some useful lemmas needed later.

Lemma 1. *Let $s_0 s_1 \cdots s_{t-1}$ be a binary sequence of length $t \leq n$, if*

$$\sum_{i=0}^{t-1} 2^{n-1-i} \left(\bigoplus_{j=0}^{i} s_j \right) \leq B(\mathbf{x}) \leq \sum_{i=0}^{t-1} 2^{n-1-i} \left(\bigoplus_{j=0}^{i} s_j \right) + 2^{n-t} - 1, \tag{2}$$

then $x_0 = s_0, x_1 = s_1, x_2 = s_2, \ldots, x_{t-1} = s_{t-1}$.

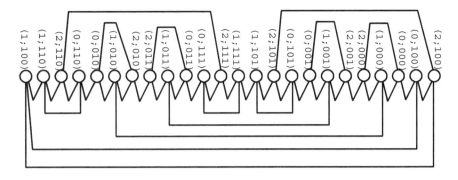

Fig. 2. A 2-page book embedding of $CCC(3)$.

Proof. We prove the statement by its contraposition. Namely, if there exists an integer $0 \le e < t$ such that $x_e \ne s_e$, then $B(\mathbf{x})$ cannot be in that range. Let e be such a minimum integer. Then,

$$\sum_{i=0}^{e-1} 2^{n-1-i} \left(\bigoplus_{j=0}^{i} s_j \right) = \sum_{i=0}^{e-1} 2^{n-1-i} \left(\bigoplus_{j=0}^{i} x_j \right). \tag{3}$$

If $\bigoplus_{j=0}^{e} x_j = 0$, then $\bigoplus_{j=0}^{e} s_j = 1$ and

$$B(\mathbf{x}) = \sum_{i=0}^{e-1} 2^{n-1-i} \left(\bigoplus_{j=0}^{i} x_j \right) + \sum_{i=e+1}^{n-1} 2^{n-1-i} \left(\bigoplus_{j=0}^{i} x_j \right) \tag{4}$$

$$< \sum_{i=0}^{e-1} 2^{n-1-i} \left(\bigoplus_{j=0}^{i} s_j \right) + 2^{n-e-1} \le \sum_{i=0}^{t-1} 2^{n-1-i} \left(\bigoplus_{j=0}^{i} s_j \right). \tag{5}$$

If $\bigoplus_{j=0}^{e} x_j = 1$, then $\bigoplus_{j=0}^{e} s_j = 0$ and

$$B(\mathbf{x}) = \sum_{i=0}^{e-1} 2^{n-1-i} \left(\bigoplus_{j=0}^{i} x_j \right) + 2^{n-1-e} + \sum_{i=e+1}^{n-1} 2^{n-1-i} \left(\bigoplus_{j=0}^{i} x_j \right) \tag{6}$$

$$> \sum_{i=0}^{e-1} 2^{n-1-i} \left(\bigoplus_{j=0}^{i} s_j \right) + \sum_{i=e+1}^{t-1} 2^{n-1-i} \left(\bigoplus_{j=0}^{i} s_j \right) + 2^{n-t} - 1 \tag{7}$$

$$= \sum_{i=0}^{t-1} 2^{n-1-i} \left(\bigoplus_{j=0}^{i} s_j \right) + 2^{n-t} - 1. \tag{8}$$

In both cases, $B(\mathbf{x})$ cannot be in that range and we conclude the proof. $\qquad\square$

In the above lemma, we obtain the following corollary when we consider the case $t = n$.

Corollary 1. *A function B is bijective.*

Lemma 2. *For four binary sequences $\mathbf{x}, \mathbf{y}, \mathbf{x}^t, \mathbf{y}^t$ of the same length n, without loss of generality let $B(\mathbf{x}) < B(\mathbf{x}^t)$ and $B(\mathbf{y}) < B(\mathbf{y}^t)$. If $B(\mathbf{x}) < B(\mathbf{y}) < B(\mathbf{x}^t)$, then $B(\mathbf{x}) < B(\mathbf{y}) < B(\mathbf{y}^t) < B(\mathbf{x}^t)$ for any $0 \le t \le n - 1$.*

Proof. By the definition of B,

$$\sum_{i=0}^{t-1} 2^{n-1-i} \left(\bigoplus_{j=0}^{i} x_j \right) \le B(\mathbf{x}) < B(\mathbf{x}^t) \le \sum_{i=0}^{t-1} 2^{n-1-i} \left(\bigoplus_{j=0}^{i} x_j \right) + 2^{n-t} - 1. \quad (9)$$

We can denote $B(\mathbf{y})$ as

$$B(\mathbf{y}) = \sum_{i=0}^{t-1} 2^{n-1-i} \left(\bigoplus_{j=0}^{i} y_j \right) + \sum_{i=t}^{n-1} 2^{n-1-i} \left(\bigoplus_{j=0}^{i} y_j \right), \quad (10)$$

We assume $B(\mathbf{x}) < B(\mathbf{y}) < B(\mathbf{x}^t)$ and from Lemma 1, we get $\sum_{i=0}^{t-1} 2^{n-1-i} \left(\bigoplus_{j=0}^{i} x_j \right) = \sum_{i=0}^{t-1} 2^{n-1-i} \left(\bigoplus_{j=0}^{i} y_j \right)$. Therefore, we can obtain

$$\sum_{i=t}^{n-1} 2^{n-1-i} \left(\bigoplus_{j=0}^{i} x_j \right) < \sum_{i=t}^{n-1} 2^{n-1-i} \left(\bigoplus_{j=0}^{i} y_j \right), \quad (11)$$

and as a result, we conclude the proof with the following inequality

$$B(\mathbf{y}^t) = \sum_{i=0}^{t-1} 2^{n-1-i} \left(\bigoplus_{j=0}^{i} y_j \right) + 2^{n-t} - 1 - \sum_{i=t}^{n-1} 2^{n-1-i} \left(\bigoplus_{j=0}^{i} y_j \right) < B(\mathbf{x}^t). \quad (12)$$

$\square$

For $n \ge 4$, our solution is subject to the parity of the dimension. First, we show that the case when the dimension is even.

Theorem 1. *The pagenumber of $CCC(n)$ is three for even $n \ge 4$.*

Proof. A mapping $\psi_e : V(CCC(n)) \to \mathbb{Z}_{n2^n}$ is defined as follows:

$$\psi_e((t; \mathbf{x})) = t2^n + ((t \pmod 2)(2^n - 1)) + (-1)^t B(\mathbf{x}). \quad (13)$$

First, we show the mapping ψ_e gives a total ordering to the vertex set of $CCC(n)$. For two vertices $u = (s; \mathbf{x}), v = (t; \mathbf{y})$ in $CCC(n)$, we assume that $s < t$. Then, two vertices are mapped in such a way that $s2^n < \psi_e(u) < (s+1)2^n \le t2^n < \psi_e(v) < (t+1)2^n$. Therefore, it is sufficient to show that the mapping is injective when $s = t$. When $s = t$, the injective property of ψ_e depends on whether the function B is injective or not. By Corollary 1, ψ_e is injective. The vertex set of $CCC(n)$ and $\mathbb{Z}_{n2^n}$ have the same cardinality, it concludes that ψ_e is bijective.

An assignment of edges to pages is given as follows:

- All S-edges: Page 0,
- C-edges such that $((s; \mathbf{x}), (s+1 \pmod n); \mathbf{x}))$ for some sequence $\mathbf{x}$ and some odd $s \in \mathbb{Z}_n$: Page 1,
- C-edges such that $((s; \mathbf{x}), (s+1 \pmod n); \mathbf{x}))$ for some sequence $\mathbf{x}$ and some even $s \in \mathbb{Z}_n$: Page 2.

An assignment of S-edges into one page For two S-edges $e = (x, x') = ((s; \mathbf{x}), (s; \mathbf{x}^s))$ and $f = (y, y') = ((t; \mathbf{y}), (t; \mathbf{y}^t))$, without loss of generality, we may assume $\psi_e(x) < \psi_e(y)$, $\psi_e(x) < \psi_e(x')$ and $\psi_e(y) < \psi_e(y')$. If $\psi_e(x') < \psi_e(y)$, then edges e and f can be embedded in the same page without crossing. Next we consider the case $\psi_e(x) < \psi_e(y) < \psi_e(x')$. Then, $s2^n \le \psi_e(x) < \psi_e(x') < (s+1)2^n$ and $s2^n < \psi_e(y) < (s+1)2^n$, and therefore $s = t$. When $s = t$ is odd, the magnitude correlation of $\psi_e(x)$ and $\psi_e(y)$ inverts that of $B(\mathbf{x})$ and $B(\mathbf{y})$. Similarly, the magnitude correlation of $B(\mathbf{x}')$ and $B(\mathbf{y}')$ can be obtained. From the above fact and Lemma 2, we obtain $\psi_e(x) < \psi_e(y) < \psi_e(y') < \psi_e(x')$. Therefore, edges e and f can be embedded into the same page without crossing. When $s = t$ is even, the similar discussion can be applied and all S-edges can be embedded into the page 0 without crossing.

An assignment of C-edges into two pages Firstly, an edge $e = (x, x') = ((0; \mathbf{x}), (n-1; \mathbf{x}))$ can be embedded in the same page as edges $f = (y, y') = ((t; \mathbf{y}), (t+1; \mathbf{y}))$ for some odd $1 \le t \le n-3$. The vertex ordering ψ_e of four vertices mentioned above is

$$\psi_e(x) = B(\mathbf{x}) < 2^n, \tag{14}$$
$$\psi_e(x') = n2^n - 1 - B(\mathbf{x}) \ge (n-1)2^n, \tag{15}$$
$$\psi_e(y) = (t+1)2^n - 1 - B(\mathbf{x}) \ge t2^n \ge 2^n, \tag{16}$$
$$\psi_e(y') = t2^n + B(\mathbf{x}) < (t+1)2^n < (n-1)2^n. \tag{17}$$

Therefore, e can be embedded in the same page as f without crossing. Secondly, for arbitrary sequences $\mathbf{x}$ and $\mathbf{y}$, we show that edges $e = (x, x') = ((0; \mathbf{x}), (n-1; \mathbf{x}))$ and $f = (y, y') = ((0; \mathbf{y}), (n-1; \mathbf{y}))$ can be embedded to the same page. Without loss of generality, we may assume that $\psi_e(x) < \psi_e(y)$. Then, $B(\mathbf{x}) < B(\mathbf{y})$ and $\psi_e(y') < \psi_e(x')$. Thus, edges e and f can be embedded to the same page without crossing and therefore all C-edges that connects the vertices of level 0 and level $n-1$ can be embedded in one page. Thirdly, all edges that connects vertices which do not belong to level 0 and level $n-1$ can be embedded in at most two pages. For two edges $e = (x, x') = ((s; \mathbf{x}), (s+1; \mathbf{x}))$ and $f = (y, y') = ((t; \mathbf{y}), (t+1; \mathbf{y}))$ for some $s, t \in \mathbb{Z}_{n-1} \setminus \{n-1\}$, without loss of generality, we may assume that $\psi_e(x) < \psi_e(x')$, $\psi_e(x) < \psi_e(y)$ and $\psi_e(y) < \psi_e(y')$. If the parity of s and t are different, the pages to which these edges are assigned are different. We consider the case the parity of s and t are the same. Edges e and f can be embedded into the same page when $\psi_e(x') < \psi_e(y)$, then we consider the case $\psi_e(x) < \psi_e(y) < \psi_e(x')$. Similar to the case of S-edge, we obtain $s = t$ from the inequality. When $s = t$ is even, the magnitude correlation of $\psi_e(x)$ and $\psi_e(y)$ is the same

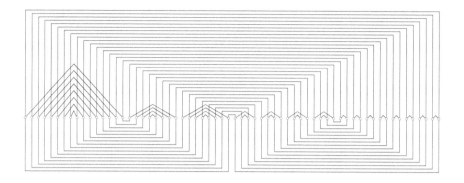

Fig. 3. A 3-page book embedding of $CCC(4)$.

as that of $B(\mathbf{x})$ and $B(\mathbf{y})$. Then, $\psi_e(y') < \psi(x')$ and edges e and f can be embedded in the same page. When $s = t$ is odd, the magnitude correlation of $\psi_e(x)$ and $\psi_e(y)$ inverts that of $B(\mathbf{x})$ and $B(\mathbf{y})$. In this case, we obtain an inequality $\psi_e(y') < \psi(x')$ and edges e and f can be embedded in the same page. $\qquad\square$

A 3-page book embedding of $CCC(4)$ is shown in Figure 3. Angled edges are assigned to page 0, rectangle edges are assigned to page 1 (upper of vertices) and page 2 (lower of vertices).

Theorem 2. *The pagenumber of $CCC(n)$ is three for odd $n \geq 5$.*

Proof. A mapping $\psi_o : V(CCC(n)) \rightarrow \mathbb{Z}_{n2^n}$ is defined as follows:

for $0 \leq t \leq n - 3$
$$\psi_o((t; \mathbf{x})) = t2^n + ((t \pmod 2))(2^n - 1) + (-1)^t B(\mathbf{x}), \qquad (18)$$
for $t = n - 2$
$$\psi_o((n - 2; \mathbf{x})) = n2^n - 1 - \left(2B(\mathbf{x}) + \bigoplus_{i=0}^{n-1} x_i\right), \qquad (19)$$
for $t = n - 1$
$$\psi_o((n - 1; \mathbf{x})) = n2^n - 1 - \left(2B(\mathbf{x}) + \neg\bigoplus_{i=0}^{n-1} x_i\right). \qquad (20)$$

First, we show that ψ_o is a bijection, that is, it gives a total ordering to the vertex set of $CCC(n)$. The mapping ψ_o is the same as ψ_e for vertices of level is less than $n - 2$ and for a vertex v that has a level less than $n - 2$, it is clear that $\psi_o(v) \leq (n - 2)2^n - 1$. For a vertex v that has a level $n - 2$ or $n - 1$, $(n - 2)2^n \leq \psi_o(v) < n2^n$. Therefore it is sufficient to show that the mapping ψ_o is bijective by showing that the mapping is injective for vertices with the level

$n-2$ or $n-1$. For vertices $u = (t; \mathbf{x})$, $v = (s; \mathbf{y})$ where s and t are $n-2$ or $n-1$, if $\mathbf{x} \neq \mathbf{y}$, $|2B(\mathbf{x}) - 2B(\mathbf{y})| \geq 2$ and $\bigoplus_{i=0}^{n-1} x_i$ is at most one, then $\psi_o(u) \neq \psi_o(v)$. Otherwise, that is, if $\mathbf{x} = \mathbf{y}$,

$$\psi_o((n-2; \mathbf{x})) - \psi_o((n-1; \mathbf{x})) = -\bigoplus_{i=0}^{n-1} x_i + \neg \bigoplus_{i=0}^{n-1} x_i. \tag{21}$$

Either $\bigoplus_{i=0}^{n-1} x_i$ or $\neg \bigoplus_{i=0}^{n-1} x_i$ becomes 1 and we obtain $\psi_o((n-2; \mathbf{x})) \neq \psi_o((n-1; \mathbf{x}))$. Therefore, the mapping ψ_o is bijectiive.

Next, we give an assignment of edges to three pages. This assignment is similar to the case n is even.

- All S-edges: Page 0,
- C-edges $((s; \mathbf{x}), (s+1; \mathbf{x}))$ for some sequence $\mathbf{x}$ and even $s \in \mathbb{Z}_n \setminus \{n-1\}$: Page 1,
- C-edges $((s; \mathbf{x}), (s+1; \mathbf{x}))$ for some sequence $\mathbf{x}$ and odd $s \in \mathbb{Z}_n$: Page 2,
- C-edges $((n-1; \mathbf{x}), (0; \mathbf{x}))$ for some sequence $\mathbf{x}$: Page 2.

An assignment of S-edges into one page. We show that all S-edges can be embedded into one page without crossing based on the vertex ordering ψ_o. Similar to the proof of Theorem 1, S-edge that connects vertices whose levels are less than $n-2$ can be embedded to one page. For S-edges that connects the vertices whose levels are $n-1$ or $n-2$, we show the following claims:

Claim.

$$|\psi_o((n-1; \mathbf{x})) - \psi_o((n-1; \mathbf{x}^{n-1}))| = 1. \tag{22}$$

Proof of the Claim :

$$\psi_o((n-1; \mathbf{x})) - \psi_o((n-1; \mathbf{x}^{n-1})) \tag{23}$$

$$= -\left(2B(\mathbf{x}) + \neg \bigoplus_{i=0}^{n-1} x_i\right) + \left(2B(\mathbf{x}^{n-1}) + \neg \left(\bigoplus_{i=0}^{n-2} x_i \oplus \overline{x_{n-1}}\right)\right) \tag{24}$$

$$= -2\bigoplus_{i=0}^{n-1} x_i - \neg \bigoplus_{i=0}^{n-1} x_i + 2\bigoplus_{i=0}^{n-2} x_i \oplus \overline{x_{n-1}} + \neg \left(\bigoplus_{i=0}^{n-2} x_i \oplus \overline{x_{n-1}}\right) \tag{25}$$

$$= -2\bigoplus_{i=0}^{n-1} x_i - \neg \bigoplus_{i=0}^{n-1} x_i + \neg 2 \bigoplus_{i=0}^{n-1} x_i + \bigoplus_{i=0}^{n-1} x_i \tag{26}$$

$$= -\bigoplus_{i=0}^{n-1} x_i + \neg \bigoplus_{i=0}^{n-1} x_i. \tag{27}$$

In the last equation, either $\bigoplus_{i=0}^{n-1} x_i$ or $\neg \bigoplus_{i=0}^{n-1} x_i$ becomes 1 and the desired result follows. □

Next, we show a claim with respect to the S-edges that connect two vertices whose levels are $n-2$.

Claim. Based on the ordering ψ_o, all S-edges that connect vertices whose levels are $n - 2$ can be embedded in one page.

Proof of the Claim : For two S-edges $e = (x, x') = ((n - 2; \mathbf{x}), (n - 2; \mathbf{x}^{n-2}))$ and $f = (y, y') = ((n - 2; \mathbf{y}), (n - 2; \mathbf{y}^{n-2}))$, without loss of generality, we may assume $\psi_o(x) < \psi_o(y)$, $\psi_o(x) < \psi_o(x')$ and $\psi_o(y) < \psi_o(y')$. If $\psi_o(x') < \psi_o(y)$, it is clear that two edges can be embedded into the same page, we consider the case $\psi_o(x) < \psi_o(y) < \psi_o(x')$. The vertex ordering of four vertices is

$$\psi_o(x) = n2^n - 1 - \left(2B(\mathbf{x}) + \bigoplus_{i=0}^{n-1} x_i \right), \tag{28}$$

$$\psi_o(x') = n2^n - 1 - \left(2B(\mathbf{x}^{n-2}) + \bigoplus_{i=0}^{n-3} x_i \oplus \overline{x_{n-2}} \oplus x_{n-1} \right), \tag{29}$$

$$\psi_o(y) = n2^n - 1 - \left(2B(\mathbf{y}) + \bigoplus_{i=0}^{n-1} y_i \right), \tag{30}$$

$$\psi_o(y') = n2^n - 1 - \left(2B(\mathbf{y}^{n-2}) + \bigoplus_{i=0}^{n-3} y_i \oplus \overline{y_{n-2}} \oplus y_{n-1} \right). \tag{31}$$

Then,

$$2B(\mathbf{x}) + \bigoplus_{i=0}^{n-1} x_i > 2B(\mathbf{y}) + \bigoplus_{i=0}^{n-1} y_i > 2B(\mathbf{x}^{n-2}) + \bigoplus_{i=0}^{n-3} x_i \oplus \overline{x_{n-2}} \oplus x_{n-1}. \tag{32}$$

Since each exclusive-or term is either zero or one and the function B is bijective, then the ordering of four vertices depends on each B term. With Lemma 2, we obtain the ordering $\psi_o(x) < \psi_o(y) < \psi_o(y') < \psi_o(x')$. Since vertices which have the level $n - 2$ are assigned between $(n - 2)2^n$ and $n2^n - 1$ and there is no S-edge that connects vertices whose levels are less than $n - 2$. Therefore, those edges can be embedded into the page 0 without edge crossing. □

From above claims, all S-edges can be embedded to one page without edge crossing.

An assignment of C-edges into two pages From the equation (21), all C-edges that connect vertices whose levels are $n-2$ and $n-1$ can be embedded into arbitrary page. Analogous to Theorem 1, it is sufficient to prove the following statements to complete the proof:

- Edges that connect vertices whose levels are $n - 1$ and 0 do not cross with other edges assigned to page 2.
- Edges that connect vertices whose levels are $n - 3$ and $n - 2$ do not cross with other edges assigned to page 1.

Firstly, we show that some C-edge $e = ((0; \mathbf{x}), (n - 1; \mathbf{x}))$ does not cross with any other edges $f = ((t; \mathbf{y}), (t + 1; \mathbf{y}))$ assigned to page 2 where $t \neq n - 1$. By the assignment of edges into page 2, t is odd and $t + 1 \leq n - 1$. Just mentioned above, some edge with $t + 1 = n - 1$ can be embedded into any page without crossing, we consider the case $t + 1 \leq n - 3$. The ordering of vertices that is incident to either an edge e or an edge f is as follows:

$$\psi_o((0; \mathbf{x})) = B(\mathbf{x}) < 2^n, \tag{33}$$

$$\psi_o((n - 1; \mathbf{x})) = n2^n - 1 - \left(2B(\mathbf{x}) + \neg \bigoplus_{i=0}^{n-1} x_i \right) \geq (n - 2)2^n, \tag{34}$$

$$\psi_o((t; \mathbf{y})) = (t + 1)2^n - 1 - B(\mathbf{y}) \geq 2^n, \tag{35}$$

$$\psi_o((t + 1; \mathbf{y})) = (t + 1)2^n + B(\mathbf{y}) < (n - 2)2^n. \tag{36}$$

Therefore, some C-edge that connects vertices whose levels are 0 and $n - 1$ do not cross with other C-edge that is assigned to page 2.

Secondly, we show that two C-edges $e = ((0; \mathbf{x}), (n-1; \mathbf{x}))$ and $f = ((0; \mathbf{y}), (n-1; \mathbf{y}))$ do not cross. Without loss of generality, we may assume $\psi_o((0, \mathbf{x})) < \psi_o((0, \mathbf{y}))$. The ordering of such vertices is

$$\psi_o((0; \mathbf{x})) = B(\mathbf{x}), \tag{37}$$

$$\psi_o((n - 1; \mathbf{x})) = n2^n - 1 - \left(2B(\mathbf{x}) + \neg \bigoplus_{i=0}^{n-1} x_i \right), \tag{38}$$

$$\psi_o((0; \mathbf{y})) = B(\mathbf{y}), \tag{39}$$

$$\psi_o((n - 1; \mathbf{y})) = n2^n - 1 - \left(2B(\mathbf{y}) + \neg \bigoplus_{i=0}^{n-1} y_i \right). \tag{40}$$

Since $B(\mathbf{x}) < B(\mathbf{y})$, $\psi_o((n - 1; \mathbf{y})) < \psi_o((n - 1; \mathbf{x}))$ and edges e and f can be embedded in the same page. From above two results, all C-edges that connect vertices whose levels are 0 and $n - 1$ can be embedded in the page 2 without any crossing.

Thirdly, we consider a C-edge $e = ((n-3; \mathbf{x}), (n-2; \mathbf{x}))$. From the assignment of edges, that edge is assigned to the page 1. Let $f = ((t; \mathbf{y}), (t + 1; \mathbf{y}))$ be an edge where even $t < n - 3$. The ordering of the vertices that are incident to either edges e or f is:

$$\psi_o((n - 3; \mathbf{x})) = (n - 3)2^n + B(\mathbf{x}) \geq (n - 3)2^n, \tag{41}$$

$$\psi_o((n - 2; \mathbf{x})) = n2^n - 1 - \left(2B(\mathbf{x}) + \bigoplus_{i=0}^{n-1} x_i \right) \geq (n - 2)2^n, \tag{42}$$

$$\psi_o((t; \mathbf{y})) = t2^n + B(\mathbf{y}) < (t + 1)2^n \leq (n - 4)2^n, \tag{43}$$

$$\psi_o((t + 1; \mathbf{y})) = (t + 2)2^n - 1 - B(\mathbf{y}) < (n - 3)2^n. \tag{44}$$

Therefore, edges that connect vertices whose levels are $n - 3$ and $n - 2$ can be embedded to the page 1 without crossing to other edges that connect vertices whose levels are neither $n - 3$ nor $n - 2$ assigned to page 1.

Finally, we prove the edges $e = ((n-3; \mathbf{x}), (n-2; \mathbf{x}))$ and $f = ((n-3; \mathbf{y}), (n-2; \mathbf{y}))$ do not cross in page 1. Without loss of generality, we may assume that

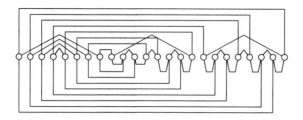

Fig. 4. A 3-page book embedding of $CCC(3)$.

$\psi_o((n-3;\mathbf{x})) < \psi_o((n-3;\mathbf{y}))$. The ordering of vertices with respect to those edges is:

$$\psi_o((n-3;\mathbf{x})) = (n-3)2^n + B(\mathbf{x}), \tag{45}$$

$$\psi_o((n-2;\mathbf{x})) = n2^n - 1 - \left(2B(\mathbf{x}) + \bigoplus_{i=0}^{n-1} x_i\right), \tag{46}$$

$$\psi_o((n-3;\mathbf{y})) = (n-3)2^n + B(\mathbf{y}), \tag{47}$$

$$\psi_o((n-2;\mathbf{y})) = n2^n - 1 - \left(2B(\mathbf{y}) + \bigoplus_{i=0}^{n-1} y_i\right). \tag{48}$$

Since $B(\mathbf{x}) < B(\mathbf{y})$, we obtain $\psi_o((n-2;\mathbf{y})) < \psi_o((n-2;\mathbf{x}))$ and edges e and f, that is, any pair of edges that connect level $n-3$ vertex and $n-2$ vertex do not cross on page 1. □

A 3-page book embedding of $CCC(3)$ according to the ordering ψ_o is shown in Figure 4. Angled edges are assigned to page 0, rectangle edges are assigned to page 1 (upper of vertices) and 2 (lower of vertices). Although the minimum odd number greater than 3 is 5, we illustrate the 3-page embedding of $CCC(3)$ since the $CCC(5)$ has 160 vertices and 240 edges, so we cannot draw in this manuscript.

References

1. G. Chartrand and L. Lesniak, Graphs & Digraphs, fourth ed., Chapman&Hall/CRC, 2004.
2. F. R. K. Chung, F. T. Leighton and A. L. Rosenberg, Embedding graphs in books: A layout problem with application to VLSI design , SIAM J. Alg. Discr. Methods, 8(1987) 33–58.
3. V. Dujmović and D. R. Wood, Stacks, Queues and Tracks: Layouts of Graph subdivisions, Discrete Math, and Theoretical Comput. Sci., Vol.7, pp.155–202, 2005.
4. T. Hasunuma, Embedding iterated line digraphs in books, Networks Vol.40, No.2, pp.51–62, 2002.
5. M. Konoe, K. Hagiwara and N. Tokura, On the pagenumber of hypercubes and cube-connected cycles, IEICE Trans., J71-D(3)(1988) 490–500 (in Japanese) .

6. D. J. Muder, M. L. Weaver, and D. B. West, Pagenumber of complete bipartite graphs, J. Graph Theory, 12 pp.469–489, 1988.
7. F. P. Preparata and J. Vuillemin, The Cube-Connected Cycles: A versatile network for parallel computation, Commun. ACM, vol.24, No.5, pp.300–309, May 1981.
8. Y. Tanaka and Y. Shibata, On the pagenumber of trivalent Cayley graphs, Disc. Appl. Math. 154, pp.1279–1292, 2006.
9. M. Yannakakis, Four pages are necessary and sufficient for planar graphs, Proc. 18th ACM Symp. on Theory of Computing pp.104–108, 1986.

$L(2,1)$-labellings of integer distance graphs[*]

Peter Che Bor Lam[1,2,**], Tao-Ming Wang[3] and Guo Hua Gu[4]

[1] Division of Science and Technology
and Division of Business Management
BNU-HKBU United International College,
Zhuhai, P. R. China
cblam@uic.edu.hk
[2] Visiting Professor
Department of Mathematics
Tunghai University
Taichung, 40704, Taiwan
[3] Department of Mathematics
Tunghai University
Taichung, 40704, Taiwan
wang@thu.edu.tw
[4] Department of Mathematics
Southeast University,
Nanjing , P. R. China

Abstract. Let D be a set of positive integers. The (integer) distance graph $G(Z, D)$ with distance set D is the graph with vertex set Z, in which two vertices x, y are adjacent if and only if $|x - y| \in D$. An $L(2,1)$-labelling of a graph G is an assignment of nonnegative integers to the vertices of G such that labels of any two adjacent vertices differ by at least 2, and labels of any two vertices that are at distance two apart are distinct. The minimum range of labels over all $L(2,1)$-labellings of a graph G is called the $L(2,1)-$labelling number, or simply the λ-number of G, and is denoted by $\lambda(G)$. We use $\lambda(D)$ to denote the λ-number of $G(Z, D)$. In this paper, some bounds for $\lambda(D)$ are established. It is also shown that distance graphs satisfy the conjecture $\lambda(G) \le \Delta^2$. For some special distance sets D, better upper bounds for $\lambda(D)$ are obtained. We shall also determine the exact values of $\lambda(D)$ for some two element set D.

1 Introduction

The channel assignment problem (FAP) is is the assignment of frequencies to television and radio transmitters subject to restrictions imposed by the distance between transmitters. This problem was first formulated as a graph coloring problem by Hale [5]. In 1988, Roberts (in a private communication to Griggs)

[*] Supported partially by the National Science Council under Grants NSC 96-2115-M-029-001 and NSC 96-2115-M-029-007
[**] The corresponding author

introduced a variation of this problem, where "close" transmitters must receive different channels and "very close" transmitters must receive channels at least two apart. Motivated by this variation, Griggs and Yeh [4] first proposed and studied the $L(2,1)$-labelling of a simple graph with a condition at distance two. This is followed by many other works. For examples, see [1–3, 6–9, 11, 12].

An $L(2,1)$-labelling f of G is a function $f : V(G) \to [0, k]$, such that $|f(u) - f(v)| \geq 2$ if $uv \in E(G)$; and $|f(u) - f(v)| \geq 1$ if $d_G(u, v) = 2$, where $d_G(u, v)$ is the length (number of edges) of a shortest path between u and v in G. Elements of the image under f are called *labels*, and the *span* of f, denoted by $\mathrm{span}(f)$, is the difference between the maximum and minimum labels of f. Without loss of generality, we assume that the minimum label of $L(2,1)$-labellings of G is 0, and so $\mathrm{span}(f)$ is the maximum label. The $L(2,1)$-*labelling number*, or λ-*number* of G, $\lambda(G)$, is the minimum span over all such labellings.

Griggs and Yeh showed that the $L(2,1)$-labelling problem, similar to the coloring problem, is NP-complete [4]. They also proved that for any graph G, $\lambda(G) \leq \Delta^2 + 2\Delta$, where Δ is the maximun degree of G, and they conjectured that for any graph G with maximum degree $\Delta \geq 2$, $\lambda(G) \leq \Delta^2$ (the Δ^2-Conjecture). Although this upper bound by Griggs and Yeh was later improved to $\Delta^2 + \Delta$ by Chang and Kuo [2], the Conjecture remains open.

To study the conjecture, we focus on the integer distance graphs. From now on, a set of positive integers will be called a *distance set*, denoted by D. The *integer distance graph* (simply called *distance graph*) $G(Z, D)$ with distance set D is the graph with vertex set Z, the set of all integers, and two vertices u and v are adjacent if and only if $|u - v| \in D$. For simplicity, we denote $G(Z, D)$ and its λ-number by $G(D)$ and $\lambda(D)$ respectively. If $g = \gcd(D)$, each component of $G(D)$ is isomorphic to $G(D')$, where $D' = \{d' : gd' \in D\}$. So we may assume that that $\gcd(D) = 1$.

In this paper, we show that $G(D)$ satisfies the Δ^2-Conjecture. We also improve the upper bounds of $\lambda(D)$ for some special k-element distance sets and for some general 2-element distance sets. Lastly, we determine $\lambda(D)$ for some special 2-element set D.

2   Basic definitions and results

In this section, we give some definitions and establish some basic results. For a finite distance set $D = \{d_1, d_2, \ldots, d_r\} \subset N$ with $d_1 < d_2 < \cdots < d_r$, define

$$D^2 = \{2d_i | 1 \leq i \leq r\} \cup \{d_j \pm d_i | 1 \leq i < j \leq r \text{ and } i \neq j\} \setminus D.$$

Clearly, $|D^2| \leq r^2 = |D|^2$. We can also verify that for $u, v \in Z$, $d_{G(D)}(u, v) = 2$ if and only if $|u - v| \in D^2$.

For any finite D, the distance graph $G(D)$ is an infinite $2|D|$-regular graph. Therefore, $\lambda(D) \geq 2|D|$. In [10], an upper bound for $\lambda(D)$ was obtained by the following theorem.

Theorem 1. [10]   *If D is finite, then* $2|D| + 2 \leq \lambda(D) \leq |D^2| + 3|D|$.

It follows that if $|D| = 1$, then $\lambda(D) = 4$. In this case each component of $G(D)$ is a path of infinite length. In [3], Griggs and Yeh proved that $\lambda(P_n) = 4$, for $n \geq 5$. Thus when $|D| = 1$, Theorem 1 is equivalent to their result on paths.

If $\lambda(D) = 2|D| + 2$, then we say the distance set D is λ-*minimal*. In Sections 4 and 5, we can see that there are several special distance sets D, $|D| \geq 2$, that are λ-minimal. On the other hand, unfortunately, we do not know whether the upper bound is attainable or not. In fact, the actual value or upper bound of $\lambda(D)$ is much smaller than this upper bound for most special distance sets D. So we shall attempt to improve the bound for some special D.

An immediate consequence of Theorem 1 is that the Δ^2-conjecture holds for distance graphs with finite distance sets.

Theorem 2. *Let $G(D)$ be a distance graph with finite D and maximum degree Δ. Then $\lambda(D) \leq \Delta^2$.*

Proof. Since $G(D)$ is $2|D|$−regular, $\Delta = 2|D| \geq 2$. It follows from Theorem 1 that $\lambda(D) \leq |D^2| + 3|D| \leq |D|^2 + 3|D| \leq \Delta^2$. ∎

Let $f : Z \to \{0, 1, 2, \ldots\}$ be a labelling. f is called *periodic with period p* if $f(i) = f(i + p)$ for all $i \in Z$ and is denoted by f_p. It is called *D-consistent* if for all $u, v \in Z$,
  (1) $|f(u) - f(v)| \geq 2$, if $|u - v| \in D$; and
  (2) $|f(u) - f(v)| \geq 1$, if $|u - v| \in D^2$.
The definition above implies that f is D-consistent if and only if it is an $L(2, 1)$-labelling of $G(D)$. Proof of the following proposition is straight-forward.

Proposition 1. *If f_p is D-consistent with period p, where $D = \{d_1, d_2, \ldots, d_r\}$ and $p > d_i$ for $i = 1, \ldots, r$, then f_p is also $\{k_1 p \pm d_1, k_2 p \pm d_2, \ldots, k_r p \pm d_r\}$-consistent for all $k_i \in N (i = 1, \ldots, r)$.*

3 Periodic labellings of distance graphs

Similar to the periodic coloring of distance graphs, we consider periodic labellings of distance graphs in the Section. We shall prove that any distance graph $G(D)$ with finite distance set D, has a periodic $L(2, 1)$-labelling using only $\lambda(D)$ labels. Its proof implies an explicit upper bound on the period for such labellings. As a consequence, it is practical to determine the λ-number of any distance graph whose distance set D is finite by a bounded search.

Theorem 3. *Suppose $D \subset N$ is finite. If $G(D)$ has a k-$L(2, 1)$-labelling, then it has a periodic k-$L(2, 1)$-labelling.*

Proof. Let $q = \max\{d : d \in D\}$. Suppose $f : Z \longrightarrow \{0, 1, 2, \cdots, k\}$ is a k-$L(2, 1)$-labelling of $G(D)$, which is a two-way infinite sequence of $(k + 1)$ labels

$$\cdots, f(-2), f(-1), f(0), f(1), f(2), \cdots.$$

In this two-way infinite sequence of labels, there are at most $(k+1)^{2q}$ possible blocks of $2q$ consecutive terms, so there exist two disjoint equal blocks of $2q$ consecutive terms, say $(f(i), f(i+1)\cdots, f(i+2q-1))$ and $(f(j), f(j+1)\cdots, f(j+2q-1))$ with $i < j$. Now we define a periodic labelling f' of the distance graph $G(D)$ with period $j - i$ as follows:

For $i \leq x \leq j+2q-1$, let $f'(x) = f(x)$; for $x \geq j+2q$, let $f'(x) = f(x+i-j)$ and; for $x \leq i-1$, let $f'(x) = f(x+j-i)$. By recursively using this rule, f' can be defined for all integers $x \in Z$.

To show that f' is an $L(2,1)$-labelling of $G(D)$, we first show that any two vertices at distance two receive different labels under f'. For otherwise, $f'(x+d) = f('x)$ for some integer $x \in Z$ and some $d \in D^2$. Since f' coincides with f on the set $\{i, i+1, \cdots, j+2q-1\}$ and $d \leq 2q$, it follows that $x \geq j$ or $x \leq i-1$. In the first case, let x be the smallest integer such that there exists $d \in D^2$ with $f'(x) = f'(x+d)$. Therefore $f'(x-(j-i)) = f'(x-(j-i)+d)$. However, by the choice of x, $f'(x-(j-i)) \neq f'(x-(j-i)+d)$. This is a contradiction. By similar argument, we get a contradiction in the second case.

We now show that any two adjacent vertices get labels differ by at least two under f'. Assume to the contrary that there exist some $x \in Z$ and some $d \in D$ such that $|f'(x) - f'(x+d)| \leq 1$. Since $d \leq q$, $x \geq j+q-1$ or $x \leq i-1$. We only need to consider the latter case because the former case is similar. Let $x(\leq i-1)$ be the largest integer such that there exists $d \in D$ such that $|f'(x)-f'(x+d)| \leq 1$. Since $f'(x) = f'(x+j-i)$ and $f'(x+d) = f'(x+d+j-i)$, then $f'(x+j-i) = f'(x+d+j-i)$, which contradicts the choice of x. Hence, f' is a periodic k-$L(2,1)$-labelling of $G(D)$. ∎

Corollary 1. *Suppose $D \subset N$ is finite and $q = max\{d : d \in D\}$. If the subgraph of $G(D)$ induced by the set $\{0, 1, 2, \cdots, 2q(k+1)^{2q}+2q-1\}$ has a k-$L(2,1)$-labelling, then $G(D)$ has a periodic k-$L(2,1)$-labelling with period not greater than $2q(k+1)^{2q}$.*

Proof. The proof of Theorem 3 shows that if f is a k-$L(2,1)$-labelling of the subgraph of $G(D)$ induced by the set $\{0, 1, 2, \cdots, 2q(k+1)^{2q}+2q-1\}$, then there are two integers $0 \leq i < j \leq 2q(k+1)^{2q}$ such that the restriction of f to the segment $\{i, i+1, \cdots, j+2q-1\}$ can be extended to a periodic k-$L(2,1)$-labelling of $G(D)$ with period $j - i \leq 2q(k+1)^{2q}$. ∎

Theorem 4. *There exists an algorithm to determine λ-number of distance graph $G(Z, D)$ for a finite set D of positive integers.*

Proof. Let $r = |D|$ and $q = max\{d : d \in D\}$. By Theorem , $\lambda(D) \leq r^2 + 3r$. For $k \leq r^2 + 3r$, we consider the $L(2,1)$-labellings of the subgraph of $G(D)$ induced by the set $\{0, 1, 2, \cdots, 2q(k+1)^{2q}+2q-1\}$. By Corollary , if this subgraph has a k-$L(2,1)$-labelling, then $G(D)$ has a periodic k-$L(2,1)$-labelling. Thus $\lambda(D) \leq k$. Therefore, $\lambda(D)$ can be determined by a bounded search. ∎

By Theorem 4 $\lambda(D) \leq \lambda(G')$, where G' is some finite subgraph of $G(D)$. So a bounded search algorithm to determine $\lambda(D)$ exists, although we do not know whether there is an efficient algorithm.

4 Finite distance sets

In this section, we will determine the exact value $\lambda(D)$ or improve the upper bound of $\lambda(D)$ in Theorem 1 for some special D.

Theorem 5. *Let k and i be two positive integer with $1 \leq i \leq k - 1$ and $\gcd(k, i) = 1$. If $nk \pm i \notin D$ for all $n \in N$ and $D \cup D^2$ contains no multiple of k, then $\lambda(D) \leq k - 1$.*

Proof. Let $Z_k = \{0, 1, \cdots, k - 1\}$ and $S = \{x : x = |iy|_k, \; y \in Z_k\}$, where $|z|_k$ denotes the integer $z' \in Z_k$ such that $z \equiv z'$ mod k. Since $\gcd(k, i) = 1$, therefore given $y_1, y_2 \in Z_k$, $|iy_1|_k = i|y_2|_k$ if and only if $y_1 = y_2$. It follows that $S = Z_k$. Now we define $f' : Z_k \rightarrow Z_k$ in the following manner. Let $x \in Z_k$. Choose $y \in Z_k$ such that $|iy|_k = x$. Define $f'(x) = y$. We then extend f' to a labelling f of Z by putting $f(z) = f'(z')$ for all $z \in Z$, where $z' \in Z_k$ and $|z'|_k = |z|_k$. Clearly f is periodic with period k and span$(f) \leq k - 1$. We shall prove that f is D-consistent:

Suppose $u, v \in Z$ and $u \neq v$. Let $u' \in Z_k$ and $v' \in Z_k$ be integers such that $|iu'|_k = |u|_k$ and $|iv'|_k = |v|_k$. Then $|f(u) - f(v)| = |u' - v'|$.

If $u' = v'$, then $|u|_k = |v|_k$, and consequently $u - v$ is a multiple of k. It follows that $u - v \notin D \cup D^2$. So if $d_{G(D)}(u, v) \leq 2$, then $|u - v| \in D \cup D^2$ and $|f(u) \neq f(v)|$.

If $|u' - v'| = 1$, then $|u - v|_k = |iu' - iv'|_k = i$. Since $nk \pm i \notin D$ for all $n \in N$, therefore $uv \notin E(G(D))$. It follows that f is D-consistent. ∎

This result can be used to determine $\lambda(G(D))$ of some specific D efficiently. Consider the set $D = \{9, 11\}$. Since $D \cup D^2 = \{2, 9, 11, 18, 20, 22\}$ contains no multiple of 7 and $7n \pm 1$ is not in D for all $n \in N$, $\lambda(D) \leq 6$. By Theorem 1, $\lambda(D) \geq 6$ and so $\lambda(9, 11) = 6$. More generally, we have the following corollary:

Corollary 2. *Suppose $D = \{a, b\}$, $1 \leq a < b$. If a, b and $b \pm a$ are not divisible by 7, then D is λ-minimal.*

Proof. We first set $k = 7$ in Theorem 5. Let $a \equiv \pm p$ (mod 7) and $b \equiv \pm q$ (mod 7), where $0 \leq p, q \leq 3$. The conditions on a, b ensure that $1 \leq p \neq q \leq 3$. Therefore there are six possible cases for (p, q), that is, $(1,2)$, $(1,3)$, $(2,1)$, $(2,3)$, $(3,1)$ and $(3,2)$. In each case, there exists an integer $i \in \{1, 2, 3\} \setminus \{p, q\}$ such that $nk \pm i \notin D$ for all $n \in N$. For example, $i = 3$ when $(p, q) = (1, 2)$. Clearly, $D \cup D^2$ contains no multiple of k, $1 \leq i \leq k - 1$ and $\gcd(k, i) = 1$. So $\lambda(D) \leq k - 1 = 6$ by Theroem 5 , $\lambda(D) = 6$ by Theorem 1, and consequently D is λ-minimal. ∎

Similarly, by setting $k = 8$ in Theorem 5, a weaker result can be obtained.

Corollary 3. *Suppose $D = \{a, b\}$, $1 \leq a < b$. Then $6 \leq \lambda(D) \leq 7$ if either (1) $a \equiv \pm 1 (mod\; 8)$, $b \equiv \pm 2 (mod\; 8)$, or (2) $a \equiv \pm 2 (mod\; 8)$, $b \equiv \pm 3 (mod\; 8)$.*

Proof. We can complete the proof by setting $i = 3$ in (1) and $i = 1$ in (2). ∎

We now turn to some special k-element distance sets D. If D contains the first k positive integers or odd integers, $\lambda(D)$ can be determined completely by next two results:

Theorem 6. *If $D = \{1, 2, ..., k\}$, then $\lambda(D) = 2k + 2$.*

Proof. By Lemma 2.1, $\lambda(D) \geq 2|D| + 2 = 2k + 2$. It suffices to show $\lambda(D) \leq 2k + 2$. Consider the periodic labelling $P_{2k+3} = 0, 2, 4, ..., 2k+2, 1, 3, 5, ..., 2k+1$. It is straightforward to check P_{2k+3} is $\{1, 2, ..., k\}$-consistent. So $\lambda(D) \leq 2k + 2$. ∎

Let $r \geq 1$ be an integer. The r-power path on n vertices, denoted by P_n^r, is the graph with the vertex set $\{v_1, v_2, ..., v_n\}$ and the edge set $\{v_i v_j \mid 1 \leq |i-j| \leq r\}$. In [2], the authors studied the $L(2,1)$-labelling of P_n^r. They obtained the exact values of $\lambda(P_n^r)$:

Proposition 2. *Let $r \geq 2$ be an integers, then*

$$\lambda(P_n^r) = \begin{cases} 2(n-1), & n \leq r+1, \\ 2r+1, & r+1 < n \leq 2r+2, \\ 2r+2, & n > 2r+2. \end{cases}$$

If $D = \{1, 2, ..., r\}$, then $G(D)$ is an r-power path on infinite vertices. Hence Proposition 2 is a special case of Theorem 6 with large n.

Theorem 7. *If $D = \{1, 3, 5, ..., 2k - 1\}$, $k \geq 2$, then $\lambda(D) = 2k + 2$.*

Proof. $\lambda(D) \geq 2k + 2$ follows from Lemma 2.1. On the other hand, we define a labelling

$$P_{2k+3} = 2k+2, k, 2k+1, k-1, 2k, k-2, ..., k+5, 3, k+4, 2, k+3, 1, k+2, 0, k+1.$$

This is a $(2k+2) - L(2,1)$-labelling of $G(D)$. So $\lambda(D) \leq 2k + 2$. ∎

By the previous two theorems, we see that $G(D)$ when $D = \{1, 2, ..., k\}$ or $\{1, 3, 5, ..., 2k - 1\}$ are the examples where the lower bound $2|D| + 2$ in Lemma 2.1 is attainable.

If D consists of arbitrary k consecutive integers, we have the following theorem.

Theorem 8. *If $D = \{a, a+1, ..., a+k-1\}$, $a, k \geq 2$, then*

$$\lambda(D) \leq \min\{2(a+k-1), 6k-2\}.$$

Proof. First define a periodic labelling $P_{2a+2k-1} = 0, 1, 2, ..., 2(a+k-1)$. It is straightforward to check $P_{2a+2k-1}$ is D-consistent. So $\lambda(D) \leq 2(a+k-1)$. Now we prove $\lambda(D) \leq 6k - 2$. By Lemma 2.1, $\lambda(D) \leq |D^2| + 3|D|$. Here $D^2 = \{1, 2, ..., k-1, 2a, 2a+2, ..., 2a+2k-2, 2a+1, 2a+3, 2a+2k-3\}$, hence $|D^2| \leq 3k - 2$. It follows that $\lambda(D) \leq 3k - 2 + 3k = 6k - 2$. ∎

In particular, for the case when $k \geq a$ we find the exact value of $\lambda(D)$.

Theorem 9. *If $D = \{a, a+1, ..., a+k-1\}$, $k \geq a \geq 2$, then $\lambda(D) = 2(a+k-1)$.*

Proof. Recall the definition of the set D^2 at the beginning of section 2. Obviously, for all $u, v \in Z$, $d_G(u, v) \leq 2$ if and only if $|u - v| \in D^2 \cup D$. Thus if $|u - v| \in D^2 \cup D$, then u and v must have distinct labels. Note that $D = \{a, a+1, ..., a+k-1\}$ and $k \geq a$, so

$$D^2 \cup D = \{1, 2, ..., a-1, a, a+1, ..., a+k-1, a+k, a+k+1, ..., 2(a+k-1)\}.$$

For all distinct $u, v \in S = \{1, 2, ..., 2(a+k-1), 2(a+k-1)+1\}$, we have $1 \leq |u - v| \leq 2(a+k-1)$, i.e., $|u - v| \in D^2 \cup D$. This implies the labels of u and v must be distinct. Since S has $2(a+k-1)+1$ numbers and 0 can be used as a label, $\lambda(D) \geq 2(a+k-1)$.

On the other hand, the periodic labelling $P_{2a+2k-1} = 0, 1, 2, ..., 2(a+k-1)$ is $\{a, a+1, ..., a+k-1\}$-consistent. Hence $\lambda(D) \leq 2(a+k-1)$. ∎

References

1. G. J. Chang, W.-T. Ke, D. Kuo, D. D.-F. Liu and R. K. Yeh, On $L(d, 1)$-labellings of graphs, *Discrete Math.* **220** (2000), 57-66.
2. G. J. Chang and D. Kuo, The $L(2, 1)$-labelling problem on graphs, *SIAM J. Discrete Math.* **9** (1996), 309-316
3. J. P. Georges, D. W. Mauro and M. I. Stein, Labelling products of complete graphs with a condition at distance two, *SIAM J. Discrete Math.* **14** (2000), 28-35.
4. J. R. Griggs and R. K. Yeh, Labeling graphs with a condition at distance 2, *SIAM J. Discrete Mathematics*, **5** (1992), 586-595.
5. W. K. Hale, Frequency assignment: Theorey and applications, *Proc. IEEE*, **68** (1980), 1497-1514.
6. F. Havet, B. Reed and J-S. Sereni, L(2,1)-labelling of graphs, *Proceedings of the Nineteenth Annual ACM-SIAM Symposium on Discrete Algorithms*, (2008), 621-630.
7. J. van den Heuvel, R. A. Leese and M. A. Shepherd, Graph labelling and radio channel assignment, *J. Graph Theory* **29** (1988), 263-283.
8. D. D.-F. Liu and R. K. Yeh, On Distance Two Labellings of Graphs, *Ars Combinatoria* **47** (1997), 13-22.
9. D. Sakai, Labeling chordal graphs with a condition at distance two, *SIAM J. Discrete Math.* **7** (1994), 133-140.
10. F. Y. Tao and G. H. Gu, $L(2, 1)$-labelling problem on distance graphs, *Journal of Southeast University* (English Edition), **20** (2004), 122-125.
11. M. A. Whittlesey, J. P. Georges and D. W. Mauro, On the λ-number of Q_n and related graphs, *SIAM J. Discrete Math.* **8** (1995), 499-506.
12. K.-F. Wu and R. K. Yeh, Labelling graphs with the circular difference, *Taiwanese J. Math.* **4** (2000), 397-405.

An improved Algorithm for the Black-and-White Coloring Problem on Trees

Daniel Berend[1] and Shira Zucker[2]*

[1] Departments of Mathematics and Computer Science, Ben-Gurion University, Beer
Sheva 84105, Israel,
berend@cs.bgu.ac.il
[2] Department of Computer Science, Ben-Gurion University, Beer Sheva 84105, Israel,
zuckers@cs.bgu.ac.il

Abstract. Given a graph G and positive integers b and w, the black-and-white coloring problem asks about the existence of a partial vertex-coloring of G, with b vertices colored black and w white, such that there is no edge between a black and a white vertex. We suggest an improved algorithm for solving this problem on trees.

Keywords: Combinatorial optimization, Black-and-White coloring, tree, vertex-separator.

1 Introduction

The *Black-and-White Coloring (BWC) problem* is de ned as follows. Given an undirected graph G and positive integers b, w, determine whether there exists a partial vertex-coloring of G such that b vertices are colored in black and w vertices in white (with all other vertices left uncolored), such that no black vertex and white vertex are adjacent.

One application of the BWC problem is to the problem of storing chemical products, where certain pairs of places cannot contain di erent products. (For other applications see, for example, [1].)

We sometimes refer to the optimization version of this problem, in which we are given a graph G and a positive integer b, and have to color b of the vertices in black, so that there will remain as many vertices as possible which are non-adjacent to any of the b vertices. These latter vertices are to be colored in white, and the resulting coloring is *optimal*. Note that it may well be the case that, given an optimal BWC, we can increase the number of black vertices without decreasing the number of white vertices. Clearly, when referring to a BWC, it su ces to refer to its black vertices only.

The BWC problem has been introduced and proved to be NP-complete by Hansen *et al.* [3]. In the same paper, a polynomial algorithm for trees was given. A

---

* Partially supported by the Lynn and William Frankel Center for Computer Sciences.

polynomial algorithm for partial k-trees with a xed k, was suggested by Kobler *et al.* [5]. Yahalom [7] gave a sub-linear algorithm for the graphs obtained by the moves of a rook on a chessboard. For an $m \times n$ board, this is in fact the Cartesian product (cf. [6]) $K_m \ K_n$ of two complete graphs. In [1], we provided explicit optimal solutions for the graphs obtained by the moves of a king on a chessboard. Note that, for an $m \times n$ board, this is in fact the strong product (cf. [2]) $P_m \boxtimes P_n$ of two simple paths.

The algorithm for trees, suggested by Hansen *et al.*, has running time of $O(n^3)$. In this paper we introduce another algorithm, whose running time is $O(n^2 \lg^3 n)$. Note that here we are content to nd, for each possible value of b, the optimal w. In the full paper, we explain how to nd a corresponding BWC.

In Section 2 we present formally the problem and the main theorem. In Section 3 we nd it convenient to state a more general version of our problem. The rest of that section is devoted to the solution of the generalized problem.

The authors are grateful to A. Melkman for drawing their attention to [4].

2 The Main Result

Let T be a tree with n vertices. The attributes of a BWC of T are given by a pair (b, w), in which b is the number of black vertices and w the number of white vertices. Thus, by having an array containing the optimal w for each value of b, we identify the attributes of all optimal BWC's. We refer to a BWC with b black and w white vertices as a (b, w)-*coloring*.

Problem 1.

   Input: A rooted n-vertex tree T.

   Output: An array maxW which, for each $0 \leq b \leq n$, gives the maximal w such that there exists a (b, w)-coloring of T.

Theorem 1. *Problem 1 is solved by Algorithm 1 in time* $O(n^2 \lg^3 n)$.

3 Proof of Theorem 1

3.1 A Recursive Approach and a More Generalized Problem

We assume that the input tree T is rooted. For each vertex v, denote by T_v the subtree rooted at v.

To each vertex v of T we attach a $3 \times (|T_v| + 1)$ table, v.maxWhite, where entry (c, b) contains the maximal w for which there exists a (b, w)-coloring of T_v, with v colored in c. Here c ranges over the set {black, white, uncolored}. Thus v.maxWhite[black], for example, contains for each value b the maximal possible w, assuming that v is colored black.

Our algorithm starts by nding a vertex v, which separates the tree into several subtrees of relatively small sizes. One of these subtrees is the tree obtained from T by removing all vertices of T_v (if v is not the root of T) and the others are rooted at the children of v. (In fact, it will be more convenient for us to adjoin v to each of the latter subtrees, so that they are all rooted at v.) The algorithm works recursively on each of these smaller subtrees, and nds for each the required table. The table of T_v is calculated iteratively by taking the union of larger and larger subtrees of T_v, two at a time.

The vertex v separating the tree may be any vertex which is not a leaf. To x ideas, it will be convenient for us to select this vertex as one which separates the tree into as small as possible subtrees. (However, any non-leaf would be equally good.) The existence of such a vertex follows from the following well-known theorem [8].

Theorem A. *Any n-vertex tree can be divided in linear time into several connected components, each with at most $\frac{n}{2}$ vertices, by removing a single vertex.*

In the course of the algorithm, after the table attached to the subtree rooted at some vertex has been found, we do not need that subtree any more. Rather, the root of that subtree becomes a leaf, and all other vertices are disposed. However, this new leaf carries with it some non-trivial information, namely the table corresponding to the subtree rooted at it. Thus, at later stages we need to deal with instances of the problem in which leaves are equipped with attached data, which may be of size $O(n)$. Thus, we replace our original problem by the following, more general one. Denote by $L(T)$ the set of leaves of T.

Problem 2.

Input: A rooted n-vertex tree T, with a $3 \times (m_v + 1)$ table v.maxWhite attached to each leaf v. For $c \in \{\texttt{black}, \texttt{white}, \texttt{uncolored}\}$ and $0 \le b \le m_v$, the number v.maxWhite$[c][b]$ is the maximal w such that, by coloring v in c, one can be "awarded" with b black and w white vertices.

Output: A table root(T).maxWhite which, for each $c \in \{\texttt{black}, \texttt{white}, \texttt{uncolored}\}$ and $0 \le b \le n + \sum_{v \in L(T)}(m_v - 1)$, provides the maximal w such that there exists a (b, w)-coloring of T with root(T) colored c.

In this problem, the table v.maxWhite provides the attributes of BWC's of some (perhaps unknown to us) subtree rooted at v with m_v vertices. Thus, for example, each entry (c, b) is at most $m_v - b$. Moreover (see Lemma 2 below), we assume that $b + v$.maxWhite$[c][b] \ge m_v - \lg m_v - 2$. Note that, for each vertex v, the entries v.maxWhite$[\texttt{black}][0]$ and v.maxWhite$[c][m_v]$ for $c \ne \texttt{black}$, are meaningless and should be ignored.

Theorem 2. *Problem 2 is solved by Algorithm 2 in time $O(nm \lg^3 m)$, where $m = \sum_{v \in L(T)} m_v$.*

During the performance of the algorithm on the original problem, some subtrees are reduced to single vertices, namely their root becomes a leaf of the current tree. At the moment a subtree T_v is reduced, v is provided with a table of size $3 \times (m_v + 1)$, where $m_v = |T_v| + \sum_{y \in L(T_v)}(m_y - 1)$. This table is attached to v until some subtree containing v will also be reduced and its root will become a new leaf, at which point the table will be disposed. Notice that, for each vertex v, we have $m_v = 1 + \sum_{y \in Y} m_y$, where Y is the set of children of v. The runtime of Algorithm 2 depends, besides the number of vertices, on the size of the tables attached to the leaves. Therefore, it depends on two parameters, n and m.

In order to nd the table root(T).maxWhite, which provides the attributes of all optimal BWC's of T, we invoke Algorithm 1. This algorithm transforms the given instance of the problem into an instance of Problem 2 and then invokes Algorithm 2 to actually nd the required table.

3.2 The Algorithm

The following simple lemmas will play a crucial role in the algorithm.

Lemma 1. *For any optimal BWC of a tree T on n vertices, the number of uncolored vertices is at most* $\lg n$.

Lemma 2. *For any optimal BWC of a tree T and any vertex v, the number of uncolored vertices in the subtree T_v is at most* $\lg |T_v| + 2$.

Algorithm 1 below constructs the array maxW, providing the maximal w for each value of b.

TreeBWC(T)
Input: A tree T with n vertices, rooted at r
Output: An array, providing the maximal W for each $0 \le b \le n$

for each leaf v
   v.maxWhite[black]$\leftarrow$ [0] //the array for T_v if v is black
   v.maxWhite[white]$\leftarrow$ [1] //the array for T_v if v is white
   v.maxWhite[uncolored]$\leftarrow$ [0] //the array for T_v if v is uncolored
r.maxWhite $\leftarrow$ generateTable(T)
for $i \leftarrow 0$ **to** n
   maxW[i] $\leftarrow$ max$\{r$.maxWhite[black][i],r.maxWhite[white][i],
                                    r.maxWhite[uncolored][i]$\}$

return maxW

Algorithm 1: Finding the attributes of all BWC's of a tree

Algorithm 2 below separates the tree into smaller parts and works recursively on each of them. If the separator vertex is not the root of the tree, we deal with the subtree rooted at the separator and then continue dealing with the rest of the tree. Otherwise, we deal with the subtrees rooted at each child separately and nally merge their results. In general, after the tables for the roots of two subtrees T_1 and T_2, with a common root but otherwise disjoint sets of vertices, have been generated, the algorithm needs to generate the arrays for $T' = T_1 \cup T_2$. This can be done simply by the following computations [3]. For any tree T, let B_T (W_T and U_T, respectively) be the array root(T).maxWhite[black] (root(T).maxWhite[white] and root(T).maxWhite[uncolored], resp.). We have:

$$
\begin{aligned}
B_{T'}[b] &\leftarrow \max\{B_{T_1}[b_1] + B_{T_2}[b_2] : b_1 + b_2 = b + 1\}, \\
W_{T'}[b] &\leftarrow \max\{W_{T_1}[b_1] + W_{T_2}[b_2] - 1 : b_1 + b_2 = b\}, \\
U_{T'}[b] &\leftarrow \max\{U_{T_1}[b_1] + U_{T_2}[b_2] : b_1 + b_2 = b\}.
\end{aligned} \tag{1}
$$

In fact, to reduce the runtime, we proceed somewhat differently. As a preparation to Algorithm 3, which performs these computations, Algorithm 2 below first translates all the values w in the table attached to each vertex v into the values $u = m_v - b - w$. By Lemma 2 we shall never need to use u values exceeding $\lg m_v + 2$. Therefore, the algorithm converts the table attached to v into a $3 \times (\lg m_v + 3)$-table, whose entries are themselves lists, of varying lengths. The list at entry (c, u') of the table is composed of a sorted list of all b-values of the pairs with $u = u'$, assuming that v is colored c. This is done by Algorithm SortU, called from Algorithm 2, whose details are omitted.

Algorithm 2 performs Algorithm 3 on pairs of lists, until it merges them all to a single one. In principle, we would like to merge at each step the two shortest lists; actually, we do not do it in this most economical way, but our ordering guarantees that we usually merge lists of average sizes. After each invocation of Algorithm 3 by Algorithm 2, the latter adjusts the resulting values, i.e., it subtracts 1 from all b-values in list$_i$[black] and from all the u-values in list$_i$[uncolored], where list$_i$ is the current table of lists built. The subtraction of 1 from the u-values in list$_i$[uncolored] is equivalent to a change of the index u of each list$_i$[uncolored][u] to $u - 1$. In order that after this adjustment we still keep all b-values for each $u \leq \lg m_v + 2$, Algorithm 3 computes these values for all $u \leq \lg m_v + 3$. After Algorithm 2 finishes dealing with the subtree rooted at v, it transforms all pairs in the resulting list back to their original form, which is a table that saves for each value of b and each color of v only the maximal w.

At the end of the process, after the algorithm has determined all pairs corresponding to optimal BWC's of a subtree T_v for some vertex v, it saves them in v.maxWhite and deletes all the children of v from the tree, along with their tables. This deletion is done by a procedure named reduce(T, v).

Algorithm 3 below gets two tables of lists representing pairs (b, u) and computes their pairwise sums by performing the procedure computePairwiseSums for each possible value of u. This procedure, performed with the parameters list$_1[q]$ and list$_2[p - q]$, computes the pairwise sums of the elements of its two parameters. Seidel (cf. [4]) gave a solution which takes $O(m \lg m)$ time, where m is the total size of the two lists. In fact, replacing each list by the generating polynomial of the numbers in the list, the set of pairwise sums is the set of powers for which the product of the two polynomial has a non-zero coefficient. Since two polynomials of degree m can be multiplied in time $O(m \lg m)$, the set of pairwise sums can be found also in time $O(m \lg m)$. For each sum (p, u), entry

```
generateTable(T)
Input: A tree T rooted at r, with a 3 × (m_v + 1) table attached to each leaf v
Output: The table r.maxWhite

T' ← T
if |T'| ≤ 2
    solve the problem directly and return the resulting table
s ← ndSeparator(T')
if s = r
    v_1, v_2, ..., v_d ← all children of s
    for i = 1 to d
        maxWhite_i ← generateTable(T_{v_i} ∪ {s})
        for b = 0 to m_{v_i}
            maxWhite_i[b] ← m_{v_i} − b− maxWhite_i[b]  //transform w's to u's
        list_i ← SortU(maxWhite_i)
    for k ← 0 to ⌈lg d⌉
        for i ← 1 to d − 2^{k+1} + 1 by 2^{k+1}
            list_i[black] ← Fusion(list_i[black],list_{i+2^k}[black])
            list_i[white] ← Fusion(list_i[white],list_{i+2^k}[white])
            list_i[uncolored] ← Fusion(list_i[uncolored],list_{i+2^k}[uncolored])
            adjust the b values in list_i[black] and the u values in list_i[uncolored]
    s.maxWhite ← list_1 transformed back to its original form
if s ≠ r
    s.maxWhite ← generateTable(T_s)
    T' ← reduce(T', s)  //delete all descendants of s to obtain a reduced tree
    r.maxWhite ← generateTable(T')
return r.maxWhite
```

Algorithm 2: Solution of Problem 2

(p, u) of a temporary boolean table becomes true. At the end of the process, the resulting list contains all the pairs.

Recall that Algorithm 2 translates the pairs in each array from (b, w) to (b, u). In principle, we could have used the pairs (b, u) from the beginning of Algorithm 1. We prefer having the original kind of pairs (b, w) during the process of the algorithm, so that the values we obtain are more meaningful to the problem.

3.3 Time Complexity

Runtime of Algorithm 3 The procedure computePairwiseSums(list$_1[q]$, list$_2[p-q]$) is performed exactly once for each p, q with $0 \leq q \leq p \leq \lg m_v + 3$, where m_v is the total size of the two lists. The runtime of this procedure, suggested by Seidel (cf. [4]), is $O(m_v \lg m_v)$. Therefore, the runtime of Algorithm 3 is $O(m_v \lg^3 m_v)$.

```
Fusion(list₁,list₂)
Input: Two (lg n + 3)-size arrays of lists – list₁ and list₂ – containing u-values
Output: An array of lists containing all maximal pairwise sums of elements of
        list₁ and list₂

tmp ← new boolean[n + 1][lg n + 4] // All entries are initialized to false.
       // tmp[b][u] =true i  b_i ∈ list_i[u_i], i = 1, 2, where ∑_i b_i = b, ∑_i u_i = u.
for p ← 0 to lg n + 3
  for q ← 0 to p
    for each x ∈ computePairwiseSums(list₁[q],list₂[p − q])
      tmp[x][p] ← true
for p ← 0 to lg n + 3
  for i ← 0 to n
    if tmp[i][p]
      outlist[p] ← outlist[p] ∪{i}
return outlist
```

Algorithm 3: Merging subtrees rooted at siblings

Runtime of Algorithm 2 Finding a separator vertex, at the beginning of the algorithm, takes $O(n)$ time. The transformation of pairs from the original form (b, w) in each maxWhite$_i$ to (b, u) and in the inverse direction in s.maxWhite are done in $O(m)$ time.

Saving the arrays in a 2-dimensional table, sorted by the value of u, and for each u sorted by the value of b, is done by Algorithm SortU in linear time in m.

The most time-consuming part of Algorithm 2 is the invocation of Algorithm 3.

Lemma 3. *For each vertex v having d_v children y_i, $1 \le i \le d_v$, attached with the tables y_i.maxWhite, Algorithm 2 finds the table v.maxWhite in $O(m_v \lg^3 m_v \lg d_v)$ time.*

In principle, the computations described in this lemma are the most time-consuming part of Algorithm 2, and therefore we could have estimated here the runtime of the algorithm. However, we prefer estimating the runtime in a way which is more suitable to the recursive nature of the algorithm.

The runtime R of Algorithm 2 on a tree $T_{n,m}$, with n vertices and m-size data, satis es the recursion

$$
R(T_{n,m}) \le
\begin{cases}
R(T_{n_1,m}) + R(T_{n-n_1+1,m}) + c_1 n, & (2 \le n_1 < n), \\
& n > 2, \ s \ne r, \\
\sum_{i=1}^{d} R(T_{n_i,m_i}) + c_1 m + c_1 m \lg^3 m \lg d, & \\
& (\sum m_i = m, \sum n_i = n - 1), \\
& n > 2, \ s = r, \\
c_2 m, & n = 2.
\end{cases}
$$

for some global constants $c_2 \geq 2c_1 \geq 2$, where the trees on the right-hand side of the formula are those attained by the separation, s is a separator vertex, r the root of $T_{n,m}$ and d the number of children of r. Recall that $m \geq n \geq 2$. For the case where $n > 2$ and $s \neq r$, the algorithm first finds the separator in $O(n)$ time, then deals with the subtree rooted at s, turns s into a leaf, and then deals with the reduced tree $T_{n-|T_s|+1,m}$. For the case where $n > 2$ and $s = r$, the algorithm first finds the separator in $O(n)$ time, then deals with all trees T_{n_i,m_i} rooted at the children of r, then adds r as a new root to each T_{n_i,m_i} in m_i time, and finally performs Algorithm 3 to merge all subtrees, which, by Lemma 3, takes $O(m \lg^3 m \lg d)$ time.

Conclusion of the Proof of Theorem 2 It suffices to prove $R(T_{n,m}) = O(nm \lg^3 m)$. We do this by induction on n. We actually prove that $R(T_{n,m}) \leq (c_2 n - 2c_2)m \lg^3 m$. For $n = 2$, the correctness is trivial. Assume the inequality holds for $2 \leq n' < n$. If $s \neq r$, then

$$
\begin{aligned}
R(T_{n,m}) &\leq R(T_{n_1,m}) + R(T_{n-n_1+1,m}) + c_1 n \\
&\leq (c_2 \cdot n_1 - 2c_2)m \lg^3 m + (c_2(n - n_1 + 1) - 2c_2)m \lg^3 m + c_1 n \\
&= (c_2 n - 2c_2)m \lg^3 m - c_2 m \lg^3 m + c_1 n \\
&\leq (c_2 n - 2c_2)m \lg^3 m.
\end{aligned}
$$

If $s = r$, then $d \geq 2$ and

$$
\begin{aligned}
R(T_{n,m}) &\leq \sum_{i=1}^{d} R(T_{n_i,m_i}) + c_1 m + c_1 m \lg^3 m \lg d \\
&\leq \sum_{i=1}^{d} (c_2 n_i - 2c_2)m_i \lg^3 m_i + 2c_1 m \lg^3 m \lg d \\
&\leq \sum_{i=1}^{d} (c_2 n_i - 2c_2)m \lg^3 m + 2c_1 m \lg^3 m \lg d \\
&= m \lg^3 m (c_2(n-1) - 2c_2 d + 2c_1 \lg d) \\
&\leq m \lg^3 m (c_2 n - 2c_2).
\end{aligned}
$$

This completes the proof of Theorem 2.

Conclusion of the Proof of Theorem 1 Algorithm 1 runs in linear time, except for the invocation of Algorithm 2. Therefore, its runtime is $O(nm \lg^3 m) = O(n^2 \lg^3 n)$.

References

1. D. Berend, E. Korach and S. Zucker, Anticoloring of a family of grid graphs, *Discrete Optimization*, 5/3:647–662, 2008.
2. N. Bray, from MathWorld – A Wolfram web resource, created by E. W. Weisstein. *url:* mathworld.wolfram.com/GraphStrongProduct.html
3. P. Hansen, A. Hertz and N. Quinodoz, Splitting trees, *Disc. Math.*, 165/6:403–419, 1997.
4. J. Erickson, Lower bounds for linear satisfiability problems, *Chicago J. Theoret. Comput. Sci.*, 1999(8).

5. D. Kobler, E. Korach and A. Hertz, On black-and-white colorings, anticolorings and extensions, preprint.

6. E. W. Weisstein, from MathWorld – A Wolfram web resource. *url:* mathworld.wolfram.com/CartesianProduct.html

7. O. Yahalom, Anticoloring problems on graphs, M.Sc. Thesis, Ben-Gurion University, 2001.

8. B. Zelinka, Medians and peripherians of trees, *Arch. Math.*, 4:87–95, 1968.

Finding Paths Between 3-Colourings

Luis Cereceda[1*], Jan van den Heuvel[1], and Matthew Johnson[2**]

[1] Department of Mathematics, London School of Economics,
Houghton Street, London WC2A 2AE, U.K.
{jan,luis}@maths.lse.ac.uk
[2] Department of Computer Science, Durham University,
Science Laboratories, South Road, Durham DH1 3LE, U.K.
matthew.johnson2@dur.ac.uk

Abstract. Given a 3-colourable graph G together with two proper vertex 3-colourings α and β of G, consider the following question: is it possible to transform α into β by recolouring vertices of G one at a time, making sure that all intermediate colourings are proper 3-colourings? We prove that this question is answerable in polynomial time. We do so by characterising the instances G, α, β where the transformation is possible; the proof of this characterisation is via an algorithm that either finds a sequence of recolourings between α and β, or exhibits a structure which proves that no such sequence exists. In the case that a sequence of recolourings does exist, the algorithm uses $O(|V(G)|^2)$ recolouring steps and in many cases returns a shortest sequence of recolourings. We also exhibit a class of instances G, α, β that require $\Omega(|V(G)|^2)$ recolouring steps.

1 Introduction

In this paper graphs are finite and do not contain loops or multiple edges unless stated otherwise. We refer the reader to [5] for standard terminology and notation not defined here. A *k-colouring* of a graph $G = (V(G), E(G))$ is a function $\alpha : V(G) \rightarrow \{1, 2, \ldots k\}$ such that $\alpha(u) \neq \alpha(v)$ for any edge uv. Throughout this paper we will assume that k is large enough to guarantee the existence of k-colourings (i.e., k is at least the *chromatic number of G*).

For a positive integer k and a graph G, we define the *k-colour graph of G*, denoted $\mathcal{C}_k(G)$, as the graph that has the k-colourings of G as its node set, with two k-colourings α and β joined by an edge in $\mathcal{C}_k(G)$ if they differ in colour on just one vertex of G. In this case, we shall also say that we can *recolour G* from α to β (and if v is the unique vertex on which α and β differ, then we also say that we can *recolour v*). Note that a path in $\mathcal{C}_k(G)$ can be described by either a sequence of colourings (the vertices of the path) or a sequence of recolourings.

---

[*] Current address: *Bayes Forecast*, Gran Vía 39, 5ªplanta, 28013 Madrid, Spain. lcereceda@bayesforecast.com.
[**] Research partially supported by Nuffield grant no. NAL/32772.

In addition, other graph-theoretical notions such as distance and adjacency can now be used for colourings.

The connectedness of the k-colour graph is an issue of interest when trying to obtain efficient algorithms for almost uniform sampling of k-colourings. In particular, $\mathcal{C}_k(G)$ needs to be connected for the single-site Glauber dynamics of G (a Markov chain defined on the k-colour graph of G) to be rapidly mixing. For details, see, for example, [6, 7], and references therein. In this setting, research on the connectedness of colour graphs has concentrated on cases where k is at least the maximum degree, or on cases where G is a highly symmetric graph such as an integer grid.

Properties of the colour graph, and questions regarding the existence of a path between two colourings, also find application in the study of radio channel reassignment. Given that a channel assignment problem can often be modelled as a graph colouring problem, the task of reassigning channels in a network, while avoiding interference and ensuring no connections are lost, can initially be thought of as a graph recolouring problem. See [1] for a discussion of these ideas in the context of cellular phone networks.

In recent work, the present authors have sought to develop a more general theory of the connectedness of colour graphs. In [4], a number of initial observations on properties of colour graphs are made and it is shown that if G has chromatic number $k \in \{2, 3\}$, then $\mathcal{C}_k(G)$ is not connected, but that for $k \geq 4$, there are k-chromatic graphs for which $\mathcal{C}_k(G)$ is not connected and k-chromatic graphs for which $\mathcal{C}_k(G)$ is connected. In [3], a characterisation of bipartite graphs whose 3-colour graph is connected is given and the problem of recognising these graphs is shown to be coNP-complete; while a polynomial algorithm is given for the restriction of the problem to planar graphs. In this paper, we consider the related problem of deciding whether two 3-colourings of a graph G belong to the same component of $\mathcal{C}_3(G)$. Formally, we have the following decision problem.

3-COLOUR PATH
Instance: A connected graph G together with two 3-colourings of G, α and β.
Question: Is there a path between α and β in $\mathcal{C}_3(G)$?

We assume our 3-COLOUR PATH instance graphs G to be connected as it is clear that the problem can be solved component-wise for disconnected graphs: there is a path between 3-colourings α and β of a disconnected graph G, if and only if for every connected component H of G there is a path between the colourings induced by α and β on H.

Our main result is the following.

Theorem 1.
The decision problem 3-COLOUR PATH *is in the complexity class* P.

We will prove Theorem 1 by describing a polynomial time algorithm that decides 3-COLOUR PATH. The algorithm stems from the proof of a characterisation of instances G, α, β where α and β belong to the same component of $\mathcal{C}_3(G)$. We will describe and prove this characterisation in Section 3. First, in Section 2, we shall examine what can forbid the existence of a path in $\mathcal{C}_3(G)$ between 3-colourings α and β of a graph G. The proof of the characterisation of connected pairs of 3-colourings is via an algorithm that, given G, α, β, either finds a sequence of recolourings between α and β, or exhibits a structure (described in Section 2) which proves that no such sequence exists. Thus this algorithm also decides 3-COLOUR PATH.

We will see that in the case that α and β belong to the same component of $\mathcal{C}_3(G)$, our algorithm will exhibit a path of length $O(|V(G)|^2)$ between them. This proves the following.

Theorem 2.
Let G be a 3-colourable graph with n vertices. Then the diameter of any component of $\mathcal{C}_3(G)$ is $O(n^2)$.

In Section 4 we turn our attention to what else can be said about the distance between a given pair of 3-colourings. We will prove that for many instances our algorithm returns a shortest path between the given 3-colourings. We will also construct a class of instances G, α, β such that α and β are connected and at distance $\Omega(|V(G)|^2)$ in $\mathcal{C}_3(G)$.

The computational complexity of the general problem k-COLOUR PATH, defined as 3-COLOUR PATH but for k-colourings instead of 3-colourings, is very different. In [2] it is proved that for fixed $k \geq 4$ this problem is PSPACE-complete and that in this case, and in contrast with Theorem 2, the distance between two k-colourings can be superpolynomial in the size of the graph.

Finally, the decision problem 2-COLOUR PATH is more or less trivial: two 2-colourings α, β of a bipartite graph G are connected, if and only if α and β are the same on each non-trivial component of G. And recolouring α to β can only involve changing the colours on isolated vertices, and hence requires at most $O(|V(G)|)$ steps.

2 Obstructions to Paths between 3-Colourings

In this section we examine what can stop us from being able to find a sequence of recolourings between a pair of 3-colourings α, β of a graph G. Informally, we call a structure in G, α, β forbidding the existence of a path between α and β in $\mathcal{C}_3(G)$ an *obstruction*.

For the remainder of this section we assume we are dealing with some fixed graph G.

The first and most obvious obstruction is given by what we call *fixed vertices*. For a 3-colouring α, we define a vertex v as *fixed* if there is no sequence of recolourings from α which will allow us to recolour v. In other words, a vertex v is fixed if for every colouring β in the same component of $\mathcal{C}_3(G)$ as α we have

$\beta(v) = \alpha(v)$. For example, if a cycle with 0 mod 3 vertices is coloured 1-2-3-1-2-3-$\cdots$-1-2-3, then every vertex on the cycle is fixed (as none can be the first to be recoloured). We call such a cycle a *fixed cycle* (as a subgraph of G, and with respect to the 3-colouring α). Similarly, a path coloured $\cdots$3-1-2-3-1-2-3-1-$\cdots$, both whose endvertices lie on fixed cycles, cannot be recoloured and is called a *fixed path*.

Given a 3-colouring α of G, we denote the set of fixed vertices of G by F_α. In the next section, we shall prove the following.

Proposition 1.
Let α be a 3-colouring of G. Then every $v \in F_\alpha$ belongs to a fixed cycle or to a fixed path.

The next lemma, which illustrates how fixed vertices may act as an obstruction, follows immediately from the definitions.

Lemma 1.
Let α and β be two 3-colourings of G. If α and β belong to the same component of $\mathcal{C}_3(G)$, then we must have $F_\alpha = F_\beta$, and $\alpha(v) = \beta(v)$ for each $v \in F_\alpha$.

We proceed to describe two further obstructions that will forbid the existence of a path between a given pair of 3-colourings. For this, we need a few more definitions.

To *orient* a cycle or a path means to orient each edge of the cycle or path to obtain a directed cycle or a directed path. If C is a cycle, then $\overrightarrow{C}$ denotes C with one of the two possible orientations. Similarly, $\overrightarrow{P}$ denotes one of the two possible orientations of a path P.

For a 3-colouring α of G, the *weight* of an edge $e = uv$ oriented from u to v is

$$w(\overrightarrow{uv}, \alpha) = \begin{cases} +1, \text{ if } \alpha(u)\alpha(v) \in \{12, 23, 31\}; \\ -1, \text{ if } \alpha(u)\alpha(v) \in \{21, 32, 13\}. \end{cases}$$

The *weight* $W(\overrightarrow{C}, \alpha)$ of an oriented cycle is the sum of the weights of its oriented edges; the same holds for the weight $W(\overrightarrow{P}, \alpha)$ of an oriented path. The following lemma and proof are from [4, 3].

Lemma 2.
Let α and β be 3-colourings of G, and let C be a cycle in G. If α and β are in the same component of $\mathcal{C}_3(G)$, then we must have $W(\overrightarrow{C}, \alpha) = W(\overrightarrow{C}, \beta)$.

Proof. Let α and α' be 3-colourings of G that are adjacent in $\mathcal{C}_3(G)$, and suppose the two 3-colourings differ on vertex v. If v is not on C, then we certainly have $W(\overrightarrow{C}, \alpha) = W(\overrightarrow{C}, \alpha')$.

If v is a vertex of C, then all its neighbours must have the same colour in α, otherwise we would not be able to recolour v. If we denote the in-neighbour of v on $\overrightarrow{C}$ by v_i and its out-neighbour by v_o, then this means that $w(\overrightarrow{v_i v}, \alpha)$

and $w(\overrightarrow{vv_o}, \alpha)$ have opposite sign, hence $w(\overrightarrow{v_iv}, \alpha) + w(\overrightarrow{vv_o}, \alpha) = 0$. Recolouring vertex v will change the signs of the weights of the oriented edges $\overrightarrow{v_iv}$ and $\overrightarrow{vv_o}$, but they will remain opposite. Therefore $w(\overrightarrow{v_iv}, \alpha') + w(\overrightarrow{vv_o}, \alpha') = 0$, and it follows that $W(\overrightarrow{C}, \alpha) = W(\overrightarrow{C}, \alpha')$.

From the above we immediately obtain that the weight of an oriented cycle is constant on all 3-colourings in the same component of $\mathcal{C}_3(G)$. ☐

The following lemma can be proved in the same way.

Lemma 3.
Let α and β be 3-colourings of G such that $F_\alpha = F_\beta \neq \varnothing$ and $\alpha(v) = \beta(v)$ for all $v \in F_\alpha$. Suppose G contains a path P with endvertices u and v, where $u, v \in F_\alpha$. If α and β are in the same component of $\mathcal{C}_3(G)$, we must have $W(\overrightarrow{P}, \alpha) = W(\overrightarrow{P}, \beta)$.

Lemmas 1, 2 and 3 give necessary conditions for two 3-colourings α and β to belong to the same component of $\mathcal{C}_3(G)$. From Lemmas 1 and 2 we have, respectively:

(C1) $F_\alpha = F_\beta$, and $\alpha(v) = \beta(v)$ for each $v \in F_\alpha$; and

(C2)   for every cycle C in G we have $W(\overrightarrow{C}, \alpha) = W(\overrightarrow{C}, \beta)$.

If for two 3-colourings α and β of G we take condition (C1) to be satisfied (so they have the same fixed vertices, coloured alike), Lemma 3 gives a third necessary condition for α and β to belong to the same component of $\mathcal{C}_3(G)$:

(C3)   for every path P between fixed vertices we have $W(\overrightarrow{P}, \alpha) = W(\overrightarrow{P}, \beta)$.

Bearing in mind that we are only considering condition (C3) if condition (C1) is already satisfied, let us observe that neither conditions (C1) and (C2) taken together, nor conditions (C1) and (C3) taken together, are sufficient to guarantee the existence of a path between 3-colourings α and β.

To see that conditions (C1) and (C2) are not sufficient, consider the graph and two 3-colourings shown in Figure 1. It is easy to check that (C1) and (C2) are satisfied (note that only vertices on the 3-cycles are fixed), but the two colourings are not connected: fix an orientation of the path between the two 3-cycles, and observe that the weight of this oriented path is $+3$ in one colouring and -3 in the other.

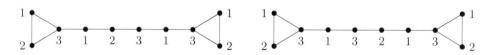

Fig. 1. Two 3-colourings of a graph G not connected in $\mathcal{C}_3(G)$.

To see that the two conditions (C1) and (C3) are not sufficient, consider two 3-colourings α and β of a 5-cycle that differ only in that the colours 1 and 2 are swapped: (C1) and (C3) are satisfied (since $F_\alpha = F_\beta = \varnothing$), but there is no path between the two colourings as the 5-cycle has different weights in the two colourings.

We prove in the next section that if all three conditions are satisfied by a pair of colourings α and β of a graph G, then these colourings are in the same component of $\mathcal{C}_3(G)$.

3 A Characterisation of Connected Pairs of 3-Colourings

In this section we prove the following characterisation of connected pairs of 3-colourings. Its proof will yield a polynomial time algorithm for 3-COLOUR PATH, proving Theorem 1. We will also prove Theorem 2 in the process.

Theorem 3.
Two 3-colourings α and β of a graph G belong to the same component of $\mathcal{C}_3(G)$, if and only if

(C1)   we have $F_\alpha = F_\beta$, and for each $v \in F_\alpha$, $\alpha(v) = \beta(v)$;

(C2)   for every cycle C in G we have $W(\overrightarrow{C}, \alpha) = W(\overrightarrow{C}, \beta)$; and

(C3)   for every path P between fixed vertices we have $W(\overrightarrow{P}, \alpha) = W(\overrightarrow{P}, \beta)$.

The necessity of the three conditions has already been established. We prove that they are sufficient by outlining an algorithm whose input is a graph G and two 3-colourings α and β of G, and whose output is either a path in $\mathcal{C}_3(G)$ from α to β, or an obstruction that shows that (C1), (C2) or (C3) is not satisfied, and hence no such path exists.

The first step of the algorithm is to find F_α and F_β. We claim that the following procedure finds the fixed vertices of a graph G with 3-colouring α.

- Let S_1, S_2, S_3 initially be the three colour classes induced by α.
- For $i \in \{1, 2, 3\}$, and for each vertex $v \in S_i$: if v has no neighbours in one or both of the other two sets, then let $S_i = S_i \setminus \{v\}$.
- Repeat the previous step until no further changes are possible. Return $S = S_1 \cup S_2 \cup S_3$.

Claim 4.
The above procedure returns $S = F_\alpha$.

Before proving the claim, let us give some more definitions. Fix a vertex v of G and set $L_0^+ = L_0^- = \{v\}$. For $i = 1, 2, \ldots$, let a vertex u belong to L_i^+ if u has a neighbour $w \in L_{i-1}^+$ and $\alpha(u) \equiv \alpha(w) + 1 \pmod 3$. (So, for example, if v is coloured 3, then L_1^+ contains all its neighbours coloured 1, L_2^+ contains all vertices coloured 2 that have a neighbour in L_1^+, and so on.) For $j = 1, 2, \ldots$, let a vertex u belong to L_j^- if u has a neighbour $w \in L_{j-1}^-$ and $\alpha(u) \equiv \alpha(w) - 1$ (mod 3). We call these sets the *levels* of v, and the sets are called *positive* or *negative* according to their superscript.

We note that if a level is empty then so are all levels with the same superscript and larger index. We observe that v lies on fixed cycle or path if and only if the number of positive and negative levels of v that are nonempty are both infinite; or, more particularly and more usefully, v lies on a fixed cycle or path if and only if both L_n^+ and L_n^- are non-empty (where, again, n is the number of vertices in G). To see this, note that if v lies on a fixed cycle then every level of v contains a vertex of this cycle, and if v lies on a fixed path then it lies between two fixed cycles; the vertices of one are in the positive levels and those of the other are in the negative levels. Let L_i^+ and L_j^- be the levels of smallest cardinality index that contain a vertex of these cycles and note that every level $L_{i'}^+$, $i' \geq i$, and $L_{j'}^-$, $j' \geq j$, contains a vertex of one of the two cycles. Next note that if there is a vertex w in L_n^+, then there is a walk from w to v that contains one vertex from each L_i^+, $0 \leq i \leq n$. As the walk contains at least one vertex twice it includes a fixed cycle. If L_n^- is also non-empty then we can find a fixed cycle among the vertices of the negative levels of v so either v lies on one of these fixed cycles (which might be the same) or lies on a fixed path between them.

Proof of Claim 4 (and Proposition 1). Suppose the procedure described above is run on G, α, and has terminated. Note that a vertex that lies on a fixed cycle or path is certainly in S. We shall show that for each vertex $v \in V(G)$:

- either v lies on a fixed cycle or path (so is both fixed and in S),

- or v is neither fixed nor in S.

This will prove that $S = F_\alpha$, and also Proposition 1.

Fix a vertex v of G and consider the levels of v. We have observed that if L_n^+ and L_n^- are both non-empty, then v lies on a fixed cycle or a fixed path. Otherwise only finitely many of either the positive or negative levels (or both) are nonempty. We show that this means we can recolour v, and hence v is not fixed. So assume that $L_t^+ = \varnothing$ or $L_t^- = \varnothing$ for some $t > 0$. Without loss of generality, let us assume $L_t^+ = \varnothing$. Thus each vertex $u \in L_{t-1}^+$ can be recoloured with $\alpha(u) + 1$ (mod 3). Then each vertex $w \in L_{t-2}^+$ can be recoloured with $\alpha(w) + 1$ (mod 3), and so on, until v is recoloured. The fact that v can be recoloured implies it is not in S: every vertex in S has a pair of differently coloured neighbours so no vertex in S can be the first to be recoloured. $\qquad\square$

Claim 4 allows us to find F_α and F_β. If $F_\alpha \neq F_\beta$, or if there is a vertex $v \in F_\alpha$ such that $\alpha(v) \neq \beta(v)$, then there is no path from α to β. The algorithm outputs F_α, F_β and, if necessary, v.

Henceforth we assume that condition (C1) is satisfied, so $F_\alpha = F_\beta$ and for all $v \in F_\alpha$, $\alpha(v) = \beta(v)$.

If $F_\alpha \neq \varnothing$, we construct, from G, a new graph G^f by identifying, for $i = 1, 2, 3$, all vertices in S_i and denoting the newly created vertex by f_i. In other words:

$$V(G^f) = (V(G) \setminus F_\alpha) \cup \{f_1, f_2, f_3\}, \quad \text{and}$$

$$E(G^f) = \bigcup_{i=1,2,3} \{ uf_i \mid u \in V(G) \setminus F_\alpha \text{ and there is a } v \in S_i \text{ with } uv \in E(G) \}$$

$$\cup \{f_1 f_2, f_1 f_3, f_2 f_3\} \cup \{ uv \in E(G) \mid u, v \in V(G) \setminus F_\alpha \}.$$

If G has no fixed vertices with respect to α, then we set $G^f = G$.

It is convenient to assume that all edges are retained so that G and G^f have the same edge set. Since S_1, S_2, S_3 are independent sets (they are subsets of the colour classes of the colouring α), this means G^f is a graph with possibly multiple edges, but no loops.

Let α^f and β^f be the colourings induced on G^f by α and β. It is easy to observe that if $F_\alpha \neq \varnothing$,

- f_1, f_2 and f_3 are the only vertices of G^f that are fixed in α^f and β^f, and
- f_1, f_2 and f_3 induce a (fixed) 3-cycle in G^f.

Note that if α and β belong to the same component of $\mathcal{C}_3(G)$, this component is isomorphic to the component of $\mathcal{C}_3(G^f)$ that contains α^f and β^f. Hence we have the following.

Claim 5.
There is a path from α to β in $\mathcal{C}_3(G)$ if and only if there is a path from α^f to β^f in $\mathcal{C}_3(G^f)$.

To prove Theorem 3, we shall prove the following claim.

Claim 6.
Two 3-colourings α^f and β^f of a graph G^f belong to the same component of $\mathcal{C}_3(G^f)$, if and only if

(C2′) for every cycle C in G^f we have $W(\overrightarrow{C}, \alpha^f) = W(\overrightarrow{C}, \beta^f)$.

Let us first establish that the claim implies the theorem, recalling that we are assuming condition (C1). Let $\overrightarrow{C}$ be an oriented cycle in G. In G^f, the oriented edges of $\overrightarrow{C}$ form a set of edge-disjoint oriented cycles. (Here we use the convention that all edges from G are retained in G^f.) Since these cycles contain the same edges as $\overrightarrow{C}$, similarly oriented, it is easy to see that the sum of the weights of these cycles is equal to $W(\overrightarrow{C}, \alpha)$. Thus if G^f, α^f, β^f satisfy (C2′), then G, α, β satisfy (C2).

Now, let $\overrightarrow{P}$ be an oriented path between fixed vertices in G. If the endvertices of P have the same colour, then the oriented edges of $\overrightarrow{P}$ again form a set of edge-disjoint oriented cycles in G^f, and (C2′) implies that $W(\overrightarrow{P}, \alpha) = W(\overrightarrow{P}, \beta)$. If the endvertices of P have a different colour, then we can suppose, without loss of generality, that the endvertices of P are coloured 1 and 2 and that $\overrightarrow{P}$ is oriented from the endvertex coloured 1 towards the endvertex coloured 2. That means that the union of the oriented edges of $\overrightarrow{P}$ and the edge $\overrightarrow{f_2 f_1}$ forms a set of oriented cycles in G^f. Since we have $w(\overrightarrow{f_2 f_1}, \alpha^f) = w(\overrightarrow{f_2 f_1}, \beta^f)$, (C2′) again implies that $W(\overrightarrow{P}, \alpha) = W(\overrightarrow{P}, \beta)$. We have shown that if G^f, α^f, β^f satisfy (C2′), then G, α, β satisfy (C3).

Conversely, if there is a cycle C in G^f such that $W(\overrightarrow{C}, \alpha^f) \neq W(\overrightarrow{C}, \beta^f)$, then this same cycle can be found in G or, if C intersects $\{f_1, f_2, f_3\}$, then there is a path between fixed vertices in G that has different weights under α and β. This shows that if G^f, α^f, β^f do not satisfy (C2′), then one of (C2) or (C3) fails for G, α, β.

Proof of Claim 6. To prove the claim we describe an algorithm that either finds a path from α^f to β^f in $\mathcal{C}_3(G^f)$, or finds a cycle C in G^f such that $W(\overrightarrow{C}, \alpha^f) \neq W(\overrightarrow{C}, \beta^f)$. The algorithm attempts to find a sequence of recolourings that transforms α^f into β^f. It maintains a set $F \subseteq V(G^f)$ such that the subgraph induced by F is connected and for each $v \in F$, the current colouring of v is $\beta^f(v)$. Initially, if F_α and F_β were not empty, we let $F = \{f_1, f_2, f_3\}$. Otherwise, we set $F = \varnothing$. We then try to increase the size of F one vertex at a time.

We show how to extend F if $F \neq V(G^f)$. If $F \neq \varnothing$, then choose a $v \notin F$ such that v is adjacent to a vertex $u \in F$. (This is possible by the assumption that G (and thus G^f) is connected.) If $F = \varnothing$, then we choose an arbitrary vertex v, and u does not exist. Suppose the current colouring is α'. If $\alpha'(v) = \beta^f(v)$, we can extend F to include v immediately. Otherwise, let us assume that $\alpha'(v) = 2$ and $\beta^f(v) = 3$. Note that this means that $\alpha'(u) = 1$ (if u exists), since $\alpha'(u) = \beta^f(u)$ and u is adjacent to v.

Now we attempt to find the positive levels of v (where the levels are defined with respect to α'). This is easily done algorithmically: $L_1^+(v)$ contains those neighbours of v coloured 3; $L_2^+(v)$ contains neighbours of vertices in $L_1^+(v)$ coloured 1, and so on. We stop if either

(L1)   we reach a level L_i^+ that is empty, or

(L2)   we find a level that contains a vertex $w \in F$.

Note that one of (L1) or (L2) must occur since v has only finitely many non-empty levels unless the levels contain a fixed cycle. Hence we eventually reach either a level that contains a vertex $w \in F$, or an empty level. If F is empty, then, of course, (L1) must occur.

If (L1) occurs, then we can recolour each vertex z in L_j^+, $j = i-1, i-2, \ldots, 0$, with $\alpha'(z) + 1$ (mod 3), starting with the highest level and working down. Thus, ultimately, v is recoloured 3 and we can now add v to F. If there are still vertices not in F, we repeat the procedure.

Suppose (L2) occurs. Then there is a path P from u to w coloured 1-2-3-1-2-3-$\cdots$-$\alpha'(w)$. Moreover, no internal vertex of P is in F. As u and w are in F, and F induces a connected subgraph, we can extend P to a cycle C using a path $Q = w \cdots u$ in F. We claim that $W(\overrightarrow{C}, \alpha') \neq W(\overrightarrow{C}, \beta^f)$, and hence the cycle C is an obstruction that shows that α' and β^f do not belong to the same component of $\mathcal{C}_3(G)$. Because α^f and α' belong to the same component of $\mathcal{C}_3(G)$, this cycle is also an obstruction showing that α^f and β^f do not belong to the same component of $\mathcal{C}_3(G)$.

To see that $W(\overrightarrow{C}, \alpha') \neq W(\overrightarrow{C}, \beta^f)$, choose the orientation $\overrightarrow{C}$ so that the edge uv is oriented from u to v. The weight of $\overrightarrow{C}$ is the sum of the weights of $\overrightarrow{P}$ and $\overrightarrow{Q}$ (taking $\overrightarrow{P}$ and $\overrightarrow{Q}$ to have the same orientation as $\overrightarrow{C}$). Let $W(\overrightarrow{Q}, \alpha') = k$. As vertices in F are coloured alike in α' and β^f, $W(\overrightarrow{Q}, \beta^f) = k$. Let p be the number of edges in P. Then $W(\overrightarrow{P}, \alpha') = p$, since each edge has weight $+1$. But $W(\overrightarrow{Q}, \beta^f) < p$, since $w(\overrightarrow{uv}, \beta^f) = -1$. Thus we find $W(\overrightarrow{C}, \beta^f) < k + p = W(\overrightarrow{C}, \alpha')$.

All the above was done under the assumption that $\alpha'(v) = 2$ and $\beta^f(v) = 3$. In the cases $\alpha'(v) = 3$, $\beta^f(v) = 1$ and $\alpha'(v) = 1$, $\beta^f(v) = 2$ we do exactly the same, again using the positive levels $L_i^+(v)$. In the other three cases, we follow the same steps, but now using the negative levels $L_i^-(v)$ of v. This completes the proof of the claim. □

This completes the proof of Theorem 3. □

Note that if α and β are in the same component of $\mathcal{C}_3(G)$ and G has n vertices, the algorithm in the proof of Claim 6 will use at most $\frac{1}{2}n(n+1)$ recolouring steps: each time a vertex is added to F, we may have to recolour all vertices not in F at most once. This proves Theorem 2. □

Note also that the procedure which finds the fixed vertices of a given 3-colouring, the construction of G^f from G, and the algorithm in the proof of Claim 6 can clearly be performed in polynomial time. This proves Theorem 1. □

Using Theorem 3, it is in fact possible to give an alternative proof of Theorem 1. We describe a modification of the algorithm that proves Theorem 3 which, given a graph G together with two 3-colourings α and β as input, decides whether or not α and β belong to the same component of $\mathcal{C}_3(G)$ by simply checking conditions (C1), (C2) and (C3).

As before, we first check whether condition (C1) is satisfied. We proceed by assuming it (else the algorithm terminates), and then transform the instance G, α, β into the instance G^f, α^f, β^f. We have already observed that these operations can be performed in polynomial time.

Having seen that condition (C2') is equivalent to conditions (C2) and (C3), we now claim that condition (C2') can be verified in polynomial time. (Note that without having first proved Theorem 1 this is not immediately obvious since the graph G^f may contain an exponential number of cycles.) In order to prove this claim, we need to recall some definitions.

Let H be a connected graph with n vertices and m edges. It is well-known that (the edge sets of) the cycles of H form a vector space over the field $\mathbb{F}_2 = \{0, 1\}$, where addition is symmetric difference. This vector space is known as the *cycle space* of H. Given any spanning tree T of H, adding any of the $m - n + 1$ edges $e \in E(H) \setminus E(T)$ to T yields a unique cycle C_e of H. These $m - n + 1$ cycles are called the *fundamental cycles* of T, and they form a basis of the cycle space of H known as a *cycle basis*. In fact, it is easy to prove that for every cycle C,

$$C = \sum_{e \in E(C) \setminus E(T)} C_e,$$

where addition is as in the vector space $(\mathbb{F}_2)^m$. We refer the reader to [5, Section 1.9] for further details.

Lemma 4.
Let H be a connected graph with n vertices and m edges. Let α be a 3-colouring of H, T a spanning tree of H, and $\{ C_e \mid e \in E(H) \setminus E(T) \}$ the set of fundamental cycles of T. Then for any cycle C in H, $W(\overrightarrow{C}, \alpha)$ is determined by the values of $W(\overrightarrow{C_e}, \alpha)$, for all $e \in E(H) \setminus E(T)$.

Proof. Let C be any cycle in H, and write $C = \sum\limits_{e \in E(C) \setminus E(T)} C_e$, with addition as in the vector space $(\mathbb{F}_2)^m$. Choose an orientation $\overrightarrow{C}$ for C. For each $e \in E(C) \setminus E(T)$, orient the fundamental cycle C_e so that e has the same orientation in $\overrightarrow{C}$ and in $\overrightarrow{C_e}$. We claim that

$$W(\overrightarrow{C}, \alpha) = \sum_{e \in E(C) \setminus E(T)} W(\overrightarrow{C_e}, \alpha), \tag{1}$$

where now addition is the normal addition of integers. We prove (1) by counting edge-weight contributions to both sides of the equation.

Let $e = uv$ be an edge of C, with orientation $\overrightarrow{uv}$ on $\overrightarrow{C}$. Clearly, $w(\overrightarrow{uv}, \alpha)$ is counted exactly once on the left-hand side (LHS) of (1). To count the contributions that e makes to the right-hand side (RHS) of (1), we distinguish two cases, according to whether or not e is an edge of T. If $e \notin E(T)$, then the definition of C_e and the choice of the orientation $\overrightarrow{C_e}$ immediately gives that e contributes exactly the weight $w(\overrightarrow{uv}, \alpha)$ to the RHS. If $e = uv \in E(T)$, we claim that it appears oriented as $\overrightarrow{uv}$ exactly one more time than it appears oriented as $\overleftarrow{uv}$ in the cycle expansion of $\overrightarrow{C}$. Note that uv is a cut-edge of T and, as such, its removal splits T into two subtrees T_u and T_v, with $u \in V(T_u)$ and $v \in V(T_v)$. We also have $V(T_u) \cup V(T_v) = V(H)$. Let $f \in E(C) \setminus E(T)$ with $uv \in E(C_f)$. Then, in fact, we can take $f = xy$ with $x \in V(T_u)$ and $y \in V(T_v)$. If f has the orientation $\overrightarrow{xy}$ in $\overrightarrow{C}$, then it has the same orientation in $\overrightarrow{C_f}$, and hence the edge uv has the orientation $\overleftarrow{uv}$ in C_f. The reverse is the case if f has the orientation $\overleftarrow{xy}$ in $\overrightarrow{C}$. Going along the oriented edges of the cycle $\overrightarrow{C}$, we have the same number of edges $\overrightarrow{xy}$ with $x \in V(T_u)$ and $y \in V(T_v)$, as we have edges between $V(T_u)$ and $V(T_v)$ going in the other direction. But since uv is one of the edges of the first count, we get exactly one more edge $xy \neq uv$ of $\overrightarrow{C}$ with $x \in V(T_u)$ and $y \in V(T_v)$ oriented as $\overleftarrow{xy}$ than oriented the other way round. That means that in the sum on the RHS of (1) we have exactly one more contribution of the form $w(\overrightarrow{uv}, \alpha)$ than of the form $w(\overleftarrow{uv}, \alpha)$.

Now suppose that $e = uv$ is not an edge of C. Clearly this edge makes no contribution to the LHS of the equation. Again, to count the contributions of this edge to the RHS of the expression, we distinguish the cases where e is an edge of T and where it is not. If $e = uv \in E(T)$, we can argue as in the preceding paragraph, to see that this time, in the RHS we have exactly the same times a contribution of the form $w(\overrightarrow{uv}, \alpha)$ as of the form $w(\overleftarrow{uv}, \alpha)$. Hence the net contribution to the RHS is zero. Lastly, if $e \notin E(T)$, it makes no contribution either, since the fundamental cycle C_e to which it corresponds does not appear in the cycle expansion of C.

This completes the proof of the lemma. $\qquad\square$

Lemma 4 gives an obvious algorithm to check if the (reduced) instance G^f, α^f, β^f satisfies condition (C2'), running in polynomial time.

4   Shortest Paths between 3-Colourings

Once again, throughout this section we assume that G is some fixed connected graph. We use the notation and terminology from the previous section.

We have seen that if α and β are 3-colourings of G that are in the same component of $\mathcal{C}_3(G)$, then they are at distance $O(|V(G)|^2)$. In this section we show that this bound on the distance between 3-colourings is of the right order. More precisely, we prove that there exists a class of instances G', α, β such that α and β are connected and at distance $\Omega(|V(G')|^2)$ in $\mathcal{C}_3(G')$.

Before doing so, we prove that in the case that α and β are connected and $F_\alpha \neq \varnothing$ (so $F_\beta = F_\alpha$, and for all $v \in F_\alpha$ we have $\alpha(v) = \beta(v)$), the algorithm described in the previous section finds a shortest path from α to β in $\mathcal{C}_3(G)$.

Theorem 7.
Let α and β be two 3-colourings of a connected graph G that are in the same component of $\mathcal{C}_3(G)$, and suppose that $F_\alpha \neq \varnothing$. Then the algorithm described in Section 3 finds a shortest path between α and β.

Proof. Our algorithm in fact finds a path from α^f to β^f in G^f, but, as we observed earlier, the relevant components of the two colour graphs are isomorphic. For a 3-colouring γ of G^f, denote by $\mathcal{C}_\gamma$ the component of $\mathcal{C}_3(G^f)$ containing γ. Note that, by assumption of connectedness, $\mathcal{C}_{\alpha^f} = \mathcal{C}_{\beta^f}$.

Recall that G^f has exactly three fixed vertices f_1, f_2, f_3 for the colourings α^f and β^f.

Let γ be any 3-colouring in $\mathcal{C}_{\beta^f}$. For any vertex v of G^f, let $\overrightarrow{P}$ be an oriented path from f_1 to v. Then the *height of v in γ* is defined as

$$h(v, \gamma) = |W(\overrightarrow{P}, \gamma) - W(\overrightarrow{P}, \beta^f)|.$$

We need to prove that this definition is independent of the choice of P. If there are two oriented paths $\overrightarrow{P_1}$ and $\overrightarrow{P_2}$ from f_1 to v, then, noting that their union is a set of oriented cycles and applying Lemma 2, we have $W(\overrightarrow{P_1}, \gamma) - W(\overrightarrow{P_2}, \gamma) = W(\overrightarrow{P_1}, \beta^f) - W(\overrightarrow{P_2}, \beta^f)$. A little rearrangement gives $|W(\overrightarrow{P_1}, \gamma) - W(\overrightarrow{P_1}, \beta^f)| = |W(\overrightarrow{P_2}, \gamma) - W(\overrightarrow{P_2}, \beta^f)|$.

Now let γ and δ be adjacent 3-colourings in $\mathcal{C}_{\beta^f}$ and let w be the unique vertex on which they differ. Note that this means that all neighbours of w are coloured the same as one another, and all these neighbours are coloured the same in both γ and δ. Let $\overrightarrow{P}$ be an oriented path from f_1 to some vertex v and let us consider how the height of v changes as γ is recoloured to δ. If w is not on $\overrightarrow{P}$, then clearly $h(v, \gamma) = h(v, \delta)$. We know $w \neq f_1$, as f_1 is fixed. If w is an internal vertex of $\overrightarrow{P}$, then the sum of the weights of the two edges of $\overrightarrow{P}$ incident with w is zero for both γ and δ, so again $h(v, \gamma) = h(v, \delta)$. If $w = v$, then the sign of the weight of the edge of $\overrightarrow{P}$ incident with v changes as we recolour. So in this last case we have $|h(v, \gamma) - h(v, \delta)| = 2$.

Note that finding a path from α^f to β^f is equivalent to finding a sequence of recolourings that reduces the height of every vertex v from $h(v, \alpha^f)$ to zero. In the previous paragraph we saw that each time we recolour, only the height of the vertex being recoloured changes, and it either increases or decreases by 2. So if we can find a sequence of recolourings that always reduces the height of the vertex being recoloured, we will have found a shortest path. We show that this is indeed what the algorithm of Claim 6 does.

Recall that the algorithm starts with a set $F = \{f_1, f_2, f_3\}$, and then it repeatedly adds vertices v to F, where v has a neighbour $u \in F$. To add v to F, the vertices in either all its positive levels or all its negative levels are recoloured before v itself is recoloured. Assume that we are in the case that to recolour v all positive levels need to be recoloured; the other case is proved in the same way. Let y be a vertex that is about to be recoloured at some stage in this process (this can be v itself, or any of the vertices in the positive levels of v). We must show that its height will be reduced. Let γ and δ be the colourings before and after y is recoloured. Let $\overrightarrow{Q}$ be an oriented path from u to y that contains one vertex from each nonnegative level of v. So if there are k edges in $\overrightarrow{Q}$, then $W(\overrightarrow{Q}, \gamma) = k$. Thus $W(\overrightarrow{Q}, \delta) = k - 2$ since the edge of $\overrightarrow{Q}$ incident with y has its weight changed from 1 to -1 when y is recoloured. Let $\overrightarrow{R}$ be an oriented path from f_1 to u containing only vertices in F, and let $\overrightarrow{P}$ be the union of $\overrightarrow{R}$ and $\overrightarrow{Q}$.

Since the colourings β^f, γ, δ agree on F, we have $W(\overrightarrow{R}, \beta^f) = W(\overrightarrow{R}, \gamma) = W(\overrightarrow{R}, \delta)$. We also know that $w(\overrightarrow{uv}, \beta^f) = -1$, and since $\overrightarrow{Q}$ has k edges, this means

$$W(\overrightarrow{Q}, \beta^f) \leq k - 2 = W(\overrightarrow{Q}, \delta) < k = W(\overrightarrow{Q}, \gamma).$$

From this we can derive

$$h(y,\gamma) = |W(\overrightarrow{P},\gamma) - W(\overrightarrow{P},\beta^f)| = |W(\overrightarrow{Q},\gamma) - W(\overrightarrow{Q},\beta^f)|$$
$$= W(\overrightarrow{Q},\gamma) - W(\overrightarrow{Q},\beta^f) = k - W(\overrightarrow{Q},\beta^f)$$

and, similarly,

$$h(y,\delta) = k - 2 - W(\overrightarrow{Q},\beta^f).$$

So indeed, every recolouring according to the lemma, reduces the height of the vertex being recoloured, completing the proof of the theorem. □

Next let us observe that if there are no fixed vertices, the algorithm may find a much longer path. For example, consider two colourings of a path that differ only on an endvertex v and its neighbour: $\alpha = $ 1-2-3-1-2-3-1-$\cdots$-1-2-3 and $\beta = $ 2-1-3-1-2-3-1-$\cdots$-1-2-3. The algorithm starts by setting $F = \varnothing$, and then chooses an arbitrary first vertex to start the recolouring. If that first vertex is v, then the algorithm will start by recolouring every vertex on the path. But clearly it is possible to get from α to β via only three recolourings. The reader should check that this shortest number of recolourings would be obtained if the first choice of the algorithm were any vertex other than v.

We believe that the algorithm from Section 3 will also be able to find a shortest path between two 3-colourings without fixed vertices.

Conjecture 8.
Let α and β be two 3-colourings of a connected graph G that are in the same component of $\mathcal{C}_3(G)$, and suppose that $F_\alpha = F_\beta = \varnothing$. For $v \in V(G)$, let $T(v)$ be the number of recolourings required by the algorithm in Section 3 when the algorithm starts by adding v to $F = \varnothing$. Then the length of the shortest path between α and β is equal to $\min\limits_{v \in V(G)} T(v)$.

We now proceed to the construction of a class of instances G, α, β where, for each G, α and β are connected and at distance $\Omega(|V(G)|^2)$ in $\mathcal{C}_3(G)$. For $N \in \mathbb{N}$, define the graph G_N as the graph consisting of a 3-cycle with an attached path of length N. More precisely, let

$$V(G_N) = \{f_1, f_2, f_3\} \cup \{v_1, v_2, \ldots, v_N\}, \quad \text{and}$$
$$E(G_N) = \{f_1 f_2, f_2 f_3, f_1 f_3\} \cup \{f_3 v_1, v_1 v_2, v_2 v_3, \ldots, v_{N-1} v_N\}$$

Let α_N be the 3-colouring of G_N given by $\alpha_N(f_i) = i$, for $i = 1, 2, 3$, and where the vertices $v_1, v_2, \ldots, v_N$ are coloured 1-2-3-1-2-3-$\cdots$. Similarly, let β_N be the 3-colouring of G_N given by $\beta_N(f_i) = i$, for $i = 1, 2, 3$, and where the vertices $v_1, v_2, \ldots, v_N$ are coloured 2-1-3-2-1-3-$\cdots$.

Theorem 9.
Let $N \in \mathbb{N}$ and let G_N, α_N, β_N be as described above. Then the 3-colourings α_N and β_N of G_N are connected and at distance $\frac{1}{2} N(N+1) = \Omega(|V(G_N)|^2)$ in $\mathcal{C}_3(G_N)$.

Proof. It is clear that G_N, α_N and β_N satisfy conditions (C1), (C2) and (C3). Therefore, by Theorem 3, α_N and β_N are connected in $\mathcal{C}_3(G_N)$.

As in the proof of Theorem 7, we consider heights of vertices. For any vertex v of G_N, let $\overrightarrow{P}$ be an oriented path from f_3 to v, noting that $f_3 \in F_{\alpha_N}$. Define the height of v in α_N as $h(v, \alpha_N) = |W(\overrightarrow{P}, \alpha_N) - W(\overrightarrow{P}, \beta_N)|$.

We have seen in the proof of Theorem 7 that finding a shortest path from α_N to β_N is equivalent to finding a sequence of recolourings that reduces the height of every vertex to zero, and that, with each recolouring, we reduce the height of the recoloured vertex by 2, while the height of all other vertices remains the same. This enables us to calculate the distance between α_N and β_N: we just need to calculate the height of all vertices in α_N.

First observe that $h(f_i, \alpha_N) = 0$, for $i = 1, 2, 3$. For $i = 1, \ldots, N$, let $\overrightarrow{P_i}$ be the oriented path from f_3 to v_i, and observe that $W(\overrightarrow{P_i}, \alpha_N) = i$ while $W(\overrightarrow{P_i}, \beta_N) = -i$. This means $h(v_i, \alpha_N) = |W(\overrightarrow{P_i}, \alpha_N) - W(\overrightarrow{P_i}, \beta_N)| = 2\,i$. We find that the distance between α_N and β_N is equal to $\frac{1}{2} \sum_{i=1}^{N} h(v_i, \alpha_N) = \sum_{i=1}^{N} i = \frac{1}{2} N (N + 1)$. Since G_N has $N+3$ vertices, we obtain that this distance is indeed $\Omega(|V(G_N)|^2)$.

$\square$

References

1. J. Billingham, R. Leese, and H. Rajaniemi, *Frequency reassignment in cellular phone networks*. Smith Institute Study Group Report (2005). Available via `http://www.smithinst.ac.uk/Projects/ESGI53/ESGI53-Motorola/Report/`.

2. P. Bonsma and L. Cereceda, *Finding paths between graph colourings: PSPACE-completeness and superpolynomial distances*. CDAM Research Report LSE-CDAM-2007-12 (2007). Available via `http://www.cdam.lse.ac.uk/Reports/Abstracts/cdam-2007-12.html`.
   Extended abstract in: *Mathematical Foundations of Computer Science 2007. Proceedings of 32nd International Symposium MFCS 2007*. Lect. Notes Comput. Sc. **4708** (2007), 738–749.

3. L. Cereceda, J. van den Heuvel, and M. Johnson, *Mixing 3-colourings in bipartite graphs*. CDAM Research Report LSE-CDAM-2007-06 (2007). Available via `http://www.cdam.lse.ac.uk/Reports/Abstracts/cdam-2007-06.html`.
   Extended abstract in: *Graph-Theoretic Concepts in Computer Science. Proceedings of 33rd International Workshop WG 2007*. Lect. Notes Comput. Sc. **4769** (2007), 166–177.

4. L. Cereceda, J. van den Heuvel, and M. Johnson, *Connectedness of the graph of vertex-colourings*. Discrete Math. **308** (2008), 913–919.

5. R. Diestel, *Graph Theory*. Springer-Verlag, Berlin, 3rd edition, 2005.

6. M. Jerrum, *A very simple algorithm for estimating the number of k-colourings of a low degree graph*. Random Structures Algorithms **7** (1995), 157–165.

7. M. Jerrum, *Counting, Sampling and Integrating: Algorithms and Complexity*. Birkhäuser Verlag, Basel, 2003.

An optimal algorithm for the k-fixed-endpoint path cover on proper interval graphs

George B. Mertzios and Walter Unger

Department of Computer Science
RWTH Aachen, Germany
{mertzios, quax}@cs.rwth-aachen.de

Abstract. In this paper we consider the k-fixed-endpoint path cover problem on proper interval graphs, which is a generalization of the path cover problem. Given a graph G and a set T of k vertices, a k-fixed-endpoint path cover of G with respect to T is a set of vertex-disjoint paths that covers the vertices of G, such that the vertices of T are all endpoints of these paths. The goal is to compute a k-fixed-endpoint path cover of G with minimum cardinality. We propose an optimal algorithm for this problem with runtime $O(n)$, where n is the number of intervals in G. This algorithm is based on the *Stair Normal Interval Representation (SNIR) matrix* that characterizes proper interval graphs. In this characterization, every maximal clique of the graph is represented by one matrix element; the proposed algorithm uses this structural property, in order to determine directly the paths in an optimal solution.

Keywords: proper interval graph, perfect graph, path cover, SNIR matrix, linear-time algorithm.

1 Introduction

A graph G is called an *interval graph*, if its vertices can be assigned to intervals on the real line, such that two vertices of G are adjacent if and only if the corresponding intervals intersect. The set of intervals assigned to the vertices of G is called a *realization* of G. If G has a realization, in which no interval contains another one properly, then G is called a *proper interval graph*. Proper interval graphs arise naturally in biological applications such as the physical mapping of DNA [1]. Several linear-time recognition algorithms have been presented for both graph classes in the literature [2,3,4,5]. These classes of graphs have numerous applications to scheduling problems, biology, VLSI circuit design, as well as to psychology and social sciences [6,7].

Several difficult optimization problems, which are NP-hard for general graphs [8], are solvable in polynomial time on interval and proper interval graphs. Some of them are the maximum clique, the maximum independent set [9,10], the Hamiltonian cycle (HC) and the Hamiltonian path (HP) problem [11]. A generalization of the HP problem is the path cover (PC) problem. That is, given a graph G, the goal is to find the minimum number of vertex-disjoint simple paths that cover all vertices of G. Except graph theory, the PC problem finds many applications in the area of database design, networks, code optimization and mapping parallel programs to parallel arcitectures [12,13,14,15].

The PC problem is known to be NP-complete even on the classes of planar graphs [16], bipartite graphs, chordal graphs [17], chordal bipartite graphs, strongly chordal graphs [18], as well as in several classes of inter-section graphs [19]. On the other hand, it is solvable in linear $O(n + m)$ time on interval graphs with n vertices and m edges [12]. For the greater class of circular-arc graphs there is an optimal $O(n)$-time approximation algorithm, given a set of n arcs with endpoints sorted [20]. The cardinal-ity of the path cover found by this approximation algorithm is at most one more than the optimal one. Several variants of the HP and the PC problems are of great interest. The simplest of them are the 1HP and 2HP problems, where the goal is to decide whether G has a Hamiltonian path with one, or two fixed endpoints respectively. Both problems are NP-hard for general graphs, as a generalization of the HP problem, while their complexity status remains open for interval graphs [21,22,23].

In this paper, we consider the k-fixed-endpoint path cover (kPC) prob-lem, which generalizes the PC problem in the following way. Given a graph G and a set T of k vertices, the goal is to find a path cover of G with minimum cardinality, such that the elements of T are endpoints of these paths. Note that the vertices of $V \setminus T$ are allowed to be endpoints of these paths as well. For $k = 1, 2$, the kPC problem constitutes a direct gener-alization of the 1HP and 2HP problems respectively. For the case, where the input graph is a cograph on n vertices and m edges, a linear $O(n+m)$ time algorithm has been recently presented, while its complexity status remains open for interval graphs [22].

We propose a runtime-optimal algorithm for the kPC problem on proper interval graphs with runtime $O(n)$, based on the zero-one *Stair Normal Interval Representation (SNIR) matrix* H_G that characterizes a proper interval graph G on n vertices [24]. In this characterization, every maximal clique is represented by one matrix element. It provides insight and may be useful for the efficient formulation and solution of several optimization problems. In particular, it has been used to solve in polynomial time the k-cluster problem on interval graphs [25], while this problem is NP-hard for chordal graphs [26]. In most of the practical applications, the intervals of the input are sorted by their right or left endpoints. Given such an interval realization of G, the proposed algorithm constructs at first in $O(n)$ time a particular perfect ordering of the vertices of G, which complies with the ordering of the vertices in the SNIR matrix H_G [24].

We introduce the notion of a *singular point* in a proper interval graph G on n vertices. An arbitrary vertex of G is called singular point, if it is the unique common vertex of two consecutive maximal cliques. Due to the special structure of H_G, we need to compute only $O(n)$ of its entries, in order to capture the complete information of this matrix. Based on this structure, the algorithm detects the singular points of G in time $O(n)$ and then it determines *directly* the paths in an optimal solution, using only the positions of the singular points. Namely, it turns out that every such path is a Hamiltonian path of a particular subgraph $G_{i,j}$ of G with two specific endpoints. Here, $G_{i,j}$ denotes the induced subgraph of the vertices $\{i, \ldots, j\}$ in the ordering of H_G. Since any algorithm for this problem has to visit at least all n vertices of G, this runtime is optimal.

Recently, while writing this paper, it has been drawn to our attention that another algorithm has been independently presented for the kPC problem on proper interval graphs with runtime $O(n + m)$ [23], where m is the number of edges of the input graph. This algorithm uses a greedy approach to augment the already constructed paths with connect/insert operations, by distinguishing whether these paths have already none, one, or two endpoints in T. The main advantage of the here proposed algorithm, besides its runtime optimality, is that an optimal solution is constructed directly by the positions of the singular points, which is a structural property of the investigated graph. Given an interval realization of the input graph G, we do not need to visit all its edges, exploiting the special structure of the SNIR matrix. Additionally, the representation of proper interval (resp. interval) graphs by the SNIR (resp. NIR) matrix [24] may lead to efficient algorithms for other open optimization problems, such as the 1HP, 2HP, or even kPC problem on interval graphs [21,22].

The paper is organized as follows. In Section 2 we recall the SNIR matrix of a proper interval graph. Furthermore, in Section 3 we present an algorithm for the 2HP, based on the SNIR matrix. This algorithm is used in Section 4, in order to derive an algorithm for the kPC problem on proper interval graphs with runtime $O(n)$. Finally, we discuss some conclusions and open questions for further research in Section 5.

2 The SNIR matrix

An arbitrary proper interval graph G on n vertices can be characterized by the *SNIR matrix* H_G, which has been introduced in [24]. This is the lower portion of the adjacency matrix of G, which uses a particular ordering of its vertices. In this ordering, the vertex with index i corresponds to the ith diagonal element of H_G. Every diagonal element has a chain of consecutive ones immediately below it, while the remaining entries of this column are zero. These chains are ordered in such a way that H_G has a stair-shape, as it is illustrated in Figure 2(a). We recall now the definitions of a stair and a pick of the SNIR matrix H_G [24].

Definition 1. *Consider the SNIR matrix H_G of the proper interval graph G. The matrix element $H_G(i,j)$ is called a* pick *of H_G, iff:*

1. *$i \geq j$,*
2. *if $i > j$ then $H_G(i,j) = 1$,*
3. *$H_G(i,k) = 0$, for every $k \in \{1, 2, \ldots, j-1\}$ and*
4. *$H_G(l,j) = 0$, for every $l \in \{i+1, i+2, \ldots, n\}$.*

Definition 2. *Given the pick $H_G(i,j)$ of H_G, the set*

$$S = \{H_G(k,\ell) : j \leq \ell \leq k \leq i\} \tag{1}$$

of matrix entries is called the stair *of H_G, which corresponds to this pick.*

Lemma 1 ([24]). *Any stair of H_G corresponds bijectively to a maximal clique of G.*

A stair of H_G can be recognized in Figure 2(a), where the corresponding pick is marked with a circle. Given an interval realization of G, the ordering of vertices in H_G can be computed in time $O(n)$ [24]. Furthermore, the picks of H_G can be also computed in time $O(n)$ during the construction of the ordering of the vertices, since every pick corresponds to the right endpoint of an interval in G [24]. Due to its stair-shape, the matrix H_G is uniquely determined by its $O(n)$ picks.

For an arbitrary vertex w of G, denote by $s(w)$ and $e(w)$ the adjacent vertices of w with the smallest and greatest index in this ordering respectively. Due to the stair-shape of H_G, the vertices $s(w)$ and $e(w)$ are the uppermost and lowermost diagonal elements of H_G, which belong to a common stair with w. Denote now the maximal cliques of G by $Q_1, Q_2, \ldots, Q_m$, $m \leq n$ and suppose that the corresponding pick to Q_i

is the matrix element $H_G(a_i, b_i)$, where $i \in \{1, \ldots, m\}$. Since the maximal cliques of G, i.e. the stairs of H_G, are linearly ordered, it holds that $1 \leq a_1 \leq \ldots \leq a_m \leq n$ and $1 \leq b_1 \leq \ldots \leq b_m \leq n$. Denote for simplicity $a_0 = b_0 = 0$ and $a_{m+1} = b_{m+1} = n+1$. Then, Algorithm 1 computes the values $s(w)$ and $e(w)$ for all vertices $w \in \{1, \ldots, n\}$, as it is illustrated in Figure 1. Since $m \leq n$, the runtime of Algorithm 1 is $O(n)$.

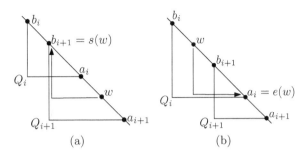

(a)   (b)

Fig. 1. The computation of $s(w)$ and $e(w)$.

Algorithm 1 Compute $s(w)$ and $e(w)$ for all vertices w

1: **for** $i = 0$ to m **do**
2:     **for** $w = a_i + 1$ to a_{i+1} **do**
3:         $s(w) \leftarrow b_{i+1}$
4:     **for** $w = b_i$ to $b_{i+1} - 1$ **do**
5:         $e(w) \leftarrow a_i$

The vertices $\{i, \ldots, j\}$ of G, where $i \leq j$, constitute a submatrix $H_{i,j}$ of H_G that is equivalent to the induced subgraph $G_{i,j}$ of these vertices. Since the proper interval graphs are hereditary, this subgraph remains a proper interval graph as well. In particular, $H_{1,n} = H_G$ is equivalent to $G_{1,n} = G$.

Definition 3. *A vertex w of $G_{i,j}$ is called* singular point *of $G_{i,j}$, if there exist two consecutive cliques Q, Q' of $G_{i,j}$, such that*

$$|Q \cap Q'| = \{w\} \tag{2}$$

Otherwise, w is called regular point *of $G_{i,j}$. The set of all singular points of $G_{i,j}$ is denoted by $S(G_{i,j})$.*

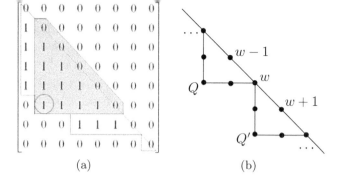

Fig. 2. (a) The SNIR matrix H_G, (b) a singular point w of $G_{i,j}$.

Proposition 1. *For every singular point w of $G_{i,j}$, it holds $i+1 \le w \le j-1$.*

Proof. Since w is a singular point of $G_{i,j}$, there exist two consecutive maximal cliques Q, Q' of $G_{i,j}$ with $Q \cap Q' = \{w\}$. Then, as it is illustrated in Figure 2(b), both Q and Q' contain at least another vertex than w, since otherwise one of them would be included in the other, which is a contradiction. It follows that $i+1 \le w \le j-1$.

Definition 4. *Consider a connected proper interval graph G and two indices $i \le j \in \{1, \ldots, n\}$. The submatrix $H_{i,j}$ of H_G is called two-way matrix, if all vertices of $G_{i,j}$ are regular points of it. Otherwise, $H_{i,j}$ is called one-way matrix.*

The intuition resulting from Definition 4 is the following. If $H_{i,j}$ is an one-way matrix, then $G_{i,j}$ has at least one singular point w. In this case, no vertex among $\{i, \ldots, w-1\}$ is connected to any vertex among $\{w+1, \ldots, j\}$, as it is illustrated in Figure 2(b). Thus, every Hamiltonian path of $G_{i,j}$ passes only once from the vertices $\{i, \ldots, w-1\}$ to the vertices $\{w+1, \ldots, j\}$, through vertex w. Otherwise, if $H_{i,j}$ is a two-way matrix, a Hamiltonian path may pass more than once from $\{i, \ldots, w-1\}$ to $\{w+1, \ldots, j\}$ and backwards, where w is an arbitrary vertex of $G_{i,j}$. The next corollary follows directly from Proposition 1.

Corollary 1. *An arbitrary vertex w of G is a regular point of the subgraphs $G_{i,w}$ and $G_{w,j}$, for every $i \le w$ and $j \ge w$.*

3 The 2HP problem on proper interval graphs

3.1 Necessary and sufficient conditions

In this section we solve the 2HP problem on proper interval graphs. In particular, given two fixed vertices u, v of a proper interval graph G, we provide necessary and sufficient conditions for the existence of a Hamiltonian path in G with endpoints u and v. An algorithm with runtime $O(n)$ follows directly from these conditions, where n is the number of vertices of G.

Denote by $2HP(G, u, v)$ this particular instance of 2HP on G. Since G is equivalent to the SNIR matrix H_G and since this matrix specifies a particular ordering of its vertices, we identify w.l.o.g. the vertices of G with their indices in this ordering. Observe at first that if G is not connected, then there is no Hamiltonian path at all in G. On the other hand, if G has only two vertices u, v and G is connected, there exists trivially a Hamiltonian path with u and v as endpoints. Thus, we assume in the following that G is connected and $n \geq 3$. The next Theorems 1 and 2 provide necessary and sufficient conditions for the existence of a Hamiltonian path with endpoints u, v in a connected proper interval graph G.

Theorem 1. *Let G be a connected proper interval graph and u, v be two vertices of G, with $v \geq u + 2$. There is a Hamiltonian path in G with u, v as endpoints if and only if the submatrices $H_{1,u+1}$ and $H_{v-1,n}$ of H_G are two-way matrices.*

Proof. Suppose that $H_{1,u+1}$ is an one-way matrix. Then, due to Definition 4, $G_{1,u+1}$ has at least one singular point w. Since $G_{1,u+1}$ is connected as an induced subgraph of G, Proposition 1 implies that $2 \leq w \leq u$.

In order to obtain a contradiction, let P be a Hamiltonian path in G with u and v as its endpoints. Suppose first that for the singular point w it holds $w < u$. Then, due to the stair-shape of H_G, the path P has to visit w in order to reach the vertices $\{1, \ldots, w-1\}$. On the other hand, P has to visit w again in order to reach v, since $w < v$. This is a contradiction, since P visits w exactly once. Suppose now that $w = u$. The stair-shape of H_G implies that u has to be connected in P with at least one vertex of $\{1, \ldots, u-1\}$ and with at least one vertex of $\{u+1, \ldots, n\}$. This is also a contradiction, since u is an endpoint of P. Therefore, there exists no such path P in G, if $H_{i,u+1}$ is an one-way matrix. Similarly, we obtain that

Algorithm 2 Construct a Hamiltonian path P in G with u, v as endpoints

---

1: $t \leftarrow u; P \leftarrow \{u\}$
2: **while** $t > 1$ **do**
3:     $p \leftarrow s(t)$ {the adjacent vertex of t with the smallest index}
4:     $P \leftarrow P \circ p; t \leftarrow p$
5: **while** $t < v - 1$ **do**
6:     **if** $(t + 1) \notin P$ **then**
7:         $P \leftarrow P \circ (t + 1); t \leftarrow t + 1$
8:     **else**
9:         **if** $(t + 2) \notin P \cup \{v\}$ **then**
10:         $P \leftarrow P \circ (t + 2); t \leftarrow t + 2$
11: **while** $t < n$ **do**
12:     $p \leftarrow e(t)$ {the adjacent vertex of t with the greatest index}
13:     $P \leftarrow P \circ p; t \leftarrow p$
14: **while** $t > v$ **do**
15:     **if** $(t - 1) \notin P$ **then**
16:         $P \leftarrow P \circ (t - 1); t \leftarrow t - 1$
17:     **else**
18:         $P \leftarrow P \circ (t - 2); t \leftarrow t - 2$
19: **return** P

---

there exists again no such path P in G, if $H_{v-1,n}$ is an one-way matrix. This completes the necessity part of the proof.

For the sufficiency part, suppose that both $H_{1,u+1}$ and $H_{v-1,n}$ are two-way matrices. Then, Algorithm 2 constructs a Hamiltonian path P in G having u and v as endpoints, as follows. In the while-loop of the lines 2-4 of Algorithm 2, P starts from vertex u and reaches vertex 1 using sequentially the uppermost diagonal elements, i.e. vertices, of the visited stairs of H_G. Since $H_{1,u+1}$ is a two-way matrix, P does not visit any two neighbored diagonal elements until it reaches vertex 1. In the while-loop of the lines 5-10, P continues visiting all unvisited vertices until vertex $v - 1$. Let t be the actual visited vertex of P during these lines. Since P did not visit any two neighbored diagonal elements in lines 2-4, at least one of the vertices $t+1$, $t+2$ have been not visited yet. Thus, always one of the lines 7 and 10 is executed.

Next, in the while-loop of the lines 11-13, P starts from vertex $v - 1$ and reaches vertex n using sequentially the lowermost diagonal elements of the visited stairs of H_G. During the execution of lines 11-13, since $H_{v-1,n}$ is a two-way matrix, P does not visit any two neighbored diagonal elements until it reaches vertex n. Finally, in the while-loop of the lines 14-18, P continues visiting all unvisited vertices until v. Similarly to the lines 5-10, let t be the actual visited vertex of P. Since P did not visit any two neighbored diagonal elements in lines 11-13, at least one of the vertices $t-1$, $t-2$ have been not visited yet. Thus, always one of the lines 16 and 18 is executed. Figure 3(a) illustrates the construction of such a Hamiltonian path by Algorithm 2 in a small example.

Theorem 2. *Let G be a connected proper interval graph with $n \geq 3$ vertices and u be a vertex of G. There is a Hamiltonian path in G with $u, u + 1$ as endpoints if and only if H_G is a two-way matrix and either $u \in \{1, n\}$ or the vertices $u - 1$ and $u + 2$ are adjacent.*

If the conditions of Theorems 1 and 2 are satisfied, Algorithm 2 constructs a Hamiltonian path with endpoints u, v. Algorithm 2 operates on every vertex of G at most twice. Thus, since all values $s(t)$ and $e(t)$ can be computed in time $O(n)$, its runtime is $O(n)$ as well. Figure 3 illustrates the construction of such a Hamiltonian path by Algorithm 2 in a small example, for both cases $v \geq u + 2$ and $v = u + 1$.

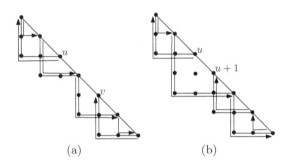

(a)   (b)

Fig. 3. The construction of the HP with endpoints u, v, where (a) $v \geq u + 2$, (b) $v = u + 1$.

3.2 The decision of 2HP in time $O(n)$

We can use now the results of Section 3.1 in order to decide in time $O(n)$ whether a given proper interval graph G has a Hamiltonian path P with two specific endpoints u, v and to construct P, if it exists. The values $s(w)$ and $e(w)$ for all vertices $w \in \{1, \ldots, n\}$ can be computed in time $O(n)$. Due to the stair-shape of H_G, the graph G is not connected if and only if there is a vertex $w \in \{1, \ldots, n-1\}$, for which it holds $e(w) = w$ and thus, we can check the connectivity of G in time $O(n)$. If G is not connected, then it has no Hamiltonian path at all. Finally, a vertex w is singular if and only if $e(w - 1) = s(w + 1) = w$ and thus, the singular points of G can be computed in $O(n)$.

Since the proper interval graphs are hereditary, the subgraphs $G_{1,u+1}$ and $G_{v-1,n}$ of G remain proper interval graphs as well. Thus, if G is connected, we can check in time $O(n)$ whether these graphs have singular points, or equivalently, whether $H_{1,u+1}$ and $H_{v-1,n}$ are two-way matrices. On the other hand, we can check in constant time whether the vertices $u - 1$ and $u + 2$ are adjacent. Thus, we can decide in time $O(n)$ whether G has a Hamiltonian path with endpoints u, v, due to Theorems 1 and 2. In the case of existence, Algorithm 2 constructs in time $O(n)$ the desired Hamiltonian path.

4 The kPC problem on proper interval graphs

4.1 The algorithm

In this section we propose Algorithm 3, which solves in $O(n)$ the k-fixed-endpoint path cover (kPC) problem on a proper interval graph G with n vertices, for any $k \leq n$. This algorithm uses the characterization of the 2HP problem of the previous section. We assume that for the given set $T = \{t_1, t_2, \ldots, t_k\}$ it holds $t_1 < t_2 < \ldots < t_k$. Denote also for simplicity $t_{k+1} = n + 1$.

Algorithm 3 Compute $C(G, T)$ for a proper interval graph G

1: **if** $G = \emptyset$ **then**
2:     **return** $\emptyset$
3: Compute the values $s(w)$ and $e(w)$ for every vertex w by Algorithm 1
4: $w \leftarrow 1$
5: **while** $w < n$ **do**
6:     **if** $e(w) = w$ **then** {G is not connected}
7:         $T_1 \leftarrow T \cap \{1, 2, \ldots, w\}$; $T_2 \leftarrow T \setminus T_1$
8:         **return** $C(G_{1,w}, T_1) \cup C(G_{w+1,n}, T_2)$
9:     $w \leftarrow w + 1$
10: **if** $k \leq 1$ **then**
11:     Call Algorithm 4
12: **if** $t_1 \in S(G)$ **then**
13:     $P_1 \leftarrow 1 \circ \ldots \circ t_1$
14:     **return** $\{P_1\} \cup C(G_{t_1+1,n}, T \setminus \{t_1\})$
15: Call Algorithm 5

Algorithm 3 computes an optimal path cover $C(G, T)$ of G. In lines 4-9, it checks the connectivity of G. If it is not connected, the algorithm computes in lines 7-8 recursively the optimal solutions of the first connected component and of the remaining graph. It reaches line 10 only if G is connected. In the case $|T| = k \leq 1$, Algorithm 3 calls Algorithm 4 as subroutine.

In lines 12-14, Algorithm 3 considers the case that G is connected, $|T| \geq 2$ and t_1 is a singular point of G. Then, Proposition 1 implies that $2 \leq t_1 \leq n-1$. Since no vertex among $\{1, \ldots, t_1 - 1\}$ is connected to any vertex among $\{t_1 + 1, \ldots, n\}$ and since $t_1 \in T$, an optimal solution must contain at least two paths. Thus, it is always optimal to choose in line 13 a path that visits sequentially the first t_1 vertices and then to compute recursively in line 14 an optimal solution in the remaining graph $G_{t_1+1,n}$. Algorithm 3 reaches line 15 if G is connected, $|T| \geq 2$ and t_1 is a regular point of G. In this case, it calls the following Algorithm 5 as subroutine.

Both Algorithms 4 and 5 compute by Algorithm 2 a Hamiltonian path of some subgraphs of G with two specific endpoints, if such a path exists. If the conditions of Theorems 1 and 2 are not satisfied, i.e. if there is no such Hamiltonian path in these subgraphs, Algorithm 2 returns "no" instead of a Hamiltonian path.

Algorithm 4 Compute $C(G,T)$, if G is connected and $|T| \leq 1$

1: **if** $k = 0$ **then**
2:     **return** $\{1 \circ 2 \circ \ldots \circ n\}$
3: **if** $k = 1$ **then**
4:     **if** $t_1 \in \{1, n\}$ **then**
5:         **return** $\{1 \circ 2 \circ \ldots \circ n\}$
6:     **else**
7:         $P_1 \leftarrow 2HP(G, 1, t_1)$
8:         $P_2 \leftarrow 2HP(G, t_1, n)$
9:         **if** $P_1 = $"no" **then**
10:             **if** $P_2 = $"no" **then**
11:                 **return** $\{1 \circ \ldots \circ t_1\} \cup \{(t_1 + 1) \circ \ldots \circ n\}$
12:             **else**
13:                 **return** $\{P_2\}$
14:         **else**
15:             **return** $\{P_1\}$

Algorithm 4 computes an optimal path cover $C(G,T)$ of G in the case, where G is connected and $|T| = k \leq 1$. If $k = 0$, then the optimal solution includes clearly only one path that visits sequentially the vertices $1, 2, \ldots, n$, since G is connected. Let now $k = 1$. If $t_1 \in \{1, n\}$, then the optimal solution is again the single path $\{1, 2, \ldots, n\}$. Otherwise, suppose that $t_1 \in \{2, \ldots, n-1\}$. In this case, a trivial path cover is that with the paths $\{1\circ\ldots\circ t_1\}$ and $\{(t_1+1)\circ\ldots\circ n\}$. This path cover is not optimal if and only if G has a Hamiltonian path P with $u = t_1$ as one endpoint. The other endpoint v of P lies either in $\{1, \ldots, t_1 - 1\}$ or in $\{t_1 + 1, \ldots, n\}$. If $v \in \{t_1 + 1, \ldots, n\}$, then H_{1,t_1+1} and $H_{v-1,n}$ have to be two-way matrices, due to Theorems 1 and 2. However, due to Definition 1, if $H_{v-1,n}$ is a two-way matrix, then $H_{n-1,n}$ is also a two-way matrix, since $H_{n-1,n}$ is a trivial submatrix of $H_{v-1,n}$.

Thus, if such a Hamiltonian path with endpoints t_1 and v exists, then there exists also one with endpoints t_1 and n. Similarly, if there exists a Hamiltonian path with endpoints $v \in \{1, \ldots, t_1 - 1\}$ and t_1, then there exists also one with endpoints 1 and t_1. Thus, Algorithm 4 calls $P_1 = 2HP(G, 1, t_1)$ and $P_2 = 2HP(G, t_1, n)$ in lines 7 and 8 respectively. If both outputs are "no", then $\{1\circ\ldots\circ t_1\}$ and $\{(t_1+1)\circ\ldots\circ n\}$ constitute an optimal solution. Otherwise, we return one of the obtained paths P_1 or P_2 in lines 15 or 13 respectively. Since the runtime of Algorithm 2 for the 2HP problem is $O(n)$, the runtime of Algorithm 4 is $O(n)$ as well.

In lines 5-9 and 12-14, Algorithm 3 separates G in two subgraphs and computes their optimal solutions recursively. Thus, since the computation of all values $s(w)$ and $e(w)$ can be done in $O(n)$ and since the runtime of Algorithms 4 and 5 is $O(n)$, Algorithm 3 runs in time $O(n)$ as well.

4.2 Correctness of Algorithm 5

The correctness of Algorithm 5 follows from the technical Lemmas 2 and 3. In these Lemmas, we assume an optimal solution C of G. Denote by P_i the path in C, which has t_i as endpoint and let e_i be its second endpoint. Observe that, if $e_1 = t_2$, then $P_1 = P_2$.

Lemma 2. *If* $\{1, \ldots, t_1 - 1\} \cap S(G) = \emptyset$, *then w.l.o.g.* $e_1 = t_2$.

Lemma 3. *If* $\{1, \ldots, t_1 - 1\} \cap S(G) \neq \emptyset$, *then w.l.o.g.* $e_1 = 1$.

Algorithm 5 considers in lines 1-10 the case that there are no singular points of G among $\{1, \ldots, t_1 - 1\}$. The proof of Lemma 2 implies for this case that $e_1 = t_2$ and, in particular that $P_1 = 2HP(G_{1,a}, t_1, t_2)$ for some $a \in \{t_2, \ldots t_3 - 1\}$. In order to maximize P_1 as much as possible, we choose the greatest possible value of a, for which $G_{1,a}$ has a Hamiltonian path with endpoints t_1, t_2. Namely, if G_{1,t_2+1} does not have such a Hamiltonian path, we set $a = t_2$ in line 3. Suppose now that G_{1,t_2+1} has such a path. In the case that there is at least one singular point of G among $\{t_2 +$

Algorithm 5 Compute $C(G,T)$, where G is connected, $|T| \geq 2$, $t_1 \notin S(G)$.

1: **if** $\{1, ..., t_1 - 1\} \cap S(G) = \emptyset$ **then** $\{e_1 = t_2\}$
2:    **if** $2HP(G_{1,t_2+1}, t_1, t_2) =$ "no" **then**
3:      $a \leftarrow t_2$
4:    **else**
5:      **if** $\{t_2 + 1, ..., t_3 - 1\} \cap S(G) \neq \emptyset$ **then**
6:        $a \leftarrow \min\{\{t_2 + 1, ..., t_3 - 1\} \cap S(G)\}$
7:      **else**
8:        $a \leftarrow t_3 - 1$
9:        $P_1 \leftarrow 2HP(G_{1,a}, t_1, t_2)$
10:       $C_2 \leftarrow C(G_{a+1,n}, T \setminus \{t_1, t_2\})$
11: **else** $\{e_1 = 1\}$
12:    **if** $2HP(G_{1,t_1+1}, 1, t_1) =$ "no" **then**
13:      $a \leftarrow t_1$
14:    **else**
15:      **if** $\{t_1 + 1, ..., t_2 - 1\} \cap S(G) \neq \emptyset$ **then**
16:        $a \leftarrow \max\{\{t_1 + 1, ..., t_2 - 1\} \cap S(G)\}$
17:      **else**
18:        $a \leftarrow t_2 - 1$
19:        $P_1 \leftarrow 2HP(G_{1,a}, 1, t_1)$
20:       $C_2 \leftarrow C(G_{a+1,n}, T \setminus \{t_1\})$
21: **return** $\{P_1\} \cup C_2$

$1, \ldots, t_3 - 1\}$, we set a to be this one with the smallest index among them in line 6. Otherwise, we set $a = t_3 - 1$ in line 8. Denote for simplicity $G_{1,n+1} = G$. Then, in the extreme cases $t_3 = t_2 + 1$ or $t_2 = n$, the algorithm sets $a = t_2 = t_3 - 1$.

Next, in lines 11-20, Algorithm 5 considers the case that there are some singular points of G among $\{1, \ldots, t_1-1\}$. Then, the proof of Lemma 3 implies that $e_1 = 1$ and, in particular that $P_1 = 2HP(G_{1,a}, 1, t_1)$, for some $a \in \{t_1, \ldots, t_2-1\}$. In order to maximize P_1 as much as possible, we choose the greatest possible value of a, for which $G_{1,a}$ has a Hamiltonian path with endpoints 1 and t_1. Namely, if G_{1,t_1+1} does not have such a Hamiltonian path, we set $a = t_1$ in line 13. Suppose now that G_{1,t_1+1} has such a path. In the case that there is at least one singular point of G among $\{t_1+1, \ldots, t_2-1\}$, we set a to be this one with the greatest index among them in line 16. Otherwise, we set $a = t_2 - 1$ in line 18. Note that in the extreme case $t_2 = t_1 + 1$, the algorithm sets $a = t_1 = t_2 - 1$.

The algorithm computes P_1 in time $O(n)$ in lines 9 and 19 respectively. Then, it computes recursively the optimum path cover C_2 of the remaining graph in lines 10 and 20 respectively and it outputs $\{P_1\} \cup C_2$. Since the computation of a 2HP by Algorithm 2 can be done in time $O(n)$, the runtime of Algorithm 5 is $O(n)$ as well.

5 Concluding remarks

In this article we proposed a simple algorithm for the k-fixed-endpoint path cover problem on proper interval graphs with runtime $O(n)$. Since any algorithm for this problem has to visit at least all n vertices of G, this runtime is optimal. The proposed algorithm is based on the characterization of proper interval graphs by the SNIR matrix. The complexity status of the k-fixed-endpoint path cover problem, as well as of 1HP and 2HP, on the general class of interval graphs remains an interesting open question for further research.

References

1. P. Hell, R. Shamir, and R. Sharan. A fully dynamic algorithm for recognizing and representing proper interval graphs. *SIAM J. Comput.*, 31(1):289–305, 2001.
2. W.L. Hsu. A simple test for interval graphs. In *WG '92: Proceedings of the 18th International Workshop on Graph-Theoretic Concepts in Computer Science*, pages 11–16, London, 1993. Springer-Verlag.
3. D. Corneil, H. Kim, S. Natarajan, S. Olariu, and A.P. Sprague. Simple linear time recognition of unit interval graphs. *Inform. Process. Lett.*, 55:99–104, 1995.
4. D.G. Corneil, S. Olariu, and L. Stewart. The ultimate interval graph recognition algorithm? In *SODA '98: Proceedings of the ninth annual ACM-SIAM symposium on Discrete algorithms*, pages 175–180, 1998.
5. B.S. Panda and S.K. Das. A linear time recognition algorithm for proper interval graphs. *Information Processing Letters*, 87(3):153–161, 2003.
6. M.C. Golumbic and A.N. Trenk. *Tolerance graphs*. Cambridge University Press, Cambridge, 2004.
7. A.V. Carrano. Establishing the order to human chromosome-specific DNA fragments. In A. D. Woodhead and B. J. Barnhart, editors, *Biotechnology and the Human Genome*, pages 37–50. Plenum Press, New York, 1988.
8. M.R. Garey and D.S. Johnson. *Computers and intractability: a guide to the theory of NP-completeness*. W.H. Freeman, San Francisco, 1979.
9. U.I. Gupta, D.T. Lee, and J.Y.T. Leung. Efficient algorithms for interval graphs and circular-arc graphs. *Networks*, pages 459–467, 1982.
10. Ju Yuan Hsiao and Chuan Yi Tang. An efficient algorithm for finding a maximum weight 2-independent set on interval graphs. *Inf. Process. Lett.*, 43(5):229–235, 1992.
11. M.S. Chang, S.L. Peng, and J.L. Liaw. Deferred-query - an efficient approach for problems on interval and circular-arc graphs (extended abstract). In *WADS*, pages 222–233, 1993.
12. S.R. Arikati and C.P. Rangan. Linear algorithm for optimal path cover problem on interval graphs. *Information Processing Letters*, 35(3):149–153, 1990.

13. G.S. Adhar and S. Peng. Parallel algorithms for path covering, hamiltonian path and hamiltonian cycle in cographs. In *International Conference on Parallel Processing*, volume 3, pages 364–365, 1990.

14. R. Lin, S. Olariu, and G. Pruesse. An optimal path cover algorithm for cographs. *Comput. Math. Appl.*, 30:75–83, 1995.

15. R. Srikant, R. Sundaram, K.S. Singh, and C.P. Rangan. Optimal path cover problem on block graphs and bipartite permutation graphs. *Theoretical Computer Science*, 115:351–357, 1993.

16. M.R. Garey, D.S. Johnson, and R.E. Tarjan. The planar hamiltonian circuit problem is np-comlete. *SIAM J. Comput.*, 5:704–714, 1976.

17. M.C. Golumbic. *Algorithmic Graph Theory and Perfect Graphs*, volume 57. Annals of Discrete Mathematics, Amsterdam, The Netherlands, 2004.

18. H. Müller. Hamiltonian circuits in chordal bipartite graphs. *Discrete Mathematics*, 156:291–298, 1996.

19. A.A. Bertossi and M.A. Bonucelli. Finding hamiltonian circuits in interval graph generalizations. *Information Processing Letters*, 23:195–200, 1986.

20. R.W. Hung and M.S. Chang. Solving the path cover problem on circular-arc graphs by using an approximation algorithm. *Discrete Applied Mathematics*, 154(1):76–105, 2006.

21. P. Damaschke. Paths in interval graphs and circular-arc graphs. *Discrete Mathematics*, 112:49–64, 1993.

22. K. Asdre and S.D. Nikolopoulos. A linear-time algorithm for the k-fixed-endpoint path cover problem on cographs. *Networks*, 50:231–240, 2007.

23. K. Asdre and S.D. Nikolopoulos. A polynomial solution to the k-fixed-endpoint path cover problem on proper interval graphs. In *18th International Conference on Combinatorial Algorithms (IWOCA'07)*, Newcastle, Australia, 2007.

24. G.B. Mertzios. A matrix characterization of interval and proper interval graphs. *Applied Mathematics Letters*, 21(4):332–337, 2008.

25. G.B. Mertzios. A polynomial algorithm for the k-cluster problem on interval graphs. *Discrete Mathematics*, 2008. doi:10.1016/j.disc.2007.12.030.

26. D.G. Corneil and Y. Perl. Clustering and domination in perfect graphs. *Discrete Applied Mathematics*, 9:27–39, 1984.

Loose Cover of Graphs by a Local Structure

Satoshi Fujita

Department of Information Engineering
Graduate School of Engineering, Hiroshima University
E-mail: fujita@se.hiroshima-u.ac.jp

Abstract. This paper introduces a new definition of embedding a local structure to a global network, called **loose cover** of graphs. We derive several basic properties on the notion of loose cover, which includes transitivity, maximality, and the computational complexity of finding a loose cover by paths and cycles. In particular, we show that the decision problem is in P if the given local structure is a path with three or less vertices, while it is NP-complete for paths consisting of six or more vertices.

1   Introduction

Let $G = (V_G, E_G)$ and $H = (V_H, E_H)$ be undirected graphs. A **loose cover** of G by H is a function f from V_G to V_H such that

$$\forall u \in V_G, \ N_H(f(u)) \subseteq f(N_G(u))$$

where $N_G(u)$ denotes the set of closed neighbors of u in G; i.e., $N_G(u) = \{v : \{u, v\} \in E_G\} \cup \{u\}$. In the following, we denote $G \sqsubseteq H$ if G can be loosely covered by H.

The notion of loose cover, which is newly introduced in the current paper, was motivated by a resource assignment problem on graphs, which is intuitively described as follows: Suppose that graph G models a physical network such as islands connected by regular lines, cities connected by highways, train stations connected by railroads, and so on. We want to locate facilities (e.g., hospital, school, and shopping mall) to each site in the network in "exactly one facility in each site" manner in such a way that several designated pairs of facilities (e.g., school and library) are assigned to neighboring sites directly connected by a link. Obviously, such a constraint can be represented by a graph H which has the set of facilities as the set of vertices, and two vertices of the graph are connected by an edge if and only if the corresponding facilities are requested to be connected by a link. A loose cover of G by H corresponds to an assignment of facilities to the sites of G to satisfy all constraints specified by H. Here, readers should notice that it is neither an embedding problem since it requests *every* site to be assigned a facility, nor a packing problem since it does not request a partitioning of G into a collection of disjoint H's.

The notion of loose cover is a generalization of **domatic partition** of graphs [4, 5] which has been investigated extensively during the past decades[1]. In fact, a domatic partition of G of size k is equivalent to a loose cover of G by a complete graph consisting of k vertices. It is known that a domatic partition of G with a maximum cardinality can be calculated efficiently if G is circular arc [2], interval [1, 9, 10, 12], or strongly chordal graph [11], although it is NP-complete for general graphs [4, 5]. In the literature, a slightly different definition of the notion of *cover by graphs* was introduced by other researchers; i.e., Kratochvil *et al.* defined a cover f of G by H in such a way that for all $u \in V_G$, $N_H(f(u)) = f(N_G(u))$. They derived several interesting results about the notion, which are completely difference from properties on "loose" cover of graphs. See [7, 8] for the details.

In this paper, we derive several properties on the notion of loose cover, which includes transitivity, maximality, and the computational complexity of finding a loose cover of G by paths and cycles. In particular, we show that the decision problem is in P if the given local structure is a path with three or less vertices, while it is NP-complete for paths consisting of six or more vertices. The remainder of this paper is organized as follows. We first derive several basic properties of loose cover in Section 2. We then clarify the computational complexity of finding a loose cover by small cycles and paths, in Sections 3 and 4, respectively. Finally, in Section 5, we conclude the paper with open problems.

2   General Properties

By definition, we immediately have the following claim.

Remark 1. For any G, it holds $G \sqsubseteq G$. In addition, if $G \sqsubseteq H$ and if we remove edges (but not vertices) from H to form H', and we add edges (but not vertices) to G to for G', then $G' \sqsubseteq H'$.

A sufficient condition for $G \sqsubseteq H$ is that the domatic number[2] of G is greater than or equal to $|V(H)|$. Of course, this condition is not necessary, since G can always be loosely covered by itself, even if the domatic number of G is smaller than $|V(G)|$.

---

[1] Domatic partition of G is a partition of the vertex set V_G such that each subset is a dominating set for G.

[2] The domatic number of G is the cardinality of a maximum domatic partition of G.

Remark 2 (Transitivity). If $G_1 \sqsubseteq G_2$ and $G_2 \sqsubseteq G_3$, then $G_1 \sqsubseteq G_3$.

Example 1. Let Q_n be a binary hypercube. It is clear that $Q_{i+1} \sqsubseteq Q_i$ for any $i \geq 1$. Hence by the above remark, we have $Q_i \sqsubseteq Q_j$ for any i, j such that $i > j$.

Example 2. Cycle $C_{i \times j}$ can be loosely covered by C_i and C_j, respectively. Hence if $G \sqsubseteq C_{i \times j}$, then $G \sqsubseteq C_i$ and $G \sqsubseteq C_j$. Note that in general, a reverse of this claim is not true.

Definition 1 (Maximal Loose Cover Graph). $H(\neq G)$ *is said to be a maximal loose cover graph (MLCG, for short) of G if $G \sqsubseteq H$ and there is no graph $H^*(\neq H, G)$ such that $G \sqsubseteq H^*$ and $H^* \sqsubseteq H$.*

Note that any graph has at least one MLCG since it is loosely covered by a graph consisting of a single vertex.

Example 3. An MLCG of K_n is a graph which is obtained from K_n by removing an edge. Let P_i denote a path consisting of i vertices. An MLCG of P_4 is P_2 (not P_3). Both C_6 and C_4 are MLCG of C_{12} (note that neither C_6 nor C_4 is a subgraph of C_{12}).

There exists a class of graphs which has exactly one MLCG. For example,

$$P_2 \to P_3 \to P_5 \to P_9 \to \cdots \to P_{2^i+1} \to P_{2^{i+1}+1} \to \cdots$$

is a class of such graphs (the unique MLCG for $P_{2^{i+1}+1}$ is P_{2^i+1}), where $P_i \to P_j$ implies that P_i is the unique MLCG of P_j, and

$$C_3 \to C_6 \to C_{12} \to C_{24} \to \cdots \to C_{3 \times 2^i} \to C_{3 \times 2^{i+1}} \to \cdots$$

is another class of such graphs (the unique MLCG for $C_{3 \times 2^{i+1}}$ is $C_{3 \times 2^i}$).

3   Complexity of Loose Cover by Cycles

If H is a cycle with $|V(G)|$ vertices, then the problem of testing if $G \sqsubseteq H$ is equivalent to the Hamiltonian cycle problem, which is well known to be NP-hard. Hence in general, the decision problem remains NP-hard even if we restrict H to the class of cycles. In this section, we derive several NP-hardness results when the cycle has a fixed size.

3.1 Loose Cover by C_{3i}

Let $C_3 = (\{a, b, c\}, \{\{a, b\}, \{b, c\}, \{c, a\}\})$ be a cycle with three vertices. A loose cover of G by C_3 is equivalent to the domatic partition of G into three subsets. In fact, we can regard function f as a labeling of V_G with three colors a, b, c, and in any loose cover by C_3, the subset of vertices which are assigned the same color corresponds to a dominating set for G. In a previous paper [3], we have proved that the problem of testing if the given cubic graph has a domatic partition of size three is NP-complete. Hence we have the following claim.

Proposition 1. *Given (cubic) graph G, the problem of testing if $G \sqsubseteq C_3$ is NP-complete.*

Corollary 1 (Loose Cover by C_{3i}). *For any $i \geq 1$, the problem of testing if $G \sqsubseteq C_{3i}$ is NP-complete.*

Note that for any $i \geq 1$, there is a graph G such that $G \sqsubseteq C_{3i}$ but $G \not\sqsubseteq C_{3j}$ for all $j > i$. (Imagine a graph consisting of two copies of C_{3i} connected by a bridge edge.)

3.2 Loose Cover by C_{4i}

Let $C_4 = (\{a, b, c, d\}, \{\{a, b\}, \{b, c\}, \{c, d\}, \{d, a\}\})$ be a cycle with four vertices. In this subsection, we show that the problem of testing if $G \sqsubseteq C_4$ is NP-hard. In the following, we illustrate a transformation of 3SAT to the decision problem. Let $U = \{u_1, u_2, \ldots, u_n\}$ and $C = \{c_1, c_2, \ldots, c_m\}$ be sets of variables and clauses, respectively, in an instance of 3SAT.

Components corresponding to variables First, we associate each variable $u_i \in U$ with a cycle of size $4 \max\{N_i^+, N_i^-\}$ as shown in Figure 1, where N_i^+ (resp. N_i^-) denotes the number of occurrences of the positive (resp. negative) literal of variable u_i in the instance of 3SAT. Note that in this cycle, every white vertex is connected with two neighbors; i.e., one is painted deep gray and the other is painted light gray. Vertices in the cycle are numbered from 1 to $4 \max\{N_i^+, N_i^-\}$. In the following construction, we associate positive (resp. negative) literals of a variable to every $(4i + 1)$st (resp. $(4i + 3)$rd) vertex in the cycle corresponding to the variable, and linearly connect all the first vertices using "adaptors" as illustrated in the figure. In the following, without loss of generality, we assume that colors a and c represent true and false values of each variable, respectively.

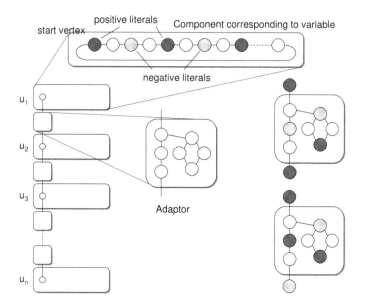

Fig. 1. A component corresponding to a variable, and the way of connecting those n components (head vertices of all components are linearly connected using "adaptors").

The right hand side of the figure illustrates two possible colorings of an adaptor; the upper coloring will be used to connect two variables assigned the same truth value (i.e., a and a, or c and c) and the lower coloring will be used to connect two variables assigned different truth values (i.e., a and c).

Components corresponding to clauses For each clause $c_j \in C$, we associate a component shown in Figure 2 (a). The "top" vertex in each component is connected to a unique vertex, called the root, as shown in Figure 3 (b). We fix the color assigned to the root to c, representing value "false." Three vertices indicated by "literals" in a component are identified with literal vertices of the corresponding variables appearing in the clause (i.e., the vertex is connected with other literal vertices of the same vertex as illustrated in Figure 1). For example, if $c_j = (u_1 + \overline{u_2} + u_3)$, then three literal vertices in the component are identified with a positive literal of variable u_1 (in a cycle corresponding to u_1), a negative literal of variable u_2 (in a cycle corresponding to u_2), and a positive literal of variable u_3 (in a cycle corresponding to u_3), respectively.

Since we are assuming that any literal vertex is assigned colors a or c, there could be four possible ways of coloring the component, as shown

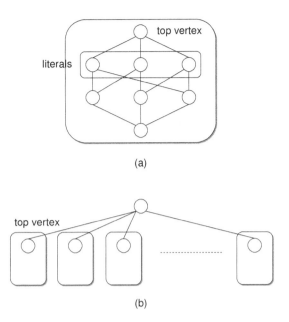

(a)

(b)

Fig. 2. (a) A component corresponding to a clause ("literal" vertices are identified with the literal vertices corresponding to the literals in the clause), and (b) the connection between those m components (top vertices are directly connected to the root vertex).

in Figure 3. Three patterns except for the lower right correspond to a satisfiable assignment for clause c_j with one, two, or three supporting literals for the "truth" of clause c_j, respectively. Note that in all of the three patterns, the condition of loose cover by C_4 is locally satisfied. The lower right pattern corresponds to an unsatisfiable assignment, which violates the loose cover by C_4. In fact, the white vertex connected to the root has four neighbors, and in the lower right pattern, those neighbors should be painted with the same color, i.e., it does not correspond to a loose cover by C_4.

Proposition 2 (Loose Cover by C_4). *The problem of testing if $G \sqsubseteq C_4$ is NP-complete.*

Corollary 2 (Loose Cover by C_{4i}). *For any $i \geq 1$, the problem of testing if $G \sqsubseteq C_{4i}$ is NP-complete.*

4 Complexity of Loose Cover by Paths

In this section, we assume that G contains no isolated vertex, without loss of generality. If $H = P_2$, i.e., if the given local structure is a path

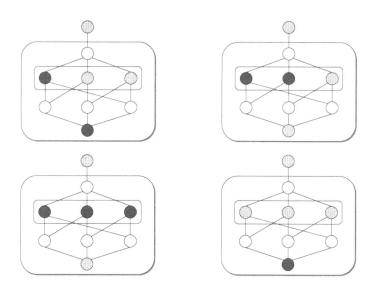

Fig. 3. Coloring of a component corresponding to a clause.

consisting of two vertices, the problem of deciding whether or not a given graph can be loosely covered by H is trivially solvable in linear time (consider a maximal independent set U of $G = (V_G, E_G)$, and assign color a to vertices in U and color b to vertices in $V_G - U$).

4.1 Loose Cover by P_3

In this section, we show that the decision problem is still contained in P if the length of the path becomes three. The proof of the NP-completeness of the case for P_6 is given in the next subsection.

Let $P_3 = (\{a, b, c\}, \{\{a, b\}, \{b, c\}\})$. In this subsection, we show that the problem of testing if $G \sqsubseteq P_3$ is in P.

We first try to remove "pendant" vertices of degree one from G. Let V_p be the set of vertices of degree one, and V_q be the set of vertices adjacent with a vertex in V_p. Note that any vertex in V_p can not be assigned color b, and any vertex in V_q must be assigned color b. Thus, if there is a vertex $u \in V_q$ such that one adjacent vertex of u is in V_p and all of the other adjacent vertices belong to V_q, then G can *never* be loosely covered by P_3. On the other hand, if there is no such vertex u in V_q, we may contract all vertices in V_q into a single vertex, and may remove vertices in V_p from G, since a "consistent" coloring of the vertices in V_p is always possible if the remaining graph is loosely covered by P_3. Let G' be the resultant graph. Note that every vertex in G' except for the contracted vertex, has a degree of two or more. To remember that the contracted vertex must be colored by b, which will be considered in later steps of the algorithm, we put a "mark" on the contracted vertex in G'. In the following, we will show that *if such a resultant G' exists, then it can always be loosely covered by P_3*.

The basic idea of the proof is to reduce G' into a graph $\hat{G}$ such that an appropriate two-edge-coloring of $\hat{G}$ corresponds to a loose cover of G by P_3, and to solve the edge coloring problem in polynomial time. Concrete procedure for the reduction is described as follows:

procedure REDUCTION

Step 1: Given $G' = (V', E')$, we try to find a collection of paths and cycles covering G'. More concretely, after letting $W := \emptyset$ and $A := \emptyset$, we repeat the following steps until $W = V'$: 1) Find a cycle consisting of vertices in $V' - W$, or find a path consisting of vertices in $V' - W$ which connects two vertices in W. 2) Add vertices (resp. edges) in the identified component to W (resp. A). Note that we do not add chord edges of the identified component to A. If the marked vertex is isolated at this time (i.e., if the degree of the marked vertex is one in G'), we reserve a minimum path connecting the marked vertex with a cycle in G', and will color the reserved path (including the marked vertex) after completing the coloring of the remaining vertics. Let $G'' = (V', A)$ be the resultant subgraph of G'. Note that any two vertices in G'' of degree three or more are not directly connected by an edge; i.e., they are connected by a path containing at least one intermediate vertex of degree two.

Step 2: Given graph G'', we repeatedly remove simple paths consisting of at least two intermediate vertices as follows: 1) Let $\{u_1, u_2, \ldots, u_k\}$ be a maximal set of vertices which induces a path in G'', where $k \geq 4$, degrees of vertices u_1 and u_k are both greater than two in G'', and the degree of u_i is two for $2 \leq i \leq k - 1$. 2) Remove vertices $u_2, u_3, \ldots, u_{k-1}$ from G''. Note that the removal of such intermediate vertices does not affect the loose cover property by P_3, since if the number of intermediate vertices of a path is at least two (i.e., if $k \geq 4$), then we can consistently assign colors to the intermediate vertices in such a way that the condition of loose cover is not violated, for any assignment of colors to the terminal vertices u_1 and u_k.

Step 3: The resultant graph G''' is a bipartite graph consisting of two vertex sets X and Y, such that X is a set of vertices of degree at least three, and Y is a set of vertices of degree two. Without loss of generality, we may assume that a marked vertex is contained in X (if any), since otherwise, we may regard a set of closed neighbors of the marked vertex (i.e., vertex set consisting of the marked vertex and its two neighbors) as an imaginary marked vertex. Let $\hat{G} = (X, E_X)$ be a graph, which is obtained from G''' by connecting any two vertices at distance two in G''' by an edge.

Note that if graph $\hat{G} = (X, E_X)$ has a two-edge-coloring such that every vertex in X is incident on at least one edge with color a and at least one edge with color c, then we immediately have a loose cover of G''' by P_3, by simply assigning color b to the vertices in X and for the vertices in Y, by assigning the same color with the corresponding edge in $\hat{G}$. A procedure to try to find such an edge coloring of $\hat{G}$ is described as follows:

procedure EDGE_COLOR

Step 1: Let u be the marked vertex (if G' contains no marked vertex, then let u be an arbitrary vertex in $\hat{G}$). For each $i \geq 0$, let X_i be the set of vertices in X at distance i from u; i.e., $X_0 = \{u\}$, X_1 is the set of vertices adjacent with u, and so on. Then, an initial edge coloring of $\hat{G}$ is determined as follows: For each edge $e \in E_X$, if it connects vertices in X_i and X_j, where $i \leq j$, and if i is even, then color it by a, and otherwise, color it by c.

Step 2: If the initial coloring satisfies the required condition, then complete the edge coloring of $\tilde{G}$, and terminate. Otherwise, go to Step 3.

Step 3: Repeat the following operation until the required condition is satisfied for all vertices in $\hat{G}$ besides root u (in the following, we use terms "parent" and "child" in a natural meaning; i.e., parent of v is an adjacent vertex of v which is closer to u than v):

1. Let v be a vertex at which the required condition is not satisfied. By construction, if $v \in X_i$, then all of its adjacent vertices in $\hat{G}$ must be contained in $X_{i\,1}$.
2. Select an arbitrary incident edge e of v (connecting to a parent vertex), and "flip" the color of the edge; i.e., color a is changed to color c, and vise versa.

3. If such a flipping violates the required condition at the parent, then conduct a similar modification towards the direction of the root vertex.

Such a propagation of modifications will always be terminated (terminate at the root vertex, in the worst case). After completing necessary updates for all vertices except for u, we will have an edge coloring such that the required condition is satisfied for all vertices except for the root.

If the required condition is not satisfied at vertex u, then we can obtain a loose cover of G from the edge coloring of $\hat{G}$, in the following manner: Without loss of generality, let us assume that all edges incident on u in $\hat{G}$ is assigned color a. If it is a marked vertex, then we may simply assign color c to the pendant vertices removed in the first step of procedure REDUCTION. If it is not a marked vertex, then we may modify the coloring of the vertices around u in the following manner (recall that in the default transformation, vertices in X are assigned color b and the other vertices are assigned color a or c): 1) Let w be a child vertex of u in $\hat{G} = (X, E_X)$ and let e be the edge connecting u and w. 2) Modify the color of the vertex in G corresponding to edge e from a to b, and assign colors a and c to vertices u and w, respectively.

The correctness of REDUCTION and EDGE_COLOR is clear from the description. In addition, the running time of the overall algorithm is bounded by $O(|V|^2)$. Thus, we have the following theorem.

Theorem 1. *The problem of testing if $G \sqsubseteq P_3$ is in P.*

4.2 NP-Completeness

Lemma 1 (Loose Covering by P_6). *The problem of testing if $G \sqsubseteq P_6$ is NP-complete.*

Proof. Let $P_6 = (\{1, 2, 3, 4, 5, 6\}, \{\{j, j+1\} : 1 \le j \le 5\})$. To prove the NP-completeness of the problem, we use a reduction from 3SAT similar to the proof for C_4. As a component corresponding to a variable, we use a cycle of size $4 \max\{N_i^+, N_i^-\}$, in which every $2j^{th}$ vertex corresponds to a positive or negative literal and is connected with two vertices connected by an edge. (Hence literal vertices of a variable are assigned colors "2 and 4" or "3 and 5".) As for a component corresponding to a clause, we use a graph consisting of six vertices $\{a_1, a_2, a_3, b_1, b_2, b_3\}$ connected by the following edges:

$$\{\{a_1, a_2\}, \{a_2, a_3\}\} \cup \{\{a_1, b_j\} : j = 1, 2, 3\}$$

where b_1, b_2, and b_3 are literal vertices, and vertices a_1, a_3 are connected with the root vertex, say r, which is connected with a leaf vertex. Without loss of generality, we associate colors 4 and 2 with values "true" and "false," respectively, and fix the color of the root to 2. In order to loose cover the resulting graph by P_5, it must hold

$$(f(a_1), f(a_2), f(a_3)) = (1, 2, 3), (3, 2, 1), (3, 6, 5), \text{ or } (5, 6, 3),$$

and in any of the cases, at least one vertex contained in $\{b_1, b_2, b_3\}$ must be assigned color 4; i.e., value "true." Hence the theorem follows.

Theorem 2 (Loose Covering by P_i for $i \geq 7$). *For any $i \geq 7$, the problem of testing if $G \sqsubseteq P_i$ is NP-complete.*

Proof. Let $P_i = (\{1, 2, \ldots, i\}, \{\{j, j+1\} : 1 \leq j \leq i-1\})$. Again, we use a reduction similar to the proof for C_4. As a component for a variable, we use a cycle of size $2(i-3) \max\{N_i^+ N_i^-\}$, in which every $(i \times j)$th vertex corresponds to a positive or negative literal and is connected with a leaf vertex. As for a component corresponding to a clause, we use a graph consisting of $i+1$ vertices $\{a_1, a_2, \ldots, a_i, b_1, b_2, b_3\}$ and with the following set of edges:

$$\{\{x, a_1\}, \{x, a_{i\ 2}\} : x \in \{b_1.b_2, b_3\}\} \cup \{\{a_j, a_{j+1}\} : 1 \leq j \leq i-3\}$$

where b_1, b_2, and b_3 are literal vertices, and vertices $a_1, a_{i\ 2}$ are connected with the root vertex, say r, which is connected with a leaf vertex. Without loss of generality, we associate color 2 with the positive value, and fix the color assigned to the root to $i - 1$. Note that since $i \geq 5$, vertex set $\{b_1, b_2, \ldots, b_{i\ 1}\}$ contains at least one vertex which has no connection with vertices in $\{r, b_1, b_2, b_3\}$. Hence in order to cover the resulting graph by P_i, each clause must contain at least one literal assigned the positive value 2. Hence the theorem follows.

5 Concluding Remarks

Open problems are listed below:

- Find a polynomial time algorithm for $H = P_4$ and P_5.
- Give an NP-completeness proof for $H = C_5$.
- Examine computational complexity of the decision problem for a special class of graphs such as perfect graphs and trees.

References

1. A. A. Bertossi. On the domatic number of interval graphs. *Information Processing Letters*, 28(6):275–280, August 1988.
2. M. A. Bonuccelli. Dominating sets and domatic number of circular arc graphs. *Discrete Applied Mathematics*, 12:203–213, 1985.
3. S. Fujita, M. Yamashita, T. Kameda, "A Study of r-Configurations – A Resource Assignment Problem on Graphs," *SIAM J. Discrete Math.*, 13(2):227–254 (2000).
4. T. W. Haynes, S. T. Hedetniemi, and P. J. Slater. *Fundamentals of Domination in Graphs*. Marcel Dekker, Inc., 1998.
5. T. W. Haynes, S. T. Hedetniemi, and P. J. Slater. *Domination in Graphs: Advanced Topics*. Marcel Dekker, Inc., 1998.
6. H. Kaplan and R. Shamir. The domatic number problem on some perfect graph families. *Information Processing Letters*. 49:51–56, 1994.
7. J. Kratochvil, A. Proskurowski, J. Telle. "Covering regular graphs," Journal of Combinatorial Theory, Series B, 71(1):1–16 (1997).
8. J. Kratochvil, A. Proskurowski, J. Telle. "Complexity of Graph Covering Problems," Nordic Journal of Computing, 5(3):173–195 (1998). *Workshop on Graph-Theoretic Concepts in Computer Science*, 1994, pages 93–105.
9. T. L. Lu, P. H. Ho, and G. J. Chang. The domatic number problem in interval graphs. *SIAM J. Disc. Math.*, 3:531–536, 1990.
10. G. K. Manacher and T. A. Mankus. Finding a Domatic Partition of an Interval Graph in Time $O(n)$. *SIAM J. Disc. Math.*, 9(2):167-172, 1996.
11. S. L. Peng and M. S. Chang. A Simple Linear Time Algorithm for the Domatic Partition Problem on Strongly Chordal Graphs, *Information Processing Letters*. 43:297–300, 1992.
12. A. Srinivasa Rao and C. P. Rangan. Linear algorithm for domatic number problem on interval graphs. *Information Processing Letters*, 33(1):29–33, October 1989.

On Conditional Covering Problem

S. Balasubramanian, S. Harini, and C. Pandu Rangan

Department of Computer Science and Engineering,
IIT Madras, Chennai India 600036
balu2901@gmail.com,harini@cse.iitm.ernet.in,
rangan@iitm.ernet.in

Abstract. The Conditional Covering Problem (CCP) aims to locate facilities on a graph, where the vertex set represents both the demand points and the potential facility locations. The problem has a constraint that each vertex can cover only those vertices that lie within its *covering radius* and no vertex can cover itself. The objective of the problem is to find a set that minimizes the sum of the facility costs required to cover all the demand points. An algorithm for CCP on paths was proposed by Horne and Smith (Networks, 46(4):177185, 2005). We show that their algorithm is incorrect and further present a correct $O(n^3)$ algorithm for the same. We also propose an $O(n^2)$ algorithm for the CCP on paths when all vertices are assigned unit costs and further extend this algorithm to interval graphs without an increase in time complexity.
Keywords: Conditional Covering Problem, Facility Location, Total Domination Problem, Paths, Interval Graphs.

1 Introduction

Consider a simple, undirected, connected graph $G = (V, E)$. The vertex set V denotes the set of sites / demand points that must be covered by some facility and facilities must be located only on the vertices of the graph. Each vertex $i \in V$ is associated with a positive cost of locating a facility at i, denoted by $c(i)$. Each edge (i, j) is associated with a positive length e_{ij}. Let D_{ij} be the shortest distance between i and j. Each vertex $i \in V$, is associated with a positive *covering radius* R_i such that, when we locate a facility at i, all the vertices within a distance of R_i from i (except i itself) are covered by i, i.e. all the vertices in the set $\{j | D_{ij} \leq R_i \text{ and } j \neq i\}$ are covered(No facility can cover itself). A set $S \subset V$ covers all the vertices in the set $\{j | \exists i \in S \text{ and } D_{ij} \leq R_i\}$.

The objective of the Conditional Covering Problem, CCP, is to minimize the sum of the costs of the facilities required to cover all the vertices in V. A *Conditional Covering set* or CCS, is a set of vertices which covers all the vertices of the graph. The cost of a set S of vertices is given by $c(S) = \sum_{i \in S} c(i)$. The CCP aims to find a CCS S, that minimizes $c(S)$.

In this paper, we give a counter example for the algorithm given in [5] for solving CCP on paths and provide a corrected $O(n^3)$ algorithm for the same in Section 2. Additionally, when all the vertices are assigned the same cost (taken to be unity without loss of generality), we give an $O(n^2)$ algorithm to solve the CCP on paths in Section 3 and show how to extend this algorithm to interval graphs without any increase in time-complexity in Section 4. Since, the algorithm proposed in [5] to solve CCP on paths is incorrect, it can be seen that the extension of the algorithm to solve the CCP in star-graphs [5] and trees given in [4] is also incorrect.

The CCP arises as an underlying graph theoretic problem in many real life facility location problems. Consider the problem of locating rescue centers in a district. Each potential site is associated with a *covering radius* and a cost of constructing a facility there. Also no rescue center can help the site at which it is located in case of a calamity at that site. Hence every facility should be covered by another facility. This problem can be modeled as CCP on a graph. CCP can similarly be used to model distribution centers and demand points where transhipment between distribution centers is required [2]. In [1], an application involving the strategic location of facilities having minimum and maximum coverage radii such as missile defense systems is discussed.

The CCP is a generalized version of the Total Dominating Set problem. Let $A \subset V$ and $B \subset V$ be two sets of vertices. A dominates B if every vertex in B -A is adjacent to at least one vertex in A; A totally dominates B if every vertex in B is adjacent to at least one vertex in A. A is called a dominating set of the graph if A dominates V and is called a totaly dominating set if it totally dominates V. The Total Dominating Set problem is a special case of the CCP when $R_i = 1$, $\forall i$ and all the edge lengths e_{ij} and facility location costs $c(i)$ are unity. Since the problem of finding the minimum Total Dominating Set in general graphs has been proved to be NP-Complete in the strong sense [9], it follows that CCP is NP-Complete on general graphs.

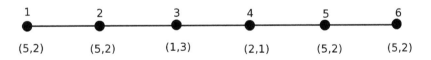

Fig. 1. Diagram to depict the counter example. $(c(i), R_i)$ for each vertex i is given below the vertex. Here we assign all edge weights to be one.

The CCP was first introduced by Moon and Chaudhry in [7] where an integer programming model for this problem was proposed and linear programming relaxation methods were applied to them. In [2], the authors consider seven greedy heuristics for solving CCP and provide computational results for the same. In [8], Moon and Papayanopoulos discuss a slight variation of CCP on tree graphs. In this problem, each demand point has a specific radius such that a facility has to be located in that radii. In [1] Smith et al. present an $O(n^2)$ algorithm for CCP on paths when the *covering radius* is uniform for all the vertices and arbitrary positive costs are assigned to vertices. They also present an $O(n)$ time algorithm when the *covering radius* is uniform and cost is unity for all vertices of the path. In [5] Horne and Smith extend the $O(n^2)$ algorithm in [1] to the case when vertices are assigned arbitrary *covering radius*.

Here, we show that the algorithm in [5] is incorrect. Let $P=123\ldots n$ be a path. The authors use the following equation to find the cost of the optimal CCS.

$$z^* = \underset{i \in V, g(i)=n}{Min} \; p(i)$$

where z^* is the cost of the optimal CCS, $g(i) = \mathrm{Max}\{k|D_{ik} \le R_i\}$ is the highest indexed vertex that i can cover i.e. the upper-reach of i and $p(i)$ is the cost required to optimally cover vertices 1 through i, by locating a facility at i and placing no facilities at vertices $i+1$, $i+2$, $\ldots n$. Consider the path in *Figure* 1. Their algorithm will consider the $p(i)$ values for all the vertices that have an upper-reach equal to n and choose the minimum among them. In this case, vertices 3, 5 and 6 have an upper-reach equal to n i.e. $g(3) = g(5) = g(6) = 6$. Hence the algorithm will compute the values $p(3)$, $p(5)$, $p(6)$ and declare the minimum among them as the cost of optimal solution to CCP. The algorithm used by them to compute $p(i)$ values, will compute these values as $p(3) = p(5) = 6$, $p(6) = 8$ and hence the cost of the optimal solution will be declared as 6. However, we can observe that the set $\{3,4\}$ is the optimal CCS with a cost of three.

Fig. 2. Diagram to depict the counter example when all the vertex costs are constant. R_i for each vertex i is given below the vertex. The value given below each edge denotes the edge weight. The optimal solution is $\{3,4\}$ but the algorithm in [5] will not even output a CCS.

Their algorithm assumes that in any optimal CCS S, the highest indexed vertex in S will always cover the vertex n. However, in the above example shown by *Figure 1*, where the optimal CCS is $\{3, 4\}$, the highest indexed vertex among $\{3, 4\}$ is 4 and still 4 does not cover 6. Instead 6 is covered by 3. Similar techniques are used by them for solving CCP in extended star and trees in [5, 4] and consequently these algorithms also fail due to similar reasons. We also note that their algorithms are incorrect even when all the vertices have unit costs as shown in *Figure 2*.

2 Conditional Covering Problem on Paths

Let P be any path graph where $V = \{1, 2, \ldots, n\}$ and $E = \{(i, i+1) | 1 \leq i \leq n - 1\}$. The vertices are assigned arbitrary positive costs and the edges are assigned arbitrary positive lengths. Let P_i denote the path $123\ldots i$.

In this section, we give a polynomial time algorithm for solving the CCP on paths.

2.1 Preliminaries

From now on, we refer any two vertices i and j satisfying $i < j$ as, i being to the *left* of j and j being to the right of i. Let (a, b) denote all the vertices in the path between the vertices indexed a and b not including a and b. Let $[a, b]$ denote all the vertices in the path between the vertices indexed a and b including a and b.

We borrow the following terminologies from [5]. For every vertex of the path, we define the *upper-reach* $g(i)$ of a vertex i as the largest indexed vertex lying within the *covering radius* of i i.e. $g(i) = \text{Max}\{k \in V | D_{ik} \leq R_i\}$ and *lower-reach* of a vertex i, given by $h(i)$, as defined as the smallest indexed vertex lying within the *covering radius* of i i.e. $h(i) = \text{Min}\{k | D_{ki} \leq R_i\}$.

The *protected cost* for vertex $i \in V$, denoted by $p(i)$, is the minimum cost to place a facility at vertex i and cover vertices 1 through i, with no facilities placed at vertices $i + 1, \ldots, n$. (The protected cost for vertex 1 is taken to be infinity as it is not possible to cover 1 as per the requirements of protected cost. Also whenever it is not possible to meet the requirements of the protected cost for a particular vertex, we take it to be infinity). The *unprotected cost* $u(i)$ for vertex $i \in V$ is the minimum cost to locate a facility at vertex i and cover vertices 1 through $i - 1$, with no facilities placed at vertices $i + 1, \ldots, n$. For the unprotected cost, vertex i is not necessarily required to be covered by smaller indexed vertices.

Further, we use the following terminologies from [5] to compute $p(i)$ and $u(i)$ $\forall i \in V$.

In order to calculate the protected cost for a vertex i, all vertices between 1 and $i-1$ that can cover vertex i are identified in two separate sets as follows.

$$GA_1(i) = \{k \in V : k+1 \leq i \leq g(k) \text{ and } h(i) > k\} \forall i \in V$$
$$GB_1(i) = \{k \in V : k+1 \leq i \leq g(k) \text{ and } h(i) \leq k\} \forall i \in V$$

The set $GA_1(i)$ includes all vertices between 1 and $i-1$ that cover vertex i, but are not covered by vertex i.
The set $GB_1(i)$ includes all vertices between 1 and $i-1$ that cover vertex i, and are covered by vertex i.

In order to calculate the unprotected cost for vertex i, all vertices $k \in V$ that can, along with i, cover vertices $k+1$ through $i-1$, but such that k does not cover i, are identified in two separate sets as follows.

$$GA_2(i) = \{k \in V : k+1 \leq h(i) \leq g(k)+1 \leq i\} \forall i \in V$$
$$GB_2(i) = \{k \in V : g(k)+1 \leq i \text{ and } h(i) \leq k\} \forall i \in V$$

The set $GA_2(i)$ includes all vertices k between 1 and $i-1$ such that i and k do not cover each other, but do cover all vertices $k+1$ through $i-1$.
The set $GB_2(i)$ includes all vertices between 1 and $i-1$ that are covered by vertex i, but do not themselves cover vertex i.

2.2 Computing the optimal CCS

We give a property of the optimal solutions to CCP using the following lemma

Lemma 1. *Consider the CCP for any path P_i. Let r be any vertex such that $g(r) = i$ and let S^* be any optimal CCS of P_i. If $r \in S^*$, then atmost one vertex $v \in (r, i]$ will be in S^*.*

Proof. Assume on the contrary that there exists an optimal CCS S' such that $r \in S'$ and, two vertices j and j' between r and i are in S'. Assume without loss of generality that $h(j') > h(j)$. All the vertices to the right of j' till i are covered by r $(g(r) = i)$ and all vertices to the left of j' till $h(j')$ are covered by j. Now, it can be clearly seen that the facility at j' does not cover any vertex that is not covered by r and j put together. Hence even with the removal of the facility at j' all the vertices of P_i continue to remain covered and consequently we have a solution of lesser cost which is a contradiction. $\square$

For computing the optimal CCS, we define the following terms:

Let $\alpha(k, i) = \text{Min}\{h(k), h(i)\}-1$. Let $Z_j^*(i)$ denote the cost of optimally covering all the vertices in $[1, i]$ by using vertices from $[1, j]$. We define $Z_j^*(i)$ only when $j \geq i$. Note that $Z_1^*(1)$ is taken to be infinity. We can use the following recursion to compute $Z_j^*(i)$:

$$Z_j^*(i) = Min\left\{Z_i^*(i), \min_{k\in[i+1,j]:\ h(k)\leq i}\{c(k) + Z_i^*(h(k)-1)\}\right\} \quad (1)$$

Similar to the argument given in Lemma 1, we can see that, atmost one vertex can be chosen from $[i+1,j]$ for optimally covering $[1,i]$ with $[1,j]$. Hence, we have just two cases, namely

1. No vertex is chosen from $[i+1,j]$: In this case, we have to cover all the vertices in $[1,i]$ by locating facilities only in $[1,i]$. The optimal cost for doing this is $Z_i^*(i)$
2. Exactly one vertex k is chosen from $[i+1,j]$: All the vertices to the right of $h(k)-1$ are covered by the facility at k. So, we have to cover the vertices in $[1,h(k)-1]$ by locating facilities only in $[1,i]$. The optimal cost for doing this is $Z_i^*(h(k)-1)$

Both the above cases have been accounted for in (1)
Clearly $Z_n^*(n)$(cost of optimally covering vertices $[1,n]$ using vertices $[1,n]$.) is the cost of the optimal solution to CCP. For computing $Z_n^*(n)$, we can use the following equation:

$$Z_n^*(n) = \min_{i\in V:\ g(i)=n}\left\{p(i), \min_{k\in(i,n+1):\ h(k)\leq i\ \text{and}\ g(k)<n}\{c(k) + u(i)\}\right\} \quad (2)$$

Consider all the vertices with an upper-reach equal to n (i.e. $g(i) = n$). Certainly, a facility must be located in one of these vertices for n to get covered. Because of Lemma 1, it follows that if we locate a facility at any vertex i for which $g(i) = n$, we can locate atmost one facility to the right of i. Hence, we have the following two cases,

1. We choose to locate no facility to the right of i : In this case, we must cover all the vertices in $[1,i]$ by locating facilities only in $[1,i-1]$ along with the facility already located at i. The optimal cost for doing this task is precisely $p(i)$.
2. We choose to locate exactly one facility k to the right of i (with $h(k) \leq i$ because otherwise there is no purpose in choosing k) : We restrict the facility k to be such that $g(k) < n$ because, the solution with $g(k) = n$ would have been accounted by the previous case i.e. the case where no facility is placed to the right of k and $g(k) = n$ and this would have been one of the solutions considered in the previous case. So, in the present case, we have to cover all vertices in $[1,\alpha(k,i)]$ by locating facilities in $[1,i-1]$ along with the facility already located in i. The optimal cost of doing this task is $Z_{i-1}^*\alpha(k,i) + c(i)$. But in this case, since $g(i) > g(k)$ and $i < k$, we have $h(i) < h(k)$ and hence $Z_{i-1}^*\alpha(k,i) + c(i)$ is the same as $u(i)$. Finally, in such a case we would incur a cost of $u(i) + c(k)$ as given in (2).

The following two *lemmas* give the equations to compute $p(i)$ and $u(i)$ recursively.

Lemma 2. *The following equation can be used to compute $p(i)$ when $p(k)$ and $u(k)$ is known $\forall k < i$ and $Z_b^*(a)$ is known $\forall$ a, b such that $a \leq b$, $b \leq i$-1.*

$$p(i) = c(i) + Min\left\{ \underset{k \in GA_1(i)}{Min}\{p(k)\},\ \underset{k \in GA_1(i)}{Min}\left\{u(k) + \underset{j \in (k,i):\ h(j) \leq k\ and\ g(j) < i}{Min}\{c(j)\}\right\},\right.$$
$$\left. \underset{k \in GB_1(i)}{Min}\{c(k) + Z_{k-1}^*\alpha(k, i)\}\right\}$$

Proof. Consider all the vertices k that have an upper-reach equal to i(i.e. $g(k) = i$). From the definition of protected cost, it follows that a facility must be located at one of these vertices for i to get covered. Because of Lemma 1, it follows that, we can place atmost one facility to the right of k. Hence, we have the following two cases.

1. We consider the first case when no facility is placed to the right of k. Vertex k covers i. Clearly, $k \in GA_1(i) \cup GB_1(i)$.
   If $k \in GA_1(i)$, the vertex k will be covered by a vertex to its left and k covers all vertices to its right till i. Hence, we should now cover vertices from 1 through k using vertices 1 through k(Note that a facility has already been placed at k). We account for all such $k \in GA_1(i)$ in the equation using the term, $c(i) + \underset{k \in GA_1(i)}{Min}\ p(k)$.
   If $k \in GB_1(i)$ then k and i cover all vertices between $min(h(i), h(k))$ and i. The vertices 1 through $\alpha(k, i)$ should be covered using vertices from 1 to $k-1$, with optimum cost, which is given by $Z_{k-1}^*(\alpha(k, i))$. We account for such $k \in GB_1(i)$ in the equation using the term $\underset{k \in GB_1(i)}{Min}\{c(k) + Z_{k-1}^*\alpha(k, i)\} + c(i)$.

2. We now consider the second case when exactly one facility namely j is placed to the right of k. Again k covers i. Clearly, $k \in GA_1(i) \cup GB_1(i)$.
   Let $k \in GA_1(i)$. Clearly, $h(j) \leq k$ (If $h(j) > k$ the purpose of j is lost) and $g(j) < i$ (If $g(j) \geq i$, such $j \in GA_1(i)$ and they would get accounted in the previous case). The only purpose of j is to cover k. Hence 1 through $k - 1$ remains to be covered optimally using 1 through k. This can be done in $u(k)$ cost. Such $k \in GA_1(i)$ are accounted in the equation using the term $\underset{k \in GA_1(i)}{Min}\{u(k) + \underset{j \in (k,i):\ h(j) \leq k\ and\ g(j) < i}{Min}\{c(j)\}\} + c(i)$.

   If $k \in GB_1(i)$, then any j placed to the right of k will be useful only when j covers at least one vertex that k does not cover. This means that $h(j) < h(k)$. But since $j > k$, it follows that $g(j) \geq g(k)$ and hence $g(j) \geq i$. So $j \in GB_1(i)$. Such a solution would have been accounted for in the previous case and hence we need not redundantly consider such solutions again.

$\square$

Lemma 3. *The following equation can be used to compute $u(i)$ when $p(k)$ and $u(k)$ is known $\forall k < i$ and $Z_b^*(a)$ is known $\forall$ a, b such that $a \leq b$, $b \leq i$-1.*

$$u(i) = Min\left\{p(i),\ \underset{k \in GA_2(i)}{Min}\{p(k)\} + c(i),\ \underset{k \in GA_2(i)}{Min}\left\{u(k) + \underset{j \in (k,i):\ h(j) \leq k\ and\ g(j) < h(i)-1}{Min}\{c(j)\}\right\} + \right.$$
$$\left. c(i),\ \underset{k \in GB_2(i)}{Min}\{c(k) + Z_{k-1}^*\alpha(k, i)\} + c(i)\right\}$$

Proof. Sometimes, it might be cheaper to cover i also when covering $[1, i-1]$ than to leave it uncovered and so we have included $p(i)$ as the first term in the equation for $u(i)$.

Now we consider the cases where i is left uncovered. Consider all the vertices k that have $g(k) < i$ and $g(k) \geq h(i) - 1$ From the definition of unprotected cost, it follows that a facility must be located at one of these vertices for every vertex between k and i to get covered without covering i. Because of Lemma 1, it follows that, we can place atmost one facility to the right of k. Hence, we have the following two cases.

1. We consider the first case when no facility is placed to the right of k. Clearly, $k \in GA_2(i) \cup GB_2(i)$.

   If $k \in GA_2(i)$, the vertex k will be covered by a vertex to its left and k along with i cover all vertices between them. We account for all such $k \in GA_2(i)$ in the equation using the term, $c(i) + \underset{k \in GA_2(i)}{Min} \, p(k)$.

   If $k \in GB_2(i)$ then k and i cover all vertices between $min\{h(i), h(k)\}$ and i. The vertices 1 through $\alpha(k, i)$ should be covered using vertices from 1 to $k - 1$. The optimum cost for doing this is given by $Z^*_{k-1}(\alpha(k, i))$. We account for such $k \in GB_2(i)$ in the equation using the term $\underset{k \in GB_2(i)}{Min} \, \{c(k) + Z^*_{k-1}\alpha(k, i)\} + c(i)$.

2. We consider the second case when exactly one facility namely j is placed to the right of k. Again k and i together cover all the vertices between them. So $k \in GA_2(i) \cup GB_2(i)$.

   Let $k \in GA_2(i)$. Clearly, $h(j) \leq k$ (If $h(j) > k$ the purpose of j is lost)and $g(j) < h(i) - 1$ (If $g(j) \geq h(i) - 1$, such j would get included in $GA_2(i)$ and they would get accounted in the previous case). The only purpose of j is to cover k. Hence 1 through $k - 1$ remains to be covered optimally using 1 through $k - 1$ along with the facility already placed at k. This can be done in $u(k)$ cost. Such $k \in GA_2(i)$ are accounted in the equation using the term $\underset{k \in GA_2(i)}{Min} \, \{u(k) + \underset{j \in (k,i): \, h(j) \leq k \text{ and } g(j) < h(i)-1}{Min} \, \{c(j)\}\} + c(i)$.

   If $k \in GB_2(i)$, then any j placed to the right of k will be useful only when j covers at least one vertex that k does not cover. This means that $h(j) < h(k)$. But since $j > k$, it follows that $g(j) \geq g(k)$ and hence $g(j) \geq h(i) - 1$. So $j \in GB_2(i)$. Such a solution would have been accounted for in the previous case and hence we need not redundantly consider such solutions again.

   □

For each i, we compute $p(i)$, $u(i)$ and $Z^*_j(i) \, \forall j \geq i$ in that order. As mentioned above, $Z^*_n(n)$ gives the optimal cost of CCP on P.

Time Complexity: For a given i and j, $Z^*_j(i)$ takes $O(j - i)$ time to be computed. For computing $Z^*_j(i) \, \forall j \geq i$, we need $\sum_{j=i}^{n}(j - i)$ operations. For computing $Z_j^*(i) \ \forall j \geq i$, $1 \leq i \leq n$ we need $\sum_{i=1}^{} \sum_{j=i}^{} (j-i) = O(n^3)$ operations.

Similarly, to compute each $p(i)$ and $u(i)$ it takes atmost $\sum_{k=1}^{i} (i-k)$ operations. For

computing $p(i)$ and $u(i)$, $1 \leq i \leq n$ we need $\sum_{i=1}^{n} \sum_{k=1}^{i} (i-k) = O(n^3)$ operations.

Thus, the algorithm takes $O(n^3)$ time.

3 Conditional Covering Problem on Paths when the cost is unity

3.1 Preliminaries

In this section, we consider the variation of CCP when each vertex is assigned a unit cost and arbitrary covering radius and we give an $O(n^2)$ algorithm for the same. Also, we show how this algorithm can be used to solve CCP on interval graphs in *Section 4*. Let $span(i) = \{k | D_{ik} \leq R_i\} \ \forall i \in V$. Note that $span(i) = \{$The set of vertices that i can cover in $V(G)\} \cup \{i\}$. A vertex i is said to be a *non-maximal* vertex if $span(i) \subset span(j)$ for some $j \in V$. Else, i is said to be a *maximal* vertex. Let M be the set of maximal vertices and M' be the set of non-maximal vertices in $V(G)$ respectively. $(V(G) = M \cup M')$

Let S be a CCS of G and let $v \in S$. A vertex v is said to *uniquely* cover a vertex u if $\nexists v' \in S$ such that v' covers u. We define a CCS S to be a *proper* CCS if every $v \in S \cap M'$ *uniquely* covers some vertex $i \in S \cap M$ which satisfies the property, $span(v) \subset span(i)$ i.e. every non-maximal vertex v in S uniquely covers some maximal vertex $i \in S$ which satisfies $span(v) \subset span(i)$. The next *Lemma* relates every CCS with a *proper* CCS.

Lemma 4. *For every CCS S, on a graph G, there exists a proper CCS S' such that $c(S') \leq c(S)$.*

Proof. Assume S is not proper. Consider every $v \in S \cap M'$ that does not *uniquely* cover any vertex $i \in M \cap S$ such that $span(v) \subset span(i)$. We can remove every such v in S and add a vertex $i \in M$ such that $span(v) \subset span(i)$ and call the resultant set S'. Since i was not uniquely covered by v in S, i remains covered in S'. Any other vertex covered by v in S is now covered by i in S'. So the set S' still remains a CCS of G and since every vertex incurs only a unit cost, $c(S') \leq c(S)$. (Note that if i is already present in S, then we would just end up removing a vertex and no new vertex which is not present in S actually gets added to S'). $\qquad \square$

3.2 Algorithm

For path graphs, we can see that, $span(i) = \{k | h(i) \leq k \leq g(i)\}$. First, we make the following remark regarding maximal vertices.

Remark 1. 1. For any $i \in M$ and $j \in M'$ if $span(j) \subset span(i)$, then i can cover all vertices in $span(j)$ except i itself.

2. For $i, j \in M$ and $i < j$, $g(i) \leq g(j)$ and $h(i) \leq h(j)$.

For any CCS S, we define $g(S) = \text{Max}\{g(i)|i \in S\}$ i.e. S cannot cover any vertex j where $j > g(S)$. Given a *proper* CCS S, we say that a vertex i is the *last maximal* vertex of S, if $i \in M \cap S$ and $\nexists j \in M \cap S$ such that $j > i$. The following lemma gives the property of $g(S)$ when S is a *proper* solution.

Lemma 5. *If S is a proper CCS, then $g(S) = g(i) = n$, where i is the last maximal vertex of S.*

Proof. Let i be the last maximal vertex of a CCS S. Let us assume on the contrary that $g(i) \neq n$ and that $g(v) = n$ where $v \in S \cap M'$. In any *proper* CCS S, if $v \in S \cap M'$ then there exists a $j \in S \cap M$ such that $span(v) \subset span(j)$ and v uniquely covers j. Clearly, given such a v and j, $g(j) \geq g(v)$. Also, since i is the last maximal vertex, $i \geq j$ and hence $g(i) \geq g(j)$ (by the second point of Remark 1). Therefore, $g(i) \geq g(j) \geq g(v)$ which is a contradiction. $\square$

We will now restrict our search for the optimal CCS for the CCP on the path P only to the set of *proper* CCS on the path P. The next *Theorem* states how the optimal solution can be obtained by searching among *proper* solutions. Let Z_i be the minimum cost *proper* CCS for path P such that i is the *last maximal* vertex of the CCS. Note that, this does not mean i is the last vertex in the CCS corresponding to Z_i. There can be a non-maximal vertex after i. Also we observe that Z_i is defined only when i is a maximal vertex and Z_i does not exist for some maximal vertices $j \in M$, when $g(j) < n$ (as a consequence of *Lemma 5*).

Theorem 1. *The cost of the optimal CCS of P, is given by $z^* = \underset{i \in M, g(i) = n}{Min} c(Z_i)$.*

Proof. We restrict our search for the optimal solution only in the set of *proper* CCS of P due to *Lemma 4*. Every *proper* CCS must have some vertex $i \in M$ to be the *last maximal* vertex. By definition, Z_i is the minimum cost *proper* CCS for path P such that i is the *last maximal* vertex of Z_i. Hence, We have to minimize $c(Z_i) \forall i \in V$. Due to *Lemma 5*, $g(Z_i) = g(i)$ and Z_i is CCS of P only when $g(i) = n$. Hence we minimize $c(Z_i)$ for all Z_i such that $g(i) = n$ to get z^*. $\square$

Calculation of $c(Z_i)$: We next describe how to find $c(Z_i)$ efficiently for $i \in M$. We define *protected* and *unprotected* costs for all the maximal vertices $i \in M$ of the path P denoted by $p(i)$ and $u(i)$ as follows. Let A_i be the minimum cost *proper* CCS for the vertices 1 through i such that i is the *last maximal* vertex of A_i. The Protected cost $p(i)$ is $c(A_i)$. Let B_i be the minimum cost *proper* CCS for the vertices 1 through $i - 1$ such that i is the *last maximal* vertex of B_i. The unprotected cost $u(i)$ is $c(B_i)$. Similar to the set Z_i, in A_i and B_i too, we can place a non-maximal vertex even to the *right* of i, as long as the solution is *proper*. The vertex i is necessarily covered by A_i (i is protected by A_i) but may not be necessarily covered by B_i(i may remain unprotected by B_i). For each $i \in M$, we define $Q_i = \{v \in M'|v \text{ covers } i \text{ and } span(v) \subset span(i)\}$. The next *Lemma* characterizes the vertices which can cover i in A_i.

Lemma 6. *In A_i, the vertex i should either be covered by a $j \in M \cap A_i$ such that $j < i$ or be uniquely covered by a $v \in Q_i$.*

Proof. In A_i, let i be covered by a maximal vertex j. By definition of A_i, i is the *last maximal* vertex of A_i and hence $j < i$.

On the other hand, if no maximal vertex covers i, then some non-maximal vertex $v \in M' \cap A_i$ must be covering i. If $v \in M' \cap A_i$ is covering i then $v \in Q_i$. (Assume $v \notin Q_i$. A_i is a *proper* CCS and it would imply the existence of a maximal vertex which is also covering i contradicting our assumption) Such a v must be *uniquely* covering i. To see this, note that A_i is a *proper* CCS. If v is not uniquely covering i, it must be uniquely covering some other maximal vertex in A_i. This maximal vertex will also cover i, contradicting our assumption. Hence the proof. □

We will now give a set of dynamic programming equations for solving $p(i)$ and $u(i)$ $\forall i \in M$. First, the following *Remark* will solve them for the base case. Let min be the least indexed maximal vertex.

Remark 2. Clearly $h(min) = 1$ due to *Remark 1*. Hence $B_{min} = \{min\}$ and $u(min) = 1$. To find A_{min}, we pick a vertex v from Q_{min} (if $Q_{min} \neq \emptyset$) and append it to B_{min}. So we have $A_{min} = B_{min} \cup \{v\}$. The cost $p(min) = 2$. If $Q_{min} = \emptyset$, we set $p(min) = \infty$ and A_{min} does not exist for the vertex min. The correctness of this remark is due to *Lemma 6*. □

For all the maximal vertices $i \in M - \{min\}$ we define the following sets. These definitions are borrowed from [5] but unlike the previous section, where the following sets are defined for all the vertices in V, we define them only on maximal vertices and further use them to calculate the $p(i)$ and $u(i)$. If these sets do not exist for any of the maximal vertex, we set them to be null.

1. $GA_1(i) = \{k \in M|k+1 \leq i \leq g(k) \text{ and } h(i) > k\}$
2. $GB_1(i) = \{k \in M|k+1 \leq i \leq g(k) \text{ and } h(i) \leq k\}$
3. $GA_2(i) = \{k \in M|k+1 \leq h(i) \leq g(k)+1 \leq i\}$
4. $GB_2(i) = \{k \in M|g(k)+1 \leq i \text{ and } h(i) \leq k\}$

Let $p'(i)$ be the cost corresponding to the minimum cost *proper* CCS for the vertices 1 through i such that i is the *last maximal* vertex of CCS and i is covered by a maximal vertex k, $k < i$. Now, we use the following *Theorem* to find $p(i)$ and $u(i)$ for $i \in M - \{min\}$. There may be some maximal vertices i for which A_i or B_i may not exist. A_i will not exist when $GA_1(i) = GB_1(i) = \emptyset$ and B_i will not exist when $(GA_2(i) = GB_2(i)) = \emptyset$ and $p'(i) = \infty$. In all such cases we will set $p(i)$ and $u(i)$ to be infinity explicitly in the algorithm. We will now consider only those cases when A_i and B_i exists and state the following *Theorem*.

Theorem 2. *The following are the dynamic programming equations to solve for* $p(i)$ *and* $u(i)$ $\forall i \in M - \{min\}$:

$$p'(i) = Min\{ \underset{k \in GA_1(i)}{Min} p(k), \underset{k \in GB_1(i)}{Min} u(k)\} + 1. \tag{3}$$

$$u(i) = Min\{p'(i), \underset{k \in GA_2(i)}{Min} p(k) + 1, \underset{k \in GB_2(i)}{Min} u(k) + 1\}. \tag{4}$$

$$p(i) = \begin{cases} Min\{p'(i), u(i)\} + 1 & p'(i) \neq u(i) \text{ and } Q_i \neq \emptyset \\ p'(i) & \text{otherwise} \end{cases} \tag{5}$$

Proof. We assume that the above equations correctly calculate $p(j)$ and $u(j)$ $\forall j \in M$ and $j < i$. We should prove that $p(i)$ and $u(i)$ are correctly calculated. By *Lemma* 6, We know that i can be covered in two possible ways. For each of the two possible ways, we analyze the terms in all the three equations.

1. i is covered by a $k \in M$ such that $k < i$:
   p'(i): Clearly, by definition $k \in GA_1(i) \cup GB_1(i)$. For a $k \in GA_1(i)$, i cannot cover k and k should be covered by some other vertex. Hence, we take the protected cost of k for calculations in equation (3). For a $k \in GB_1(i)$, i covers k. Hence, we take the unprotected cost of k for calculations in equation (3). We add a cost one due the facility placed at i.
   u(i): A vertex $k \in GA_2(i)$ is not covered by i and hence we use the protected cost $p(k)$ in (4). Similarly, a vertex $k \in GB_2(i)$ is covered by i and hence we use the unprotected cost $u(k)$ in (4). We add a cost one due to the facility placed at i. Sometimes, it may be cheaper to cover vertex i than to leave it uncovered. The optimal cost of doing this is $p'(i)$ and hence we include the term $p'(i)$ in (4).
   p(i): When the i in $p(i)$ is covered by a maximal vertex, the cost is just $p'(i)$.
2. i is uniquely covered by a vertex $v \in Q_i$:
   p'(i): By definition of $p'(i)$, i is covered by a maximal vertex and hence i cannot be uniquely covered by a vertex in $Q(i)$ in this case.
   u(i): Since it is not necessary to cover i in $u(i)$, we will not choose any non-maximal vertex v to cover i. This is because, if the solution has to be a *proper* solution, then v has to uniquely cover i. But since span$(v) \subset$ span(i), by removing v, all the vertices covered by v except i are still covered by i (i need not be covered in $u(i)$ anyway). Hence v can be removed.
   p(i): If $Q_i = \emptyset$, this case cannot be considered at all and $p(i) = p'(i)$. We assume Q_i is not empty and we use a $v \in Q_i$ to cover i uniquely. Now, vertices 1 to $i - 1$ should be covered using minimum cost and this cost is clearly given by $u(i)$. So, If Q_i is not empty, we can pick a vertex v from it and append it to B_i. The cost of this new CCS $(B_i \cup \{v\})$ of P_i is given by $u(i) + 1$. Also, if $u(i) = p'(i)$, then $p(i) = p'(i)$. This is because, $u(i) = p'(i)$ means that in B_i, i is covered by a maximal vertex and a $v \in Q_i$ will no

longer cover i *uniquely* and hence v will not cover any maximal vertices *uniquely*. Therefore for the case $u(i) = p'(i)$, we set $p(i) = p'(i)$.

Those vertices i, for which we were able to find a CCS in the second case (when $p'(i) \neq u(i)$ and $Q_i \neq \emptyset$), we finally set $p(i) = Min\{p'(i), u(i)+1\}$ as in equation (5). $\square$

Algorithm 1 : Algorithm to compute $p(i)\ \forall i \in M$ - $\{min\}$

Require: A Path $P = 123\ldots n$ along with $R_i,\ \forall i \in V(P)$.
  /* Preprocessing Step */
  Compute $g(i)$, $h(i)\ \forall i \in V(P)$.
  Find the sets M and M' such that $V = M \cup M'$.
  Find $GA_1(i)$, $GA_2(i)$, $GB_1(i)$, $GB_2(i)$ and $Q_i\ \forall i \in M$.
  /* Dynamic Programming Step */
  for $i{=}min + 1$ to n **do**
    if $i \in M$ **then**
      if $(GA_1(i) = GB_1(i)) = \emptyset$ **then**
        $p'(i) = \infty$
      else
        $p'(i) = Min\{ \underset{k \in GA_1(i)}{Min}\ p(k),\ \underset{k \in GB_1(i)}{Min}\ u(k)\} + 1.$
      end if
      if $(GA_2(i) = GB_2(i)) = \emptyset$ and $p'(i) = \infty$ **then**
        $u(i) = \infty$
      else
        $u(i) = Min\{p'(i),\ \underset{k \in GA_2(i)}{Min}\ p(k) + 1,\ \underset{k \in GB_2(i)}{Min}\ u(k) + 1\}.$
      end if
      if $u(i) \neq p'(i)$ and $Q_i \neq \emptyset$ **then**
        $p(i) = Min\{p'(i), u(i) + 1\}.$
      else
        $p(i) = p'(i).$
      end if
    end if
  end for

Remark 2 is used to find the protected and unprotected costs for min. *Algorithm* 1, uses *Theorem* 2 to find $p(i)$, $u(i)$ and hence $z(i)$(if it exists) for all $i \in M$ - $\{min\}$ and hence the CCS is always a *proper* CCS. It first does the preprocessing to calculate all the required values. Then the dynamic programming equations are used to find the protected and unprotected costs. After calculating these values, we use *Theorem* 1 to find the cost of the optimal solution for CCP on paths. The calculation of $GA_1(i)$, $GA_2(i)$, $GB_1(i)$ and $GB_2(i)$ for all i takes $O(n^2)$ time. Similarly, the dynamic programming step also takes $O(n^2)$ time. All other calculations take linear time. Thus *Algorithm* 1 runs in $O(n^2)$ time.

4 Conditional Covering Problem on Interval Graphs

A graph G is an interval graph if its vertices can be put in one-to-one correspondence with a family F of intervals on the real line such that two vertices are adjacent in G *iff* their corresponding intervals have nonempty intersection [3]. F is known as the intersection model of the graph. Let $I_i = [a_i, b_i]$ be the interval corresponding to a vertex i of the graph. We can assume that the set of $2n$ left and right end points are distinct for the graph. A graph G is a proper interval graph, *iff* no interval is properly contained within another interval of the graph. Any interval graph is a proper interval graph *iff* $a_1 < a_2 < a_3 < \ldots a_n$ and $b_1 < b_2 < b_3 < \ldots b_n$.

Consider an interval graph $G = (V, E)$ such that G has unit edge weights. But each vertex i has an arbitrary positive *covering radius* value R_i and unit cost values. We give a $O(n^2)$ algorithm to solve CCP on such a graph. But first, we make the following *Remark* about CCP on proper interval graphs.(When each vertex is assigned positive arbitrary *covering radius* and unit cost.)

Remark 3. Consider a proper interval graph. For any two vertices i and j, if $a_i < a_j$ and $b_i < b_j$ then we say that $i < j$. Such an ordering of vertices of the proper interval graph is called the proper interval graph ordering. We can observe that for any vertex i, $span(i) = \{k | h(i) \leq k \leq g(i)\}$. *Remark* 1 and hence *Lemma* 5 holds correct for proper interval graphs. As a consequence all the algorithms used for paths hold good in this case also without any modification. □

Now, we turn our attention to interval graphs. The first property stated in *Remark* 1 holds true for any graph with any ordering. However we should prove that the second property of the *Remark* 1 is valid for the maximal vertices of the interval graph.
We order all the vertices of the interval graph according to their left endpoints a_i (i.e., $i < j$ iff $a_i < a_j$).

Lemma 7. *If $i, j \in M$ such that $i < j$, then $g(i) \leq g(j)$ and $h(i) \leq h(j)$.*

Proof. We prove that we will get a contradiction otherwise.
If $g(i) \leq g(j)$ and $h(i) > h(j)$, then $span(i) \subset span(j)$ and i is not maximal anymore i.e. $i \notin M$.
If $g(i) > g(j)$ and $h(i) \leq h(j)$, then $span(j) \subset span(i)$ and j is not maximal anymore i.e. $j \notin M$
If $g(i) > g(j)$ and $h(i) > h(j)$, then we consider two cases.
Case 1: $b_i < b_j$. Since $i < j$, $a_i < a_j$. Now, $a_i < a_j$ and $h(i) > h(j)$ imply that $R_j > R_i$. Also $b_i < b_j$ and $g(i) > g(j)$ imply that $R_i > R_j$. This leads to a contradiction.
Case 2: $b_i > b_j$. Since $i < j$, $a_i < a_j$. Clearly $I_j \subset I_i$. Now, $a_i < a_j$ and $h(i) > h(j)$ imply that $R_j > R_i$. Also since $I_j \subset I_i$, we see that the upper-reach of i can be reached by j in atmost $R_i + 1$ steps. But since $R_j > R_i$, we have $R_j \geq R_i + 1$ and hence $g(j) \geq g(i)$ which is a contradiction. Hence the proof. □

As a consequence of *Lemma 7*, we see that *Remark 1* and hence *Lemma 5* hold good for interval graphs. Hence, as stated above for proper interval graphs, all the algorithms used for paths hold good for interval graphs without any modification.

5 Conclusion

In this paper, we studied the Conditional Covering Problem on some special classes of graphs such as paths and interval graphs. We showed that Horne and Smith's algorithm[5] for CCP on paths, and its extension to star and trees in [5, 4] is incorrect and gave a correct $O(n^3)$ algorithm for CCP on paths. We also presented a $O(n^2)$ algorithm for CCP on paths when unit cost is assigned to all vertices and further extended the algorithm to interval graphs without increasing the time complexity. It is an interesting open problem to see if our $O(n^3)$ algorithm for paths can be extend to stars and trees. Additionally, CCP can be examined on special classes of graphs such as series-parallel and asteroidal triple free graphs on which the total domination problem can be solved efficiently [9, 6].

References

1. Jeffrey B, Goldberg Brian, J.Lunday, and J. Cole Smith. Algorithms for solving the conditional covering problem on paths. *Naval Research Logistics*, 52(4):293–301, 2005.
2. Sohail S. Chaudhry, I. Douglas Moon, and S. Thomas McCormick. Conditional covering: Greedy heuristics and computational results. *Computers & OR*, 14(1):11–18, 1987.
3. M. C Golumbic. *Algorithmic Graph Theory and Perfect Graphs*. Academic Press New York, 1980.
4. Jennifer A. Horne and J. Cole Smith. A dynamic programming algorithm for the conditional covering problem on tree graphs. *Networks*, 46(4):186–197, 2005.
5. Jennifer A. Horne and J. Cole Smith. Dynamic programming algorithms for the conditional covering problem on path and extended star graphs. *Networks*, 46(4):177–185, 2005.
6. Dieter Kratsch. Domination and total domination on asteroidal triple-free graphs. *Discrete Applied Mathematics*, 99(1-3):111–123, 2000.
7. I. Douglas Moon and S.S. Chaudhry. An analysis of network location problems with distance constraints. *Management Science*, 30:290–307, 1984.
8. I. Douglas Moon and Lee Papayanopoulos. Facility location on a tree with maximum distance constraints. *Computers & OR*, 22(9):905–914, 1995.
9. J. Pfaff, R. Laskar, and S.T. Hedetniemi. Linear algorithms for independent domination and total domination in series-parallel graphs. *Management Science*, 30:290–307, 1984.

Range Median of Minima Queries, Super-Cartesian Trees, and Text Indexing

Johannes Fischer[1,*] and Volker Heun[2]

[1] Center for Bioinformatics (ZBIT), University of Tübingen,
Sand 14, 72076 Tübingen, Germany
`fischer@informatik.uni-tuebingen.de`
[2] Institut für Informatik, Ludwig-Maximilians-Universität München,
Amalienstr. 17, 80333 München, Germany
`Volker.Heun@bio.ifi.lmu.de`

Abstract. A *Range Minimum Query* asks for the position of a minimal element between two specified array-indices. We consider a natural extension of this, where our further constraint is that if the minimum in a query interval is not unique, then the query should return an approximation of the median position among all positions that attain this minimum. We present a succinct preprocessing scheme using only about $2.54\,n + o(n)$ bits in addition to the static input array, such that subsequent "range median of minima queries" can be answered in constant time. This data structure can be constructed in linear time, and only $o(n)$ additional bits are needed at construction time. We introduce several new combinatorial concepts such as Super-Cartesian Trees and Super-Ballot Numbers, which we believe will have other interesting applications in the future. We stress the importance of our result by giving two applications in text indexing; in particular, we show that our ideas are needed for fast construction of one component in Compressed Suffix Trees [19], a versatile tool for numerous tasks in text processing, and that they can be used for fast pattern matching in (compressed) suffix arrays [14].

Key words: Range Minimum Query; Succinct Data Structure; Cartesian Tree; Ballot Number; Text Indexing; Suffix Tree; Suffix Array.

1 Introduction

The Range Minimum Query (RMQ) problem is to preprocess an array A of n numbers (or other objects from a totally ordered universe) such that queries $\text{RMQ}_A(l, r)$ asking for the position of the minimum element in $A[l, r]$ can be answered in constant time; more formally, $\text{RMQ}_A(l, r) = \text{argmin}_{i \in \{l,\dots,r\}} \{A[i]\}$. This problem is of fundamental algorithmic importance due to its connection to many other problems, such as computing lowest common ancestors in trees [2].

* Worked on this article while at Ludwig-Maximilians-Universität München. Partially funded by the German Research Foundation (DFG, Bioinformatics Initiative).

Starting with [2], several authors have given algorithms to solve this problem with a linear-time preprocessing, leading to the currently best solution that never uses more than $2n + o(n)$ bits in addition to the input array [6].

In the case when the minimum in the query interval is not unique, RMQs can return an arbitrary minimum (although it is easy to adapt all of the above algorithms to return the leftmost or rightmost minimum). However, a natural (and useful, as we shall see!) extension of this is to require that queries return the *median* position of all positions in the query interval that attain the minimum value. We call such queries *range median of minima queries*.

We slightly relax this problem by requiring that the queries only return a *pseudo-median* of the minima — we shall see that this does not affect the algorithms that use these queries. A pseudo-median of minima is a minimum whose position has rank between $\frac{1}{C}y$ and $(1-\frac{1}{C})y$ among the y positions that attain the minimum in the query-interval (for some constant $C \geq 2$). Such a *pseudo range median of minima query* for a given interval $[l, r]$ is denoted by $\mathrm{RMQ}^{\mathrm{med}}(l, r)$. Although this is the first time that this problem is addressed, it has already interesting and useful applications, as we shall see in Sect. 4.

The main contribution of this article is to give a linear-time preprocessing scheme such that pseudo range median of minima queries can be answered in constant time. The space of our scheme is only about $2.54\,n+o(n)$ bits in addition to the input array, summarized as follows:

Theorem 1. *For a static array A of n elements from a totally ordered universe, there is a data structure with space-occupancy of $\log(3 + 2\sqrt{2})\,n + o(n) \approx 2.54311\,n + o(n)$ bits[3] that allows to answer "pseudo range median of minima queries" in $O(1)$ time, i.e., queries that for a given interval $[l, r]$ return the position of a minimum in $A[l, r]$ that has relative rank between $1/16$ and $15/16$ among all positions that attain the minimum in $A[l, r]$. This data structure can be built in $O(n)$ time, using only $o(n)$ additional bits at construction time.*

1.1 Overview of Techniques

In the sequel, we briefly sketch how to preprocess an array $A[0, n-1]$ in linear time for $\mathrm{RMQ}^{\mathrm{med}}$ (with $C = 16$). Our general idea is similar to other RMQ-algorithms, such as the one in [6]: decompose a query into several sub-queries, each of which has been precomputed; then the overall minimum inside of the query range can be obtained by taking the minimum over all precomputed intervals. There are two major deviations from previous approaches. First, instead of storing the leftmost minimum, each precomputed query stores the *perfect* median among all its minima. And second, in addition to returning the position of the minimum, each sub-query must also return the number of minima in the query-interval.

The array A to be preprocessed is (conceptually) divided into superblocks $B'_0, \ldots, B'_{n/s'-1}$ of size $s' := \log^{2+\epsilon} n$, where B'_i spans from $A[is']$ to $A[(i+1)s' -$

---

[3] Throughout this article, log denotes the binary logarithm.

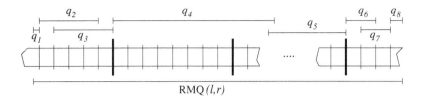

Fig. 1. The internal decomposition of $\mathrm{RMQ}^{\mathrm{med}}(l,r)$ into 8 sub-queries $q_1,\ldots,q_8$. Thin lines denote block-boundaries, thick lines superblock-boundaries.

1]. Here, $\epsilon > 0$ is an arbitrary constant. Likewise, each superblock is divided into (conceptual) blocks of size $s := \log n/(2\log\rho)$ (constant ρ to be defined later). The choice of s will become evident after Eq. (1). Call the resulting blocks $B_0,\ldots,B_{n/s-1}$. The reason for introducing the block-division is that a query $\mathrm{RMQ}^{\mathrm{med}}(l,r)$ can be divided into at most five sub-queries: one *superblock-query* that spans several superblocks, two *block-queries* that span the blocks to the left and right of the superblock-query, and two *in-block-queries* to the left and right of the block-queries.

We will preprocess the (super-)block-queries with a two-level storage scheme according to Munro [17], and the in-block-queries by the "Four-Russians-Trick" (precomputation of all answers for a sufficiently small number of possible instances).

As in previous algorithms [2,6], it suffices to precompute the superblock- and block-queries only for lengths being a power of two, as every query can be decomposed into two (possibly overlapping) sub-queries of length 2^l for some l. Thus, a general range minimum query is internally answered by decomposing it in fact into 8 (usually overlapping) sub-queries (see also Fig. 1): 2 in-block-queries q_1 and q_8 at the very ends of the query-interval, 4 block queries q_2, q_3, q_6, q_7, and 2 superblock-queries q_4 and q_5.

A key-technique for answering in-block-queries is a generalization of Cartesian Trees [22] to *Super-Cartesian Trees*, including algorithms for their construction and enumeration. Super-Cartesian Trees will be used for a unique representation of different blocks having the same results for all possible range median of minima queries within such a block. To avoid an explicit space-consuming construction of Super-Cartesian Trees, we present a sophisticated enumeration of these Super-Cartesian Trees using newly introduced Super-Ballot numbers. We believe that in particular these new concepts will turn out to have more interesting applications in the future, as Super-Cartesian Trees can be used to describe arbitrary properties about the minima in an interval.

Finally, we have to find the pseudo-median among the perfect medians of the eight sub-queries. Assume now that for each sub-query q_i ($1 \leq i \leq 8$) we know that at position p_i there is the perfect median among the y_i minima in the respective query interval. Let k be the minimum value in the complete query-interval ($k := \min_{i \in [1,8]}\{A[p_i]\}$), and $I \subseteq [1,8]$ be the set indicating which sub-query-intervals contain the minimum, $I := \{i \in [1,8] : A[p_i] = k\}$. Further, let j be an interval that contains most of these minima, i.e., $j := \text{argmax}_{i \in I}\, y_i$. The algorithm then returns the value p_j as the final answer to RMQ^{med}; this guarantees that the returned position has rank between $\frac{1}{16}y$ and $\frac{15}{16}y$ among all $y \le \sum_{i \in I} y_i \le 8y_j$ positions that attain the minimum value, as p_j has rank $\frac{y_j/2}{y} y \ge \frac{y_j/2}{8y_j} y = \frac{1}{16}y$ (and rank $\le \frac{15}{16}y$ by a similar calculation).

We thus conclude that if a query interval contains y minima, we have an algorithm which returns a pseudo-median with rank between $\frac{1}{16}y$ and $\frac{15}{16}y$, provided that the precomputed queries return the true median of minima. Section 3 explains how this goal is met.

1.2 Approximation Scheme

We just mention briefly that for any constant $\alpha > 0$ it is possible to approximate the median of y minima with an element having a rank in $[(\frac{1}{2} - \alpha)y, (\frac{1}{2} + \alpha)y]$ if we store for each block the positions of the r-quantiles instead of the perfect median position for an appropriately chosen r (depending on α). As long as r is a constant, it can be easily verified that our algorithm presented in the rest of the paper can be modified to determine for each block the positions of the r-quantiles (as well as the number of minima between each pair of consecutive r-quantiles) instead of the perfect median position. Using these r-quantiles, the number of minima in each region defined by a pair of overlapping queries q_2 and q_3, q_4 and q_5, or q_6 and q_7, respectively, can be approximated within in a factor of $1 + \frac{2}{r}$. Using these approximations for the number of minima in these 5 non-overlapping regions and the r-quantiles for the 8 queries, an element with rank within $[(\frac{1}{2} - \alpha)y, (\frac{1}{2} + \alpha)y]$ (where $\alpha = O(\frac{1}{r})$) can be selected by a more involved method than the one stated in the previous section. The details are omitted due to space limitations.

1.3 Overview of Applications

One natural application of pseudo range median of minima queries comes from text indexing. Basically, such queries are needed for finding quickly (in $O(\log|\Sigma|)$) time) the correct outgoing edge of a node in a suffix tree (Σ being the underlying alphabet), when the suffix tree is only represented implicitly via the suffix- and lcp-array, or by a parentheses sequence, as in Sadakane's Compressed Suffix Tree (CST) [19]. We will see that parts of our preprocessing scheme are necessary for the quick construction of one component of the CST. We emphasize that the original proposal [19] *does* make use of range median of minima queries, but *does not* explain how to construct the corresponding structures in linear time — a gap that we close here. A different but related application of RMQ^{med} is an $O(m \log|\Sigma|)$-time algorithm to locate a length-m-pattern in a text over the alphabet Σ, with the help of full-text- or word-suffix arrays [4, 14].

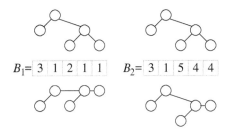

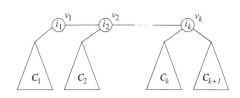

Fig. 2. Two blocks B_1 and B_2 with equal Cartesian Trees (top), but a different layout of minima. The Super-Cartesian Trees (bottom) reflect the positions of minima by horizontal edges.

Fig. 3. Illustration to the definition of Super-Cartesian Trees. The horizontal edges can be considered as right edges with a different "label"; in this sense, the Super-Cartesian Tree is in fact a Schröder Tree.

2 Preliminaries

We use the standard word-RAM model of computation, where fundamental arithmetic operations on words consisting of $\Theta(\log n)$ consecutive bits can be computed in constant time.

For an array A of n elements, we write $A[l, r]$ to denote A's sub-array from l to r. One important definition [22] that plays a central role in almost every algorithm for answering RMQs is as follows:

Definition 1. *The* Cartesian Tree *of an array $A[l, r]$ is a binary tree $\mathcal{C}(A)$ whose root is a minimum element of A (ties are broken to the left), labeled with the position i of this minimum. The left child of the root is the Cartesian Tree of $A[l, i - 1]$ if $i > l$, otherwise it has no left child. The right child is defined analogously for $A[i + 1, r]$.*

Cartesian Trees are sometimes called *treaps*, as they exhibit the behavior of both a search tree (here on the array-indices) and a min-heap (here on the array-values).

3 Preprocessing for Range Median of Minima Queries

In this section, we present the details of the preprocessing algorithm sketched in the introduction.

3.1 Preprocessing for Short Queries

Let us first consider how to precompute the perfect median of the minima inside the blocks of size s. The difficulty is that we cannot blindly adopt the usual RMQ-solutions such as the one in [6], because for normal RMQs blocks with the same Cartesian Tree are regarded as equal, totally ignoring their distribution of minima. As an example, look at the two blocks B_1 and B_2 in Fig. 2, where

Table 1. Example-arrays of length 3, their Super-Cartesian Trees, and their corresponding paths in the graph in Fig. 4. The last column shows how the algorithm in Fig. 5 calculates the index of $\mathcal{C}^{\mathrm{sup}}(A)$ in an enumeration of all Super-Cartesian Trees.

A	$\mathcal{C}^{\mathrm{sup}}(A)$	path	index in enumeration	A	$\mathcal{C}^{\mathrm{sup}}(A)$	path	index in enumeration
123			0	111			$\hat{C}_{13}+\hat{C}_{02}=6$
122			$\hat{C}_{03}=1$	221			$\hat{C}_{13}+\hat{C}_{02}+\hat{C}_{01}=7$
132			$\hat{C}_{03}+\hat{C}_{02}=2$	212			$\hat{C}_{13}+\hat{C}_{12}=8$
121			$\hat{C}_{03}+\hat{C}_{02}+\hat{C}_{02}=3$	211			$\hat{C}_{13}+\hat{C}_{12}+\hat{C}_{02}=9$
231			$\hat{C}_{03}+\hat{C}_{02}+\hat{C}_{02}+\hat{C}_{01}=4$	321			$\hat{C}_{13}+\hat{C}_{12}+\hat{C}_{02}+\hat{C}_{01}=10$
112			$\hat{C}_{13}=5$				

$\mathcal{C}(B_1) = \mathcal{C}(B_2)$. But for B_1 we want $\mathrm{RMQ}^{\mathrm{med}}(1,5)$ to return position 4 (the median position of the 3 minima), whereas for B_2 the same RMQ should return position 2, because this is the unique position of the minimum.

We overcome this problem by introducing a new kind of Cartesian Tree which is tailored to meet our special needs for this task:

Definition 2. *Let $A[l,r]$ be an array having its minima at positions $i_1, \ldots, i_k$ for some $k \geq 0$. Then the Super-Cartesian Tree $\mathcal{C}^{\mathrm{sup}}(A)$ of A is recursively constructed as follows:*

- *If $l > r$, $\mathcal{C}^{\mathrm{sup}}(A)$ is the empty tree.*
- *Otherwise, create k nodes $v_1, \ldots, v_k$, where v_j is labeled with i_j. v_1 is the root of $\mathcal{C}^{\mathrm{sup}}(A)$, and v_{i-1} is connected to v_i with a horizontal edge for $i > 1$. Recursively construct $\mathcal{C}_1 := \mathcal{C}^{\mathrm{sup}}(A[l, i_1 - 1])$, $\mathcal{C}_2 := \mathcal{C}^{\mathrm{sup}}(A[i_1 + 1, i_2 - 1])$, $\ldots$, $\mathcal{C}_{k+1} := \mathcal{C}^{\mathrm{sup}}(A[i_k + 1, r])$. For $1 \leq i < k$, the left child of v_i is the root of $\mathcal{C}_i$. Finally, the left and right children of v_k are the roots of $\mathcal{C}_k$ and $\mathcal{C}_{k+1}$, respectively. See also Fig. 3.*

Note that $\mathcal{C}^{\mathrm{sup}}$ is in fact a binary tree, where right edges may either be "horizontal" or "vertical." Fig. 2 shows $\mathcal{C}^{\mathrm{sup}}$ for our two example blocks, and the columns labeled "$\mathcal{C}^{\mathrm{sup}}(A)$" in Tbl. 1 give further examples. The reason for calling this tree the *Super-Cartesian Tree* will become clear when we analyze the number of such trees (Sect. 3.1.2).

The reason for our definition of the Super-Cartesian Tree is the following lemma, which follows immediately from Def. 2.

Lemma 1 (Relating RMQs and Super-Cartesian Trees). *Let A and B be two arrays, both of size n. Then the following two statements are equivalent:*

1. *$\forall 0 \leq i \leq j < n$: the minima in $A[i, j]$ occur at the same positions as the minima in $B[i, j]$*
2. *$\mathcal{C}^{\mathrm{sup}}(A) = \mathcal{C}^{\mathrm{sup}}(B)$.*

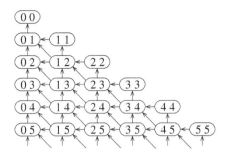

Fig. 4. The infinite graph whose vertices are (p, q) for all $0 \leq p \leq q$. There is an edge from (p, q) to $(p-1, q)$ if $p > 0$, and to $(p, q-1)$ and $(p-1, q-1)$ if $q > p$.

This lemma implies that for answering in-block-queries, we can use a global lookup-table P that stores the answers only for all possible Super-Cartesian Trees (and not for all occurring blocks). In order to index into P, each block B_i, $0 \leq i \leq n/s - 1$, has to be given a number, such that blocks with the same number have the same Super-Cartesian Tree. The computation of these block numbers will be explained in Sect. 3.1.3. Before that, in Sections 3.1.1 and 3.1.2 we take a closer look at Super-Cartesian Trees.

3.1.1 Construction of Super-Cartesian Trees. We give a linear algorithm for constructing $\mathcal{C}^{\mathrm{sup}}$. This is a straightforward extension of the algorithm for constructing the usual Cartesian Tree [8], treating the "equal"-case in a special manner, as explained next. Let $\mathcal{C}_{i-1}^{\mathrm{sup}}(A)$ be the Super-Cartesian Tree for $A[0, i-1]$. We want to construct $\mathcal{C}_i^{\mathrm{sup}}(A)$ by inserting a new node w with label i at the correct position in $\mathcal{C}_{i-1}^{\mathrm{sup}}(A)$. Let $v_1, \ldots, v_k$ be the nodes on the rightmost path in $\mathcal{C}_{i-1}^{\mathrm{sup}}(A)$ with labels $l_1, \ldots, l_k$, respectively, where v_1 is the root, and v_k is the rightmost leaf. Let x be defined such that $A[l_x] \leq A[i]$ and $A[l_{x'}] > A[i]$ for all $x < x' \leq k$. We first create a new node w labeled with i. We then remove v_x's right child v_{x+1} and append it as the left child of w. Now, if $A[l_x] = A[i]$, connect w with a *horizontal* edge to v_x. Otherwise (i.e., $A[l_x] < A[i]$), w becomes the "normal" right child of v_x (i.e., it is connected to v_x with a *vertical* edge). An easy amortized argument (as in [8]) shows that the overall running time is linear.

3.1.2 The Number of Super-Cartesian Trees. We show that there is a one-to-one correspondence between Super-Cartesian Trees and paths from $\boxed{s\ s}$ to $\boxed{0\ 0}$ in a the graph shown in Fig. 4. This bijection is obtained from the above construction algorithm for $\mathcal{C}^{\mathrm{sup}}$ (Sect. 3.1.1), where we first bijectively map the tree to a sequence of numbers $l_1, \ldots, l_n$, which can in turn be mapped bijectively to a path in the graph, explained as follows.

Consider step i of the above construction algorithm. Let l_i count the number *vertical* edges on C_{i-1}^{sup}'s rightmost path that are traversed (and therefore removed from the rightmost path) when searching for w's correct insertion point v_x. Note that the l_i's do *not* count the traversed (i.e. removed) horizontal edges on C_{i-1}^{sup}'s rightmost path. In the graph in Fig. 4, we translate this into a sequence of l_i upwards moves, and then either a *leftwards* move if $A[l_x] < A[i]$, or a *diagonal* move if $A[l_x] = A[i]$. This gives a one-to-one correspondence between Super-Cartesian Trees and paths from $\boxed{s\ s}$ to $\boxed{0\ x}$ for some x (which is then canonically continued to $\boxed{0\ 0}$), because in step i we constrain the number of upwards moves in the graph in Fig. 4 by the number of strict leftwards moves that have already been made. See again Tbl. 1 for examples.

It is well known [20] that the number of paths from $\boxed{s\ s}$ to $\boxed{0\ 0}$ is given by the s'th *Super Catalan Number* $\hat{C}_s$ (also known under the name *Little Schröder Numbers* due to their connection with Schröder Trees). These numbers are quite well understood, for our purpose it suffices to know that (e.g., [16], Thm. 3.6)

$$\hat{C}_s = \frac{\rho^s}{\sqrt{\pi s}(2s-1)}(1 + O(s^{-1})) \,, \tag{1}$$

with $\rho := 3 + 2\sqrt{2} \approx 5.8284$.

As already mentioned, this means that we do not have to precompute the in-block-queries for all n/s *occurring* blocks, but only for $O(\rho^s)$ "sample blocks" with pairwise different Super-Cartesian Trees. (Now our choice for the block-size s becomes clear.) We simply do a naive precomputation of all possible s^2 queries inside of each sample block. Because the blocks are of size s, the resulting table P needs $O(\hat{C}_s \times s^2 \cdot \log s) = O(\sqrt{n}\log^2 n \log \log n) = o(n)$ bits, and the time to compute it is $O(\hat{C}_s \times s^3) = o(n)$. (The additional factor s accounts for finding the median of the minima in each step.) As explained in the introduction, we also have to store the number of minima for each possible query. This information can be stored along with P within the same space bounds.

3.1.3 Computing the Block Types.

All that is left now is to assign a *type* to each block that can be used for indexing into table P, i.e., we wish to find a surjection

$$t : \mathcal{A}_s \rightarrow \{0, \ldots, \hat{C}_s - 1\}, \text{ and } t(B_i) = t(B_j) \text{ iff } \mathcal{C}^{\text{sup}}(B_i) = \mathcal{C}^{\text{sup}}(B_j) \,, \tag{2}$$

where $\mathcal{A}_s$ is the set of arrays of size s.

Our strategy is to simulate the construction algorithm for Super-Cartesian Trees (Sect. 3.1.1), thereby simulating a walk along the corresponding path in the graph in Fig. 4. These paths can be enumerated as follows. First observe that the number of paths from an arbitrary node $\boxed{p\ q}$ to $\boxed{0\ 0}$ in the graph in Fig. 4 is given by the recurrence

$$\hat{C}_{00} = 1, \quad \hat{C}_{pq} = \begin{cases} \hat{C}_{p(q-1)} + \hat{C}_{(p-1)q} + \hat{C}_{(p-1)(q-1)}, & \text{if } 0 \le p < q \ne 0 \\ \hat{C}_{p(q-1)} & \text{if } p = q \ne 0 \,, \end{cases} \tag{3}$$

Input: a block B_j of size s

Output: the type of B_j, as defined by Eq. (2)

```
1  let R be an array of size s + 1              // simulates rightmost path of C_{i-1}^{sup}(B_j)
2  R[1] ← −∞                                    // stopper element on stack R
3  q ← s, N ← 0, h ← 0      // h = # horizontal edges on C_{i-1}^{sup}(B_j)'s rightmost path
4  for i ← 1, ..., s do
5      while R[q + i + h − s] > B_j[i] do
6          N ← N + Ĉ_{(s−i)q} + Ĉ_{(s−i)(q−1)}   // accounts for upwards move in Fig. 4
7          while R[q + i + h − s − 1] = R[q + i + h − s] do h−−
8          q−−
9      if R[q + i + h − s] = B_j[i] then
10         N ← N + Ĉ_{(s−i)q}                    // accounts for diagonal move in Fig. 4
11         h++, q−−
12     R[q + i + h − s + 1] ← B_j[i]             // push B_j[i] on R
13 return N
```

Fig. 5. An algorithm to compute the type of a block B_j.

and $\hat{C}_{pq} = 0$ otherwise. This follows from the fact that the number of paths from node $\boxed{p\;q}$ to $\boxed{0\;0}$ is given by summing over the number of paths from each of the at most three cells that can be reached from $\boxed{p\;q}$ in a single step. Because the numbers $\hat{C}_{pq}$ generalize the Ballot Numbers in the same way as the Super-Catalan Numbers generalize the Catalan Numbers, we call them *Super-Ballot Numbers*[4]. The first few resulting Super-Ballot Numbers, laid out such that they correspond to the nodes in Fig. 4, are

$$
\begin{array}{l}
1 \\
1\;1 \\
1\;3\;3 \\
1\;5\;11\;\;11 \\
1\;7\;23\;45\;\;\;45 \\
1\;9\;39\;107\;197\;197 \;.
\end{array}
\tag{4}
$$

Due to this construction the Super Catalan Numbers appear on the rightmost diagonal of (4); in symbols, $\hat{C}_s = \hat{C}_{ss}$. Although we do not have a closed formula for our Super-Ballot Numbers we can construct the $s \times s$-array at startup, by means of (3).

We are now ready to describe the algorithm in Fig. 5 which computes a function satisfying (2). It simulates the construction algorithm for Super-Cartesian Trees (Sect. 3.1.1), *without actually constructing them*! The general idea behind this is given by the bijection from Super-Cartesian Trees to paths from $\boxed{s\;s}$

---

[4] A different generalization of the Ballot Numbers [9] also goes under the name "Super Ballot Numbers" and should not be confused with our concept.

to $\boxed{0\ 0}$ in the graph in Fig. 4, as explained in Sect. 3.1.2. By moving along this path, we count the number of paths that have been "skipped" when making an upwards (line 6) or diagonal move (line 11). At the beginning of step i of the outer for-loop, the position in the graph is $\boxed{(s-i+1)\ q}$, and h keeps the number of horizontal edges on the rightmost path in $\mathcal{C}_{i-1}^{\mathrm{sup}}(B_j)$. Stack R keeps the nodes on the rightmost path of $\mathcal{C}_{i-1}^{\mathrm{sup}}(B_j)$, with $q+i+h-s$ pointing to R's top (i.e., to $\mathcal{C}_{i-1}^{\mathrm{sup}}(B_j)$'s rightmost leaf). The loop in line 7 simulates the traversal of horizontal edges on $\mathcal{C}_{i-1}^{\mathrm{sup}}(B_j)$'s rightmost path by removing all elements that are equal to the top element on the stack. The if-statement in line 10 accounts for adding a new horizontal edge to $\mathcal{C}_{i-1}^{\mathrm{sup}}(B_j)$, which is translated into a diagonal move in Fig. 4.

In total, if we denote by l_i the number iterations of the while-loop from line 5 to 12 in step i of the outer for-loop, l_i equals the number of vertical edges in $\mathcal{C}_{i-1}^{\mathrm{sup}}(B_j)$ that are removed from the rightmost path. It follows from Sect. 3.1.2 that the algorithm in Fig. 5 correctly computes a function satisfying (2) in $O(s)$ time. Again, see Tbl. 1 for examples.

Thus, we store the type of each block (i.e., the number of its Super-Cartesian Tree in the above enumeration) in an array $T[0, n/s - 1]$. T's size is $|T| = \frac{n}{s}\lceil \log \hat{C}_s \rceil \leq \frac{n}{s}\log \rho^s = n\log \rho \approx 2.54311\,n$ bits.

3.2 Preprocessing for Long Queries

To complete the proof of Thm. 1, we now show how to precompute the perfect median of the minima in all queries that exactly span 2^j blocks or superblocks for all reasonable j (queries q_2–q_7 in Fig. 1). We note that this problem is harder than in the case of "normal" RMQs [6], as it is now not obvious how to compute the median position of all minima in an interval I from the median positions in I's first and second half, respectively. Because the techniques are similar, we only show how to preprocess the superblocks; the reader can convince himself that the space-bounds for blocks are also within $o(n)$, provided that we store the positions of minima only relative to the beginning their superblock [6].

For each superblock B_j' we define a (temporary) array $X_j'[1, n_j']$ (where $n_j' \leq s'$ denotes the number of minima in block B_j') such that $X_j'[i]$ holds the position of the i'th minimum in B_j'. The arrays X_j' are not represented explicitly; instead, we set up a *bit-vector* $D_j'[1, s']$ such that $D_j'[i]$ is 1 iff $B_j'[i]$ is a minimum in B_j. Then prepare each D_j' for constant time select_1-operations, where $\mathrm{select}_1(D_j', i)$ returns the position of the i'th 1 in D_j'. Then $X_j'[i]$ can be obtained by $\mathrm{select}_1(D_j', i)$. We already note at this point that the D_j''s and their corresponding structures for select_1 are only auxiliary structures and can hence be deleted after the preprocessing for the long queries. The D_j''s take a total of $n/s' \cdot s' = n$ bits.[5] The additional structures for constant time select_1-operations

---

[5] As we saw in Sect. 3.1, the final data structure for the short queries uses more than n bits. Therefore if we first construct the data structures for the long queries, the temporary n bits for the D_j''s constitute no extra space at construction time, as this space can later be re-used.

take additional $o(n)$ bits overall (if the D'_j's are concatenated into one bit-vector of length n) and can be computed in $O(n)$ time using standard structures for rank/select [3, 11, 17].

We now define a table $M'[0, \frac{n}{s'}][0, \log(n/s')]$, where $M'[i][j]$ stores the *perfect median position* of all minima in the sub-array $A[is', (i + 2^j)s' - 1]$, i.e., in the sub-array covering all superblocks from B'_i to B'_{i+2^j-1}. These are exactly all possible sub-queries q_4 and q_5 in Fig. 1. Thus, to answer a query q_4 or q_5, we simply look up the corresponding table entry in M'. Because we also need to know the number of minima for each of the sub-queries q_4 and q_5, we also set up a table Y' of similar dimensions as M', such that $Y'[i][j]$ stores the number of minima in $A[is', (i + 2^j)s' - 1]$.

We wish to fill tables M' and Y' in a top-down manner, i.e., filling $M'[\cdot][j]$ and $Y'[\cdot][j]$ before $M'[\cdot][j + 1]$ and $Y'[\cdot][j + 1]$. To get started, initialize the 0'th column of Y' in $O(n)$ time by setting $Y'[i][0]$ to the number of minima in superblock B'_i for all $0 \leq i \leq \frac{n}{s'}$. Then initialize the 0'th column of M' with the position of the true median in the superblocks: $M'[i][0] = X'_j[\lceil Y'[i][0]/2 \rceil]$.

We now show how to fill entry i of tables M' and Y' on level $j > 0$, i.e., how to compute the value $M'[i][j]$ as the perfect median of all minima in $A[is', (i + 2^j)s' - 1]$, and $Y'[i][j]$ as the number of minima in this sub-array. Because of the order in which M' and Y' are filled, we proceed by splitting the interval into two smaller intervals of size $\ell := 2^{j-1}s'$, as explained next.

Suppose there are $y_l := Y'[i][j - 1]$ minima in the left half $A[is', is' + \ell - 1]$, and $y_r := Y'[i + 2^{j-1}][j - 1]$ minima in the right half $A[is' + \ell, (i + 2^j)s' - 1]$. The easy case is when the overall minimum occurs only in one half, say the left one: then we can safely set $M'[i][j]$ to $M[i][j - 1]$, and $Y'[i][j]$ to y_l.

The more difficult case is when the minimum occurs in both halves. The value $Y'[i][j]$ is simply set to $y_l + y_r$. We know that the true median has rank $r := \lceil (y_l + y_r)/2 \rceil$ among all minima in the interval. If $y_l \geq y_r$, we know that the true median *must* be in the left half, and that it must have rank r in there. If, on the other hand, $y_l < y_r$, the median must be in the right half, and must have rank $r' := r - y_l$ in there. In either case, we know in which half we have to look for a new minimum with a certain rank.

We recurse in this manner "upwards," until either the minimum occurs in only one half, or until we reach level 0, where we can select the appropriate minimum from our X'_j-arrays (via D'_j). Due to the "height" of table M', the number of recursive steps is bounded by $O(\log(n/s'))$. In total, filling M' takes $O(\frac{n}{s'} \log(n/s') \log(n/s')) = O(n)$ time (recall $s' = \log^{2+\epsilon} n$).

The size of M' (and also that of Y') is $O(\frac{n}{s'} \times \log(n/s') \cdot \log n) = o(n)$ bits.

4 Applications in Compressed Suffix Trees and Arrays

4.1 Improvements in Compressed Suffix Trees

It is well-known that RMQs in general are a versatile tool for many string matching tasks. The most important application of this kind is to preprocess the array containing the lengths of the *longest common prefixes* of lexicographically adjacent suffixes (LCP-array for short) for constant-time range minimum queries. Then the longest common prefix between *arbitrary* suffixes can be found in constant time. This, in turn, can be used for many tasks in approximate and exact string matching, and also for navigational operations in (compressed) suffix trees, such as computing so-called *suffix links*, when the suffix tree is only represented *implicitly* by the suffix- and LCP-array.

The most fundamental navigational operation, however, is to locate the outgoing edge of a node v that is labeled with a given character $c \in \Sigma$, called *getChild*(v, c) (Σ is the alphabet). It can be shown [19] that one can retrieve all outgoing edges by performing subsequent queries of the form $\text{RMQ}_{\text{LCP}}(l, r)$, where the query indices l and r are the interval in LCP that represent node v, as explained by Abouelhoda et al. [1]. If the RMQs return the leftmost minimum, then *getChild*(v, c) takes $O(|\Sigma|)$ time, as in the worst case all $|\Sigma|$ minima have to be visited.

To speed up this search, Sadakane [19] proposes to use RMQ^{med} instead of plain RMQs in his Compressed Suffix Tree (CST), such that the interval $[l, r]$ can be binary-searched in $O(\log |\Sigma|)$ time. However, he does not give the details how the structures for RMQ^{med} can be constructed efficiently, and a naive approach would give roughly $O(n^2)$ construction time, or maybe $O(n \log n)$ with the help of binary search trees. As the construction times of the other structures in the CST are at most $O(n \log^\delta n)$ for any constant $0 < \delta \le 1$ [21], in order to maintain this time bound it is important to have an efficient preprocessing algorithm for pseudo range median of minima queries — a gap that we close in this paper.

Lemma 2. *There is a linear-time preprocessing scheme such that the number of search steps performed by getChild(v, c) in Compressed Suffix Trees is* $\log_{\frac{16}{15}} |\Sigma| \approx 10.74 \log |\Sigma|$ *in the worst case.*

We mention that the constant $1/12$ in Lemma 5 of [19] should also be $1/16$, as there the queries are also decomposed into 8 different sub-queries.[6] We also remark that the $2.54\,n$ bits from array T (Sect. 3.1.3) are *not* necessary in the CST, as the balanced parentheses sequence (using $4n$ bits) can be re-used to index into the table of precomputed in-block-queries.

We note two further applications of our new scheme. First, as our scheme for RMQ^{med} completely substitutes the so-called *child table* in Abouelhoda et al.'s *Enhanced Suffix Array* [1] and in Kim et al.'s Compressed Suffix Tree [13], we improve on their space consumption as well. Further, our preprocessing from Sect. 3.2 is also necessary for Fischer et al.'s entropy bounded CST [7].

4.2 Pattern Matching in (Compressed) Suffix Arrays

For a given pattern P of length m, the most common task in pattern matching is to check whether P is a substring of T. We now show that our scheme for

---

[6] Confirmed by K. Sadakane (personal communication, June 2007).

RMQ^{med} leads to a $O(m \log |\Sigma| t_{SA})$ search-method in suffix arrays, where t_{SA} is the time to access an element from the *suffix array* (see Tbl. 1 in [19]). This method is simply obtained by calling *getChild* subsequently for all characters in P, each time invoking an evaluation of the suffix array, hence the additional factor t_{SA}.

Let us first consider uncompressed suffix arrays, where $t_{SA} = O(1)$. The best results on pattern matching in uncompressed suffix arrays are due to Kim and Park [13], who give a succinct version of Kim et al.'s data structure [12] that allows $O(m \log |\Sigma|)$-time pattern searches. This table [13] requires $5n + o(n)$ bits of space; with our scheme for RMQ^{med}, this is reduced to $\approx 2.54\, n + o(n)$ bits, a space reduction by a factor of about 2. Given the practical importance of pattern matching, this is not negligible. Our new technique can also be used to get an $O(m \log |\Sigma|)$-pattern search-algorithm in (uncompressed) suffix arrays on words [4].

Naturally, our scheme for RMQ^{med} can also be combined with compressed versions of the suffix array. For example, combining Grossi and Vitter's Compressed Suffix Array (Thm. 2 in [10]) with our search strategy, matching takes $O(m \log |\Sigma| \log_{|\Sigma|}^{\alpha} n)$ time ($0 < \alpha \le 1$), for *any* alphabet size $|\Sigma|$, while needing only $\alpha^{-1} H_0 n + O(n)$ bits in total (H_0 being the empirical order-0 entropy of the input text [15]). Although this cannot be directly compared to more recent advances in compressed text indexing, such as Ferragina et al.'s Alphabet-Friendly FM-Index [5], it is interesting to note that the size of our structure for RMQ^{med} is not dependent on the size of the alphabet, which stands in contrast to all other structures in compressed suffix arrays [18].

References

1. M. I. Abouelhoda, S. Kurtz, and E. Ohlebusch. Replacing suffix trees with enhanced suffix arrays. *J. Discrete Algorithms*, 2(1):53–86, 2004.
2. O. Berkman and U. Vishkin. Recursive star-tree parallel data structure. *SIAM J. Comput.*, 22(2):221–242, 1993.
3. D. R. Clark. *Compact Pat Trees*. PhD thesis, University of Waterloo, Canada, 1996.
4. P. Ferragina and J. Fischer. Suffix arrays on words. In *Proceedings of the 18th Annual Symposium on Combinatorial Pattern Matching*, volume 4580 of *LNCS*, pages 328–339. Springer, 2007.
5. P. Ferragina, G. Manzini, V. Mäkinen, and G. Navarro. Compressed representations of sequences and full-text indexes. *ACM Transactions on Algorithms*, 3(2):Article No. 20, 2007.
6. J. Fischer and V. Heun. A new succinct representation of RMQ-information and improvements in the enhanced suffix array. In *Proceedings of the International Symposium on Combinatorics, Algorithms, Probabilistic and Experimental Methodologies*, volume 4614 of *LNCS*, pages 459–470. Springer, 2007.
7. J. Fischer, V. Mäkinen, and G. Navarro. An(other) entropy-bounded compressed suffix tree. In *Proceedings of the 19th Annual Symposium on Combinatorial Pattern Matching*, volume 5029 of *LNCS*, pages 152–165. Springer, 2008.

8. H. N. Gabow, J. L. Bentley, and R. E. Tarjan. Scaling and related techniques for geometry problems. In *Proceedings of the 16th Annual ACM Symposium on Theory of Computing*, pages 135–143. ACM Press, 1984.

9. I. M. Gessel. Super ballot numbers. *J. Symbolic Computation*, 14(2–3):179–194, 1992.

10. R. Grossi and J. S. Vitter. Compressed suffix arrays and suffix trees with applications to text indexing and string matching. *SIAM J. Comput.*, 35(2):378–407, 2005.

11. G. Jacobson. Space-efficient static trees and graphs. In *Proceedings of the 30th Annual Symposium on Foundations of Computer Science*, pages 549–554. IEEE Computer Society, 1989.

12. D. K. Kim, J. E. Jeon, and H. Park. An efficient index data structure with the capabilities of suffix trees and suffix arrays for alphabets of non-negligible size. In *Proceedings of the 11th Symposium on String Processing and Information Retrieval*, volume 3246 of *LNCS*, pages 138–149. Springer, 2004.

13. D. K. Kim and H. Park. A new compressed suffix tree supporting fast search and its construction algorithm using optimal working space. In *Proceedings of the 15th Annual Symposium on Combinatorial Pattern Matching*, volume 3537 of *LNCS*, pages 33–44. Springer, 2004.

14. U. Manber and E. W. Myers. Suffix arrays: A new method for on-line string searches. *SIAM J. Comput.*, 22(5):935–948, 1993.

15. G. Manzini. An analysis of the Burrows-Wheeler transform. *J. ACM*, 48(3):407–430, 2001.

16. D. Merlini, R. Sprugnoli, and M. C. Verri. Waiting patterns for a printer. *Discrete Applied Mathematics*, 144(3):359–373, 2004.

17. J. I. Munro. Tables. In *Proceedings of 16th Annual Conference on the Foundations of Software Technology and Theoretical Computer Science*, volume 1180 of *LNCS*, pages 37–42. Springer, 1996.

18. G. Navarro and V. Mäkinen. Compressed full-text indexes. *ACM Computing Surveys*, 39(1):Article No. 2, 2007.

19. K. Sadakane. Compressed suffix trees with full functionality. *Theory of Computing Systems*, 2007.

20. R. P. Stanley. *Enumerative Combinatorics*, volume 2. Cambridge University Press, 1999.

21. N. Välimäki, W. Gerlach, K. Dixit, and V. Mäkinen. Engineering a compressed suffix tree implementation. In *Proceedings of the 6th Workshop on Experimental Algorithms*, LNCS 4525, pages 217–228. Springer-Verlag, 2007.

22. J. Vuillemin. A unifying look at data structures. *Comm. ACM*, 23(4):229–239, 1980.

An Improved Version of Cuckoo Hashing: Average Case Analysis of Construction Cost and Search Operations

Reinhard Kutzelnigg[*]

Institute of Discrete Mathematics and Geometry
Vienna University of Technology
Wiedner Hauptstr. 8–10
A-1040 Wien, Austria
kutzelnigg@dmg.tuwien.ac.at

Abstract. Cuckoo hashing is a hash table data structure introduced in [1], that offers constant worst case search time. As a major contribution of this paper, we analyse modified versions of this algorithm with improved performance. Further, we provide an asymptotic analysis of the search costs of all this variants of cuckoo hashing and compare this results with the well known properties of double hashing and linear probing. The analysis is supported by numerical results. Finally, our analysis shows, that the expected number of steps of search operations can be reduced by using a modified version of cuckoo hashing instead of standard algorithms based on open addressing.

Keywords: Hashing; Cuckoo hashing; Open addressing; Algorithms

1 Introduction

Hash tables are frequently used data structures in computer science [2]. Their efficiency has strong influence on the performance of many programs. Standard implementations like open addressing and hashing with chaining (see, *e.g.*, [3, 4]) are well analysed, and the expected cost of an operation is low. However, as a well known fact, the worst case behaviour of hash tables is inefficient. As a consequence, new implementations have been suggested [5–10]. One of these algorithms is cuckoo hashing [1, 11]: The data structure consists of two tables of size m and possesses $n = m(1 - \varepsilon)$ keys, where $\varepsilon \in (0, 1)$ holds. The algorithm is based on two hash functions h_1 and h_2, both map a key to a unique position in the first resp. second table. These are the only allowed storage locations of this key and, hence search operations need at most two look-ups. To insert a key x, we

---

[*] The author was supported by EU FP6-NEST-Adventure Programme, Contract number 028875 (NEMO) and by the Austrian Science Foundation FWF, project S9604, that is part of the Austrian National Research Network "Analytic Combinatorics and Probabilistic Number Theory".

put it into its primary storage cell $h_1(x)$ of the first table. If this cell was empty, the insertion is complete. Otherwise, there exists a key y such that $h_1(x) = h_1(y)$. We move this key to its secondary position $h_2(y)$ in the second table. If this cell was previously occupied too, we proceed with rearranging keys in the same way until we hit an empty cell. Obviously, there are of course situations where we might enter an endless loop because the same keys are moved again and again. In such a case, the whole data structure is rebuild by using two new hash functions. As a strong point, this is a rare event [11, 12]. Figure 1 depicts the evolution of a small cuckoo hash table.

We want to emphasise that our analysis is based on the assumption, that the storage locations of the keys form a sequence of pairs of independent uniform random integers. If a rehash is necessary, we assume that all new hash values are independent from all previous attempts. This seems to be a practical unrealisable request. But we also recall that uniform hashing (using a similar independence assumption), and double hashing (using very simple hash functions) are indistinguishable for all practical purposes [3].

As a weak point, it turns out, that these simple hash functions do not work well for Cuckoo hashing. To overcome this situation, one can for instance use polynomial hash functions with pseudo random behaviour [13, 14]. In particular, our experiments show that functions of the form $ax + b \mod m$ are suitable for table sizes up to approximately 10^6, where a and b are random 32-bit numbers, and m is a prime number. An other kind of suitable hash function is based on the ideas described in [15]. Denote the 4 bit blocks of an 32-bit integer key s by $s_7, s_6, \ldots, s_0$ and assume that f denotes an array of 32-bit random integers $f[0], f[1], \ldots, f[127]$. Then, we define the hash function h as follows

$$h(s) = (f[s_0] \oplus f[s_0 + s_1 + 1] \oplus \cdots \oplus f[s_0 + \cdots + s_7 + 7]) \mod m, \quad (1.1)$$

where $\oplus$ is the bitwise exclusive or operator. This algorithm seems to work well for tables of all sizes and is therefore used in the experiments described in the further sections. Further information on hash functions suitable for practical implementation of cuckoo hashing can be also found in [16].

We continue with the definition of modified versions of cuckoo hashing. The further sections analyse the performance of both original and modified versions of the algorithm. Hereby, we count the expected number of steps (hash function evaluations resp. table cells accessed), that are necessary to insert or search a randomly selected key. In detail, we will show that at least one of the modified algorithms offers better performance in almost all aspects than the standard algorithm. Further, its expected performance of search operatons is also superior to linear probing and double hashing.

2 Asymmetric Cuckoo Hashing

This section introduces a modified cuckoo hash algorithm using tables of different size. Clearly, we choose the tables in such a way, that the first table holds more memory cells than the second one. Thus, we expect that the number of keys

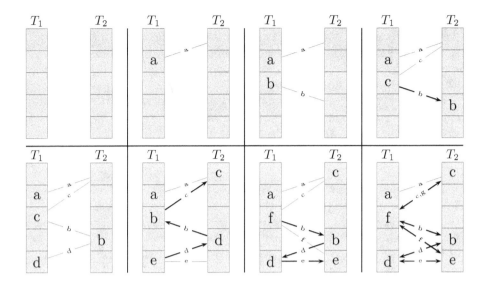

Fig. 1. An evolving cuckoo hash table. We insert the keys a to f sequentially into the previously empty data structure. Each picture depicts the status after the insertion of a single key. The lines connect the two storage locations of a key. Thus, they indicate the values of the hash functions. Arrows symbolise the movement of a key, if it has been kicked-out during the last insertion. Finally, we try to insert the key g on the middle position of T_1, which causes and endless loop and therefore is impossible.

actually stored in the first table increases, what leads to improved search and insertion performance. On the other hand, one has to examine the influence of the asymmetry on the failure probability, what will be done in this section.

This natural modification was already mentioned in [1], however no detailed analysis was known so far. By using the model discussed in the previous section, the following theorem holds.

Theorem 1. *Suppose that $c \in [0,1)$ and $\varepsilon \in (1 - \sqrt{1 - c^2}, 1)$ are fixed. Then, the probability that an asymmetric cuckoo hash of $n = \lfloor (1 - \varepsilon)m \rfloor$ data points into two tables of size $m_1 = \lfloor m(1 + c) \rfloor$ respectively $m_2 = 2m - m_1$ succeeds, is equal to*

$$1 - \frac{(1 - \varepsilon)^3(10 - 2\varepsilon^3 + 9\varepsilon^2 - 3c^2\varepsilon^2 + 9\varepsilon c^2 - 15\varepsilon + 2c^4 - 10c^2)}{12(2\varepsilon - \varepsilon^2 - c^2)^3(c^2 - 1)} \frac{1}{m} + O\left(\frac{1}{m^2}\right). \tag{2.1}$$

Note that the special case c equals 0 corrresponds to standard cuckoo hashing. Thus, this theorem is a generalisation of the analysis of the usual algorithm described in [12].

Proof (Sketch). We model asymmetric cuckoo hashing with help of a labelled bipartite multigraph, the cuckoo graph (see [11]). The two sets of labelled nodes represent the memory cells of the hash table, and each labelled edge represents a key x and connects $h_1(x)$ to $h_2(x)$. It is obviously necessary and sufficient for the success of the algorithm, that every component of the cuckoo graph has less or equal edges than nodes. Thus, each connected component of the graph must either be a tree, or contain exactly one cycle.

First, consider the set of all node and edge labelled bipartite multigraphs containing m_1 resp. m_2 nodes of first resp. second type and n edges. It is clear that the number of all such graphs equals

$$\#G_{m_1,m_2,n} = m_1^n m_2^n. \tag{2.2}$$

We call a tree bipartite if the vertices are partitioned into two classes such that no node has a neighbour of the same class. Let $t_1(x, y)$ denote the generating function of all bipartite rooted trees, where the root is contained in the first set of nodes, and define $t_2(x, y)$ analogous. Furthermore, let $\tilde{t}(x, y)$ denote the generating function of unrooted bipartite trees. Using these notations, the following relations hold:

$$t_1(x, y) = xe^{t_2(x,y)} \qquad t_2(x, y) = ye^{t_1(x,y)} \tag{2.3}$$

$$\tilde{t}(x, y) = t_1(x, y) + t_2(x, y) - t_1(x, y)t_2(x, y) \tag{2.4}$$

With help of these functions, we obtain the generating function

$$\frac{m_1! m_2! n!}{(m_1 + m_2 - n)!} \frac{\tilde{t}(x, y)^{m_1+m_2-n}}{\sqrt{1 - t_1(x, y)t_2(x, y)}} \tag{2.5}$$

counting bipartite graphs containing only tree and unicyclic components. Our next goal is to determine the coefficient of $x^{m_1} y^{m_2}$ of this function. Application of Cauchy's formula leads to an integral that can be asymptotically evaluated with help of a double saddle point method. It turns out that the saddle point is given by

$$x_0 = \frac{n}{m_2} e^{-\frac{n}{m_1}} \quad \text{and} \quad y_0 = \frac{n}{m_1} e^{-\frac{n}{m_2}}. \tag{2.6}$$

The calculation of an asymptotic expansion of the coefficients of large powers $f(x, y)^k$ for a suitable bivariate generating function f is derived in [17] (see also [18]). We use a generalisation this result to obtain an asymptotic expansion of the coefficient of $f(x, y)^k g(x, y)$ for suitable functions f and g. The calculation itself has been done with help of a computer algebra system. Due to the lack of symmetry it is more complicated than the analysis of the unmodified algorithm, altough it follows the same idea. Comparing this result with (2.2) completes the proof. □

The analysis shows a major drawback of this modification. The algorithm requires a load factor less than $\sqrt{1 - c^2}/2$ and this bound decreases if the asymmetry increases. Note that the latter bound is not strict, the algorithm might also succeed with higher load. But experiments show that the failure rate tends to one quickly if this bound is exceeded. Furthermore, even if we stay within this boundary, the success probability of an asymmetric cuckoo hash table decreases as the asymmetry increases. Additionally, we provide numerical data, given in Tab. 1, that verify this effect. The data show that the asymptotic result provides a usefull estimate of the expected number of failures for sufficiently large tables. A futher discussion of this variant of cuckoo hashing is given in Sect. 4 and 5 together with an analysis of the performace of search and insertion operations.

3   A Simplified Version of Cuckoo Hashing

In this section, we propose a further modification of cuckoo hashing: Instead of using two separate tables, we "glue" them together and use one table of double size only. Further, both hash functions address the whole table. As a result of this change, the probability that the first hash function hits an empty cell increases, hence we expect a better performance for search and insertion operations. Details will be discussed later.

   This suggestion was already made in [1], but again a detailed study of the influence of this modification is missing. Further, this suggestion was also made in the analysis of d-ary cuckoo hashing [19], that is a generalised version of the algorithm using d tables and d hash functions instead of only two.

   However, there is a slightly complication caused by this modification. Given an occupied table position, we do not know any longer if this position is the primary or secondary storage position of the key. As a solution, we can either reevaluate a hash function, or provide additional memory to store this information. Furthermore, a very clever variant to overcome this problem is given in [1]. If we change the possible storage locations in a table of size $2m$ for a key x to be $h_1(x)$ and $(h_2(x) - h_1(x)) \mod 2m$, the alternative location of a key y stored at position i equals $(h_2(y) - i) \mod 2m$. We assume henceforth that the second or third suggestion is implemented, and we do not take the cost of otherwise necessary reevaluations of hash functions into account.

   An asymptotic analysis of the success probability can be done by using similar methods as in the proof of Theorem 1.

Theorem 2. *Suppose that $\varepsilon \in (0,1)$ is fixed. Then, the probability that a simplified cuckoo hash of $n = \lfloor (1-\varepsilon)m \rfloor$ data points into a table of size $2m$ succeeds, is equal to*

$$1 - \frac{(5 - 2\varepsilon)(1 - \varepsilon)^2}{48\varepsilon^3} \frac{1}{m} + O\left(\frac{1}{m^2}\right). \tag{3.1}$$

Proof (Sketch). Similar to the analysis of the standard algorithm, we model simplified cuckoo hashing with help of a labelled (non-bipartite) multigraph. Its labelled nodes represent the memory cells of the hash table, and each labelled edge represents a key x and connects $h_1(x)$ to $h_2(x)$. Again, it is obviously necessary and sufficient for the success of the algorithm, that every component of the cuckoo graph has less or equal edges than nodes. Thus, each connected component of the graph must either be a tree, or contain exactly one cycle.

Obviously, the number of all node and edge labelled multigraphs posessing $2m$ nodes an n edges equals

$$(2m^2)^n. \tag{3.2}$$

Instead of bivariate generating functions, we make use of the well known generating functions $t(x)$ and $\tilde{t}(x)$ of rooted resp. unrooted trees that satisfy the equations

$$t(x) = xe^{t(x)}, \qquad \tilde{t}(x) = t(x) - \frac{1}{2}t(x)^2. \tag{3.3}$$

The evolution of the graph is described by the multigraph process of [20], with the only difference that we consider labelled edges too.

Thus, we obtain that the generating function counting graphs without components containing more than one cycle equals

$$\frac{(2m)!n!}{(2m-n)!} \frac{\tilde{t}(x)^{2m-n}}{\sqrt{1-t(x)}}. \tag{3.4}$$

Now, we are interested in the coefficient of x^{2m} of this function. We continue using Cauchy's formula and obtain an integral representation. Again, the coefficient can be extracted with help of the saddle point method. The required method is related to results given in [18] and [21]. □

Numerical data are given in Tab. 1. We now conclude that the success probability of simplified cuckoo hashing is slightly decreased compared to the standard algorithm, but the practical behaviour is almost identical in this aspect.

Table 1. Number of failures during the construction of $5 \cdot 10^5$ cuckoo hash tables. The table provides numerical data (data) and the expected number of failures (exp.) calculated with Theorem 1 resp. 2. We use a pseudo random generator to simulate good hash functions.

m	standard data	standard exp.	$c=0.1$ data	$c=0.1$ exp.	$c=0.2$ data	$c=0.2$ exp.	$c=0.3$ data	$c=0.3$ exp.	$c=0.4$ data	$c=0.4$ exp.	simplified data	simplified exp.
$5\cdot10^3$	656	672	653	730	858	951	1336	1575	2958	3850	710	767
10^4	308	336	384	365	456	476	725	787	1673	1925	386	383
$5\cdot10^4$	65	67	66	73	81	95	154	157	373	385	87	77
10^5	38	34	43	36	55	48	81	79	176	193	32	38
$5\cdot10^5$	7	7	3	7	10	10	8	16	29	39	7	8
$5\cdot10^3$	4963	7606	5578	8940	8165	15423	16392	51955	49422	$5\cdot10^5$	5272	8100
10^4	2928	3803	3368	4470	5185	7712	11793	25977	44773	$5\cdot10^5$	3122	4050
$5\cdot10^4$	701	761	867	894	1388	1542	3894	5195	30758	$19\cdot10^4$	737	810
10^5	385	380	435	447	683	771	2187	2598	24090	96139	417	405
$5\cdot10^5$	75	76	97	89	165	154	532	520	10627	19228	85	81

(The first five data rows correspond to $\varepsilon = 0.2$, and the last five to $\varepsilon = 0.1$.)

4 Search

Of course, we may perform a search in at most two steps. Assume that we always start a search after a key x at the position $h_1(x)$. As a consequence, we can perform a successful search in one step only, if the cell determined by the first hash function holds x. Further, it is satisfied that a search operation is unsuccessful, if the position indicated by h_1 is empty, as long as our data structure meets the following rules:

- We always try to insert a key using h_1 first, the second hash function is used only if the inspected cell is already occupied.
- If we delete an element, it is not allowed to mark the cell "empty" instead we have to use a marker "previously occupied". This is similar to the deletions in hashing with open addressing [4].

Similar to the analysis of linear probing and uniform probing in [4], our analysis considers hashing without deletions. Clearly, our results apply also to the situations where deletions are very rare. We want to emphasise that the notations are a little bit different so, we state the results in terms of the load factor $\alpha = n/(2m)$. As a consequence, the results can be directly compared.

Theorem 3 (Search in standard and asymmetric cuckoo hashing). *Under the assumptions of Theorem 1, assume that a cuckoo hash table has been constructed successfully. Then, the expected number of inspected cells of an successful search is asymptotically given by*

$$2 - \frac{1+c}{2\alpha}\left(1 - e^{-2\alpha/(1+c)}\right) + O\left(m^{-1}\right), \tag{4.1}$$

where $\alpha = n/(2m)$ denotes the load factor of the table. Furthermore, the expected number of steps of an unsuccessful search is asymptotically given by

$$2 - e^{-2\alpha/(1+c)} + O\left(m^{-1}\right). \tag{4.2}$$

Proof. Consider an arbitrary selected cell z of the first table. The probability, that none of the randomly selected values $h_1(x_1), \ldots, h_1(x_n)$ equals z is given by $p = \left(1 - \lfloor m(1+c) \rfloor^{-1}\right)^n$. Let p_s denote the probability that z is empty, conditioned on the property that the construction of the hash table succeeds, and let p_a denote the probability that z is empty, conditioned on the property that the construction of the hash table is unsuccessful. By the law of total probability, we have $p = p_s + p_a$. Due to Theorem 1, the relation $p_a = O(m^{-1})$ holds. Thus, we obtain

$$p_s = \left(1 - \lfloor m(1+c)\rfloor^{-1}\right)^n + O\left(m^{-1}\right) = e^{-n/(m(1+c))} + O\left(m^{-1}\right). \tag{4.3}$$

This equals the probability, that the first inspected cell during a search is empty. All other unsuccessful searches take exactly two steps.

Similarly, we obtain, that the expected number of occupied cells of the first table equals $\lfloor m(1+c)\rfloor(1 - e^{-2\alpha/(1+c)}) + O(1)$. That gives us the number of keys which might be found in a single step, while the search for any other key takes exactly two steps. $\qquad\square$

Theorem 4 (Search in simplified cuckoo hashing). *Under the assumptions of Theorem 2, the expected number of inspected cells C_n of an successful search satisfies the relation*

$$2 - \frac{1 - e^{-\alpha}}{\alpha} + O\left(m^{-1}\right) \le C_n \le 2 - \frac{1 - e^{-2\alpha}}{2\alpha} + O\left(m^{-1}\right), \qquad (4.4)$$

where $\alpha = n/(2m)$ denotes the load factor of the table. Furthermore, the expected number of steps of an unsuccessful search is given by $1 + \alpha$.

Proof. The number of steps of an unsuccessful search is determined by the number of empty cells which equals $2m - n$. Thus, we need only one step with probability $1 - \alpha$ and two steps otherwise.

Similar to the proof of Theorem 3, we obtain, that the probability that an arbitrary selected cell is not a primary storage location, equals

$$p = (1 - 1/(2m))^n + O\left(m^{-1}\right) = e^{-n/(2m)} + O\left(m^{-1}\right). \qquad (4.5)$$

Hence, $2m(1 - p)$ is the expected number of cells accessed by h_1 if the table holds n keys. It is for sure that each of this memory slots holds a key, because an insertion always starts using h_1. However, assume that the primary position of one of this keys y equals the secondary storage position of an other key x. If x is kicked-out, it will subsequently kick-out y. Thus, the total number of steps to find all keys increases. Figure 2 gives an example of such a situation.

Let q denote the probability, that a cell z, which is addressed by both hash functions, is occupied by a key x such that $h_1(x) = z$ holds. Then, the expected number of keys reachable with one memory access equals

$$2m \left((1 - p)p + (1 - p)^2 q\right). \qquad (4.6)$$

By setting $q = 1$ and $q = 0.5$ we get the claimed results. The latter value corresponds to a natural equilibrium. □

Note that further moments can be calculated using the same method. Hence it is straightforward to obtain the variances.

Again, we provide numerical results, which can be found in Tab. 2. Since the cost of an unsuccessful search is deterministic for the simplified version and closely related to the behaviour of the successful search otherwise, we concentrate on the successful search. From the results given in the table, we find that our asymptotic results are a good approximation, even for hash tables of small size. In particular, we notice that the simplified algorithm offers an improved performance compared to the other variants for all investigated settings. The good performace of successful searches is due to the fact, that the load is unbalanced, because the majority of keys will be usually stored using the first hash function.

Figure 3 displays the asymptotic behaviour of a successful search, depending on the load factor α. Experiments show that the actual behaviour of the simplified algorithm is closer to the lower bound of Theorem 4, than to the upper bound, especially if the load factor is small. Further, Fig. 4 shows a plot according to an unsuccessful search. Note that the algorithm possessing an asymmetry $c = 0.3$ allows a maximum load factor of approximately .477 according to Th. 1. Experiments show that the failure rate increases dramatically if this bound is exceeded.

Concerning asymmetric cuckoo hashing, all these results verify the conjecture that the performance of search operations increases as the asymmetry increases. However, the simplified algorithm offers even better performance without the drawbacks of a lower maximal fill ratio and increasing failure probability (see Sect. 2). The improved performance can be explained by the increased number of keys accessible in one step, resp. by the higher probability of hitting an empty cell. We conclude that simplified cuckoo hashing offers the best average performance over all algorithms considered in this paper, for all feasible load factors. Thus it is highly recommendable to use this variant instead of any other version of cuckoo hashing discussed here.

Compared to linear probing and double hashing, the chance of hitting a non-empty cell in the first step is identical. However, simplified cuckoo hashing needs exactly two steps in such a case, but there is a non-zero probability that the two other algorithms will need more than one additional step.

Finally, we compare simplified cuckoo hashing to modified versions of double hashing that try to reduce the average number of steps per search operation by using modified insertion algorithms. In particular, we consider Brent's variation [22] and binary tree hashing [23]. Note that there is almost no difference in the behaviour of successful searches of these two algorithms for all the load factors considered in this paper. Furthermore our numerically obtained data show that simplified cuckoo hashing offers very similar performance. However, Brent's algorithm does not influence the expected cost of unsuccessful searches compared to double hashing. Hence we conclude that simplified cuckoo hashing offers a better performance in this aspect.

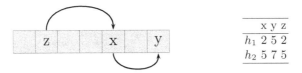

Fig. 2. Additional search costs in simplified cuckoo hashing. Two memory cells are accessible by h_1 for the current data, but only the key z can be found in a single step.

5 Insertion

Until now, no exact analysis of the insertion cost of cuckoo hashing is known. However, it is possible to establish an upper bound.

Table 2. Average number of steps of a successful search for several variants of cuckoo hashing. We use random 32-bit integer keys, the hash functions described in Sect. 1, and consider the average taken over $5 \cdot 10^4$ successful constructed tables. Further, we provide data covering linear probing, double hashing, and Brent's variation of double hashing obtained using the well known asymptotic approximations.

memory	standard $\alpha = .35$	$\alpha = .475$	asymm. $c = 0.2$ $\alpha = .35$	$\alpha = .475$	asymm. $c = 0.3$ $\alpha = .35$	$\alpha = .475$	simplified $\alpha = .35$	$\alpha = .475$
$2 \cdot 10^3$	1.2807	1.3541	1.2421	1.3085	1.2265	1.2896	1.1849	1.2695
$2 \cdot 10^4$	1.2808	1.3544	1.2423	1.3091	1.2268	1.2904	1.1851	1.2706
$2 \cdot 10^5$	1.2808	1.3545	1.2423	1.3092	1.2268	1.2905	1.1851	1.2706
$2 \cdot 10^6$	1.2808	1.3545	1.2423	1.3092	1.2268	1.2905	1.1851	1.2706
asympt. result(s)	1.2808	1.3545	1.2423	1.3092	1.2268	1.2905	1.1563 1.2808	1.2040 1.3545

	double hashing	linear probing	Brent's variation
$\alpha = .35$	1.2308	1.2692	1.1866
$\alpha = .485$	1.3565	1.4524	1.2676

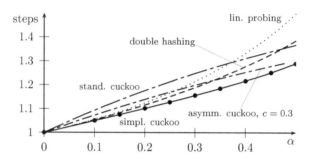

Fig. 3. Comparison of successful search. The curves are plotted from the results of Theorem 3 resp. 4 together with the well known asymptotic results of the standard hash algorithms. For simplified cuckoo hashing, the grey area shows the span between the upper and lower bound. The corresponding curve is obtained experimentally with tables containing 10^5 cells and sample size $5 \cdot 10^4$.

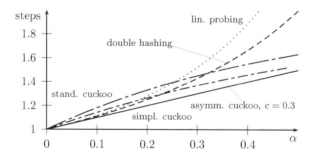

Fig. 4. Comparison of unsuccessful search. The curves are plotted from the results of Theorem 3 resp. 4 together with the well known asymptotic results of the standard hash algorithms.

Theorem 5. *Under the assumptions of Theorem 1 resp. 2, an upper bound of the number of expected memory accesses during the construction of a standard resp. simplified cuckoo hash table is given by*

$$\min\left(4, \frac{-\log(1-2\alpha)}{2\alpha}\right) n + O(1),\tag{5.1}$$

where $\alpha = (1 - \varepsilon)/2$ denotes the load factor and the constant implied by $O(1)$ depends on α.

Similar to the proofs of Theorem 1 and 2, the proof is again based on the bipartite resp. usual random graph related to the data structure. To be more precise we obtain this two bounds using two different estimators for the insertion cost in a tree component, namely the component size and the diameter. More details are omitted due to space requirements, a detailed proof will be given in [24].

We propose the usage of a slightly modified insertion algorithm for practical implementation. In contrast to the algorithm described in [1], we perform an additional test during the insertion of a key x under following circumstances. If the location $h_1(x)$ is already occupied, but $h_2(x)$ is empty, the algorithm places x in the second table. If both possible storage locations of x are already occupied, we proceed as usual and kick-out the key stored in $h_1(x)$. This modification is motivated by two observations:

- We should check if the key is already contained in the table. If we do not perform the check a priory and allow duplicates, we have to perform a check after each kick-out step. Further we have to inspect both possible storage locations to perform complete deletions. Because of this, it is not recommended to skip this test. Hence we have to inspect $h_2(x)$ anyway, and there are no negative effects on the average search time caused by this modification.
- The probability that the position $h_2(x)$ is empty is relatively high. For the simplified algorithm possessing a load α, this probability equals $1-\alpha$. Further we expect an even better behaviour for variants consisting of two tables, because of the unbalanced load. In contrast, the component that contains $h_1(x)$ possesses at most one empty cell and it usually takes time until this location is found.

Experiments show, that this modified insertion algorithm reduces the number of required steps by about 5 to 10 percent.

The attained numerical results, given in Tab. 3, show that the expected performance is by far below this upper bound. Further, we notice that the simplified version offers better average performance for all investigated settings compared to the other variants of cuckoo hashing. The average number of memory accesses of an insertion into a table using simplified cuckoo hashing is approximately equal to the expected number of steps using linear probing and thus a bit higher than using double hashing. Finally, the average number of steps for an insertion using simplified cuckoo hashing is approximately equal to the average number of cells inspected during an insertion using Brent's variation, at least for sufficiently large table sizes.

Table 3. Average number of steps per insertion for several versions of cuckoo hashing. We use random 32-bit integer keys, the hash functions described in Sect. 1, and consider the average taken over $5 \cdot 10^4$ successful constructed tables. Further, we provide data covering linear probing and double hashing obtained using the well known asymptotic approximations and numerical data for Brent's variation of double hashing.

	standard		asymm. $c = 0.2$		asymm. $c = 0.3$		simplified	
memory	$\alpha = .35$	$\alpha = .475$	$\alpha = .35$	$\alpha = .475$	$\alpha = .35$	$\alpha = .475$	$\alpha = .35$	$\alpha = .475$
$2 \cdot 10^3$	1.3236	1.7035	1.2840	1.7501	1.2724	1.9630	1.2483	1.5987
$2 \cdot 10^4$	1.3203	1.5848	1.2801	1.6531	1.2656	2.0095	1.2462	1.5119
$2 \cdot 10^5$	1.3200	1.5079	1.2796	1.5129	1.2650	1.9647	1.2459	1.4517
$2 \cdot 10^6$	1.3200	1.488	1.2796	1.4585	1.2650	1.81071	1.2459	1.4401

	bound of Th. 5	double hashing	linear probing	Brent's variation
$\alpha = .35$	1.7200	1.2308	1.2692	1.275
$\alpha = .485$	3.6150	1.3565	1.4524	1.447

6 Summary and Conclusions

The main contribution of this paper was the detailed study of modified cuckoo hash algorithms. Unlike standard cuckoo hashing, we used tables of different size or granted both hash functions with access to the whole table. This enhanced the probability that the first hash function hit an empty cell.

As an important result, the variant using one table only improved the behaviour of cuckoo hashing significantly. Thus, we obtained an algorithm that took on average less steps per search operation than linear probing and double hashing and possessed constant worst case search time and approximately the same construction cost as linear probing.

However, this does not mean that a cuckoo hash algorithm is automatically preferable for each application. For instance, it is a well known fact that an implementation of linear probing might be faster than double hashing, if the load factor is small, although the number of steps is higher. This is due to the memory architecture of modern computers. Linear probing might need more probes, but the required memory cells are with high probability already loaded into the cache and might be investigated faster than the time needed to resolve a single cache miss [25]. Similarly, a step using cuckoo hashing can be more expensive if the evaluation of the hash function takes more time than the evaluation of a "simpler" hash function. Nonetheless, simplified cuckoo hashing offers very interesting properties, and seems predestinated for applications requiring low average and worst case search time.

In the future, we suggest to extend the analysis of the performance of search and insertion operations to d-ary cuckoo hashing [19] and cuckoo hashing using buckets of capacity greater than one [26].

Acknowledgements The author would like to thank Matthias Dehmer and Michael Drmota for helpful comments on this paper.

References

1. Pagh, R., Rodler, F.F.: Cuckoo hashing. Journal of Algorithms **51**(2) (2004) 122–144
2. Cormen, T.H., Leiserson, C.E., Rivest, R.L., Stein, C.: Introduction to Algorithms. Second edn. MIT Press, Cambridge, Massachusetts London, England (2001)
3. Gonnet, G.H., Baeza-Yates, R.: Handbook of algorithms and data structures: in Pascal and C. Second edn. Addison-Wesley, Boston, MA, USA (1991)
4. Knuth, D.E.: The Art of Computer Programming, Volume III: Sorting and Searching. Second edn. Addison-Wesley, Boston (1998)
5. Azar, Y., Broder, A.Z., Karlin, A.R., Upfal, E.: Balanced allocations. SIAM J. Comput. **29**(1) (1999) 180–200
6. Broder, A.Z., Mitzenmacher, M.: Using multiple hash functions to improve ip lookups. In: INFOCOM. (2001) 1454–1463
7. Czech, Z.J., Havas, G., Majewski, B.S.: Perfect hashing. Theoretical Computer Science **182**(1-2) (1997) 1–143
8. Fredman, M.L., Komlós, J., Szemerédi, E.: Storing a sparse table with O(1) worst case access time. J. ACM **31**(3) (1984) 538–544
9. Dalal, K., Devroye, L., Malalla, E., McLeis, E.: Two-way chaining with reassignment. SIAM J. Comput. **35**(2) (2005) 327–340
10. Vöcking, B.: How asymmetry helps load balancing. J. ACM **50**(4) (2003) 568–589
11. Devroye, L., Morin, P.: Cuckoo hashing: Further analysis. Information Processing Letters **86**(4) (2003) 215–219
12. Kutzelnigg, R.: Bipartite random graphs and cuckoo hashing. In: Proc. 4th Colloquium on Mathematics and Computer Science. DMTCS (2006) 403–406
13. Dietzfelbinger, M., Gil, J., Matias, Y., Pippenger, N.: Polynomial hash functions are reliable. In: ICALP '92. Volume 623 of LNCS., Springer (1992) 235–246
14. Dietzfelbinger, M., Woelfel, P.: Almost random graphs with simple hash functions. In: STOC '03, ACM (2003) 629–638
15. Carter, L., Wegman, M.N.: Universal classes of hash functions. J. Comput. Syst. Sci. **18**(2) (1979) 143–154
16. Tran, T.N., Kittitornkun, S.: Fpga-based cuckoo hashing for pattern matching in nids/nips. In: APNOMS. (2007) 334–343
17. Good, I.J.: Saddle-point methods for the multinomial distribution. Ann. Math. Stat. **28**(4) (1957) 861–881
18. Drmota, M.: A bivariate asymptotic expansion of coefficients of powers of generating functions. European Journal of Combinatorics **15**(2) (1994) 139–152
19. Fotakis, D., Pagh, R., Sanders, P., Spirakis, P.G.: Space efficient hash tables with worst case constant access time. Theory Comput. Syst. **38**(2) (2005) 229–248
20. Janson, S., Knuth, D.E., Łuczak, T., Pittel, B.: The birth of the giant component. Random Structures and Algorithms **4**(3) (1993) 233–359
21. Gardy, D.: Some results on the asymptotic behaviour of coefficients of large powers of functions. Discrete Mathematics **139**(1-3) (1995) 189–217
22. Brent, R.P.: Reducing the retrieval time of scatter storage techniques. Commun. ACM **16**(2) (1973) 105–109
23. Gonnet, G.H., Munro, J.I.: The analysis of an improved hashing technique. In: STOC, ACM (1977) 113–121
24. Drmota, M., Kutzelnigg, R.: A precise analysis of cuckoo hashing. preprint (2008)
25. Ross, K.A.: Efficient hash probes on modern processors. IBM Research Report RC24100, IBM (2006)

26. Dietzfelbinger, M., Weidling, C.: Balanced allocation and dictionaries with tightly packed constant size bins. Theoretical Computer Science **380**(1-2) (2007) 47–68
27. Kirsch, A., Mitzenmacher, M., Wieder, U.: More robust hashing: Cuckoo hashing with a stash. In: Proceedings of the 16th Annual European Symposium on Algorithms. (2008)

Author Index

www.ingramcontent.com/pod-product-compliance
Lightning Source LLC
LaVergne TN
LVHW012328060326
832902LV00011B/1778